GENESIS

Genesis

The Evolution of Biology

JAN SAPP

2003

OXFORD
UNIVERSITY PRESS

Oxford New York
Auckland Bangkok Buenos Aires Cape Town Chennai
Dar es Salaam Delhi Hong Kong Istanbul Karachi Kolkata
Kuala Lumpur Madrid Melbourne Mexico City Mumbai Nairobi
São Paulo Shanghai Taipei Tokyo Toronto

Published by Oxford University Press, Inc.
198 Madison Avenue, New York, New York 10016

http://www.oup-usa.org

Library of Congress Cataloging-in-Publication Data
Sapp, Jan.
Genesis : the evolution of biology / by Jan Sapp.
p. cm.
Includes bibliographical references.
ISBN 0-19-515618-8; ISBN 0-19-515619-6 (pbk.)
1. Biology—History. 2. Evolution (Biology)—History. 3. Genetics—
History. I. Title.
QH305.S27 2003
570—dc21 2002152271

1 3 5 7 9 8 6 4 2

Printed in the United States of America
on acid-free paper

For Carole, Will, and Elliot

A little Learning is a dang'rous thing;
Drink deep, or taste not the Pierian spring:
There shallow draughts intoxicate the brain,
And drinking largely sobers us again.
Fir'd at first sight with what the Muse imparts,
In fearless youth we tempt the heights of Arts,
While from the bounded level of our mind,
Short views we take, nor see the lengths behind;
But more advanc'd, behold with strange surprise
New distant scenes of endless science rise!

Alexander Pope (1688–1744), "A Little Learning"

Preface

What is evolution? What is an organism? What is a gene? This book explores these concepts and the controversies that have surrounded them. It aims to provide a short history of biology, one that can be read by nonspecialists and one that incorporates new evolutionary research programs, contemporary Darwinian and non-Darwinian theory, changing concepts of the organism, and shifting concepts of the gene, all of which advance research today. A central question motivating the book is: Why did the history of biology and evolutionary thought unfold the way it did? In the book, I search for answers in the use of specific techniques, models, and analogies; financial support; institutional conditions; and sometimes larger social and intellectual movements.

Though a book such as this cannot contain all of the outstanding scholarship pertaining to the history of biology, I have selected the major historic transitions and key figures representative of them. Part I describes the emergence of evolutionary theory in France led by Lamarck and analyzes the subsequent genesis of Darwin's theory of natural selection, its philosophical and social significance, and objections to it in the nineteenth century. Part II describes parallel research on the cell in development and heredity and highlights nineteenth-century attempts to discern the processes by which animals develop from eggs. Part III follows episodes in genetics and evolutionary theory, from the rediscovery of Mendel's laws to the neo-Darwinian synthesis and the development of microbial genetics. Part IV examines the rise of molecular biology, the genetic code, its central doctrines, and their critics. It also explores research on hereditary mechanisms in addition to chromosomal genes, whose investigations were developed in the latter decades of the twentieth century.

When sketching historical changes, I have tried to underscore important themes in the history, philosophy, and social studies of biology while avoiding arcane language. I have drawn from various studies to illustrate the two-way traffic between social theory and evolutionary explanation. I explore the development of evolutionary thought from Lamarck to Darwin in the context of social change in the nineteenth century (chapters 1–4). I examine the ways evolutionists extended their views about nature to society and, inversely, how they extended their views about society to nature (chapters 4, 5, and 20). I observe the concept of the division of labor in Darwin's theory (chapter 4) and in cell theory and the concept of the organism (chapter 8). I also probe the metaphor of heredity itself (chapter 12), as well as the contentious use of human categories in concepts of parasitism, mutualism, and symbiosis (chapters 5, 19, and 20). With the rise of molecular biology, social concepts of the organism were supplemented with technological ones from communication technology (chapters 16 and 17).

I also observe the rhetorical paradigms of nineteenth-century evolutionists who emphasized the survival of the fittest and materialist philosophy against teleology and theology in debates over "man's place in nature" (chapters 1, 3, and 4). Studies of mutualism, associated with anarchism and natural theology, became a casualty of that polarity between mechanistic materialism and theology (chapter 5). When the evolution of cooperation came to the fore of neo-Darwinian concerns in the latter part of the twentieth century, it was characterized by heated debates over individualism, so-called selfish genes, and how far adaptation and evolutionary explanation could be applied to human social relations (chapter 20).

Chapters are driven by historiographic themes, the birth of specialties, and the conceptual oppositions, techniques, and controversies that have shaped the development of the life sciences. As many teachers of science have noted, scientific problems are usually much better understood from studying their history rather than their logic alone. Biologists from many specialties have investigated heredity and evolution, emphasizing different problems and techniques while often developing different perspectives and alternative theories. I explore how these specialties emerged and their relationships with each other. The history of biology is regarded as a contest over what questions are important, what answers are acceptable, what phenomena are interesting, and what techniques are most useful.

Although one model of nature is sometimes overthrown, a new one seldom, if ever, entirely wipes out the old. We shall see, for example, that comparative morphology was overshadowed by experimental embryology in the late nineteenth century (chapters 9 and 10), and it in turn by genetics in the twentieth century (chapter 12). Embryologists have long protested against geneticists' views about the centrality of chromosomal genes in heredity and development and have emphasized the organization of the egg. They did not participate in the development of the neo-Darwinian synthesis (chapters 13 and 15) but in recent decades, attempts to complete the modern synthesis with evolutionary developmental biology have been on the upswing. Protests over the centrality of DNA continue to the present day, as the definition of the gene itself has become ever more complex and abstract (chapter 17). Critics of gene theory continue to emphasize that only a cell can make a cell, and that plant and animals emerge from eggs, not genes. Thus, I also include research on non-DNA-based inheritance. In so doing, I examine current concepts of the gene and molecular biology and compare them to the central doctrines of classical molecular biology of the 1960s (chapters 16 and 17).

Traditionally, books on the history of evolutionary thought have aimed at explaining away the opposition to Darwin's theory, clearing up the confusion, and reconciling differences to reveal the "evolutionary synthesis" of the 1930s and 1940s and its subsequent articulations. Evolution is understood in terms of gene pools and population genetics. Such accounts ignore studies of non-Mendelian heredity and of other mechanisms of evolutionary change that lie outside Darwinian research traditions. As a result, for the general public and for many scientists, the expression "non-Darwinian evolution" is an oxymoron, and "Darwinian evolution" redundant.

Classical evolutionary biology has been concerned with the last 560 million years of evolution; it is essentially about animals and plants. Accordingly, its historians do not consider 85 percent of the earth's evolutionary time—the evolution of bacteria, now held by some estimates to be the largest biomass on earth, with the greatest biochemical diversity on earth—nor do they consider how protists, fungi, plants, and animals emerged from them. Indeed, whole books on the history of evolutionary biology have been written from which the word "bacteria" is virtually absent.

Techniques as much as new theories have been drivers of scientific change, and whole new approaches to evolution have arisen with the field of "molecular evolution" (chapter 18). Bringing bacteria into the evolutionary framework of biology has entailed vital changes to evolutionary theory. Theories about symbiosis in evolution, for example, that were developed at the end of the nineteenth century were recast at the end of the twentieth. Microbial evolutionists insist that symbiosis, mergers, and transfer of genes between different kinds of microbes are cardinal mechanisms of evolutionary change. We consider the concept of the "symbiome" in chapter 19, as biologists today recognize many other genomes within cells and organisms besides that of the nucleus: mitochondria, chloroplasts, viral genomes, and other symbionts inside and outside the cell. Their recognition entails dramatic alterations to neo-Darwinian theory and to our concept of the individual.

The distinguished population geneticist Theodosius Dobzhansky aptly asserted in 1973 that "Nothing in biology makes sense except in the light of evolution."[1] He did so to emphasize the importance of evolutionary thinking to nonevolutionists and to those biological researchers who simply do not consider it as much as they ought to. Here I emphasize the bidirectionality of that adage. Some readers may feel that I am going to the opposite extreme of conventional histories by devoting too little space to the articulations of neo-Darwinism and population genetics. I would argue that embracing alternative mechanisms of heredity and evolution is vital because they encompass so much of organismic biology.

I have also highlighted the significance of scientists' own storytelling and their themes about individual triumph, neglect, and rediscovery as an integral rhetorical aspect of the process of discovery. I examine myths about Lamarck (chapter 1) and reasons the foundation of the cell theory was attributed to Schleiden and Schwann, who actually opposed its central principle (chapter 7). I also investigate conflicting stories about what Mendel really discovered, why he was neglected, and how these stories functioned in debates over evolution (chapter 12). I probe an analogous story about Sir Archibald Garrod in the origins of biochemical genetics and how it was used in debates over how genes work (chapter 14).

Patronage has been crucial in shaping biology. I examine scientists' rhetoric about the relationships between pure and applied science in the early development of genetics (chapter 12), during the Cold War when Western geneticists emphasized the importance of academic freedom (chapter 15), and after the Cold War when critics decried that Western biology had become too commercialized and tied to medical interests—at the expense of fundamental problems in heredity, development, and evolution (chapters 17, 18, and 19).

Acknowledgments

I am delighted to thank many friends and colleagues who assisted me in the writing of this book. Arthur Forer read the entire manuscript when it was in its rudimentary stages, offering valuable advice and discussion. I have benefited greatly from correspondence with Joshua Lederberg about many issues discussed in this book, both historical and biological, and from his critical reading of chapters 16, 17, and 18. Carl Woese offered helpful comments on early drafts of chapters 18 and 19; Brian Hall did the same for chapters 8 and 9, and Camille Limoges for chapters 1 and 2. Lynn Margulis has been a font of stimulating conversation about symbiosis in evolution and the organizer of many inspiring symposia in which I have been fortunate to participate over the past fifteen years. I thank Douglas Allchin, Richard Burkhardt, Linnda Caporael, Francisco Carrapiço, Angela Craeger, Ford Doolittle, Michael Dietrich, Elihu Gerson, Michael Gray, Jim Griesemer, Lily Kay, Jane Maienschein, Bill Martin, Ernst Mayr, Everett, Mendelsohn, Gregg Mitman, Lynn Nyhart, Robert Olby, Michael Ruse, James Strick, David Thaler, Tom Wakeford, Fred Weizmann, Polly Winsor, and many other historians, philosophers, and biologists who are represented in the notes to chapters. The writing of this book was greatly facilitated by the excellent research conditions afforded me during my two years as Canada Research Chair in the History of the Biological Sciences at the Centre interuniversitaire de recherche sur la science et la technologie at the Université du Québec à Montréal. I especially thank Lucie Comeau for her information expertise and library assistance. I am also grateful to Kirk Jensen for his thoughtful advice and editorial help in seeing this book through to its completion. Finally, I am pleased to acknowledge financial support from the Social Sciences and Humanities Research Council of Canada.

Permission to use materials from the following articles is generously granted by Kluwer Academic Publishers: Jan Sapp, "The Struggle for Authority in the Field of Heredity, 1900–1932. New Perspectives on the Rise of Genetics," *Journal of the History of Biology* 16 (1983): 311–342. Jan Sapp, "The Nine Lives of Gregor Mendel," in H. E. Le Grand, ed., *Experimental Inquiries*. Kluwer Academic Publishers, 1990, pp. 137–166.

Contents

PART I EVOLUTION AND MORPHOLOGY

1. Evolution and Revolution 3
    - Two Worldviews 4
    - Revolution to Evolution 5
    - Lamarckian Myths 6
    - Simple to Complex 8
    - Disconnecting the Unity of Life 11
    - The Cuvier-Geoffroy Debate 13

2. The Origin 16
    - When Making Other Plans 17
    - Darwin's Bible 18
    - The *Beagle* Voyage 20
    - Natural Selection and Natural Theology 22
    - Wallace's Manuscript 24
    - Concepts of *The Origin* 26

3. Darwin's Champions 31
    - Man's Place in Nature 31
    - Natural Theology and Agnosticism 33
    - Archetype and Idealism 35
    - Ontogeny and Phylogeny 36
    - Materialism for Mysticism 40

4. Darwinism and Sociopolitical Thought 43
    - Laissez-faire 44
    - Social Darwinism Exported 46
    - War and Racism 47

Darwinism on the Left 48
Was Darwin a Social Darwinist? 49
Social Theory in Evolution 50
The Division of Labor 50
Darwin and Malthus 52

5. Mutualism 55
Anarchism 55
Between Individuals 57
Between Species 58
Roots in Natural Theology 61

6. Dissent from Darwin 63
Is the Earth Old Enough? 64
Blending Inheritance 64
What Is a Species? 65
Speciation and Isolation 66
Is Everything Adapted? 66
Holes in the Record 67
Neo-Lamarckism 68
Orthogenesis 69
Saltationism 71

PART II THE CELL IN DEVELOPMENT AND HEREDITY

7. The Myth of the Cell Theory 75
An Historical Paradox 75
Cells from Cells 77
More than Meets the Eye 78
Vitalism, Materialism, and Spontaneous Generation 80

8. The Body Politic 82
The Cell State 82
The Dawn of Protistology 85
A Cell Is Not a Cell 86
What's in a Word 87
Organisms within Organisms 90
Weismannism 91

9. Evolving Embryology 95
Technical Virtuosity 96
The Organism as a Whole 98
Epigenesis and Preformation 100

10. The Egg 103
The Body Plan in the Egg 104
Maternal Inheritance 106
Cellular Differentiation 109
Cytoplasmic Evolution 112

PART III GENETICS AND THE CLASSICAL SYNTHESIS

11. Mendel Palimpsest 117
Mendel's Laws 118
Neglect and Rediscovery 119
Making a Discoverer 121
Why Multiple Meanings? 122
Geneticists versus Statisticians 124
Mendel Made Darwinian 126
Is the Scientific Paper Fiction? 126

12. Emerging Genetics 130
The Field of Heredity 130
Genotype and Phenotype 134
Disciplinary Design 135
Biology out of Balance 138
Are Genes Real? 140

13. Darwinian Renaissance 143
Merging Mendelism 144
The Importance of Sex 146
Population Genetics 147
Random Drift and Nonadaptive Change 149
The Species Problem 151
Microevolution as Macroevolution 152
Lessons of Synthesis 154

14. Genes, Germs, and Enzymes 157
The Garrod Tale 158
Physiology and Genetics 158
Early Gene-Enzyme Associations 160
The One-Gene: One-Enzyme Hypothesis 161
Domesticating Microbes 163
The Chosen Few 164
The Rockefeller Foundation 168

15. Genetic Heresy and the Cold War 171
Non-Darwinian Development 173
Plasmon to Plasmagenes 174

The Inheritance of Acquired Characteristics 176
University Politics 179
Morgan's Smile 181

PART IV MOLECULAR BIOLOGY AND ORGANISMIC COMPLEXITY

16. Conceiving a Master Molecule 187
DNA or Protein? 188
Transformation and Transduction 189
Chromatography 191
X-Ray Crystallography 192
Digital DNA 194
Transcription and Translation 196
Turning Genes On and Off 197
Classical Doctrines of Molecular Biology 198

17. Beyond the Genome 201
Complexity and the Human Genome 203
A Genetic Plan? 205
Confronting Old Dogmas 206
Cell Architecture and Spatial Information 207
Field Heredity 211
Epinucleic Inheritance 214

18. Molecular Evolution and Microbial Phylogeny 217
Precambrian Explosion 218
Molecular Clocks 220
The Origin of the Code 221
A Code for Classification 224
A Trilogy of Life 225
Dissension and Disaffection 228
Lateral Gene Transfer 230

19. Symbiomics 234
Developmental Symbiosis 235
Symbiosis Silhouette 236
Why It Has Been Difficult to Imagine 240
Toward a Unified Theory 243
Symbiogenetic Renaissance 245
Macroevolutionary Change 247

20. The Evolution of Relationships 252
The Individual and the Group 253
Kin Selection 255

The Organism as a Beehive 255
The Lessons of Sociobiology 257
About Just-So Stories 258
Symbiotic Ties 261

Epilogue 267

Notes 273

Index 347

I

EVOLUTION AND MORPHOLOGY

1

Evolution and Revolution

> As compared to the periods which we look upon as great in our ordinary calculations, an enormous time and wide variation in successive conditions must doubtless have been required to enable nature to bring the organization of animals to that degree of complexity and development in which we see it at its perfection.
>
> —Jean-Baptiste Lamarck, 1809

THERE ARE FEW GENERALIZATIONS in biology and few universal theories. But among them, evolution is the most important. Evolutionary theory holds that the natural world is steadily changing, that organisms have diverged from common ancestors, and that they have been transformed in geological time. Every biological specialty, from genetics to ornithology, is enriched and informed by an evolutionary viewpoint. Today, biologists regard evolution as a confirmed fact, one of the greatest facts of science, as much a fact as that the earth revolves around the sun. And the influence of evolutionary thinking has extended well beyond biology. Virtually every aspect of human thought has been touched by it.

Charles Darwin convinced the scientific world of evolution in the nineteenth century. Evolutionary thinking was part of a whole new approach to investigating and understanding natural history that gradually developed in the latter part of the eighteenth century. It emerged out of a naturalistic, or mechanistic, way of understanding nature. The forces underlying the diversity of life and the relationships between species and their interactions with the physical environment would be understood in the same way as one understood a machine, a factory, or a city. There would be no unknowable mystical or magical forces. The underlying dynamics of nature could be investigated through observation and experimentation. This monumental change in worldview is made plain when Darwinian theory is compared with traditional supernatural Judeo-Christian beliefs.[1]

Two Worldviews

The biblical six days of creation were thought to have taken place only a few thousand years ago. In the seventeenth century, Archbishop James Ussher had calculated the origin of creation to the year 4004 B.C. The famed naturalist Georges Louis Leclerc, Comte de Buffon (1707–1788), estimated that the earth was about 75,000 years old, and that plants and animals arose about 37,000 years ago. Today, scientists maintain that the universe is about 10–20 billion years old, the earth is about 4.5 billion years old, and life arose about 3.5 billion years ago; hominids resembling our species appeared 4 million years ago, and our species, *Homo sapiens*, appeared about 130,000 years ago. Of course, there are believers in creation by "intelligent design" today who may assume that the earth is only several thousand years old, but even by Darwin's time, geologists and naturalists recognized that the earth was millions of years old. Geologists showed that the earth itself has experienced immense changes, and the fossil record has produced an archive of evidence of long-extinct forms. There was no evidence of the existence of humans until what amounts to a geological moment ago.

Traditional natural theology held the world to be static: God had formed all species just as they appear today; no genealogical relationship between them existed. Certainly, there had been great cataclysms such as the biblical flood, but Noah had saved all the species living today. The great philosophers of ancient Greece shared this view. In the Aristotelian and the Platonic order of things, life-forms were ordered in single file—from the most simple inanimate objects, to plants, to lower and higher animals—as a fixed plan of creation. The increasing perfection in this *scala naturae* or "great chain of being" was understood in terms of different kinds of "soul": more reason and a greater advance toward God.[2] In contrast to the great chain of being, evolutionary theory holds that all of life is related and that genealogical relations between species resemble not a ladder, but a complex branching tree (see, however, chapters 18 and 19).

Advocates of special creation have always insisted that the complexity and harmony of a plant or an animal and its place in nature have been designed by its creator. Natural processes could not have led to the production and reproduction of such complex structures as the eye or the brain, nor of coadaptations as of insects and flowers. All such complex organs, as well as the harmonious mutualistic relations between species, are evidence of the wisdom and benevolence of a Creator. This natural theology emerged in the seventeenth century and was maintained by natural theologists of the eighteenth and nineteenth centuries. That organisms and their natural relations can be explained only by purpose, by design, is known as "teleology."

According to Darwinian evolutionary theory, there is no design in the natural world, no preconceived plan. Organisms evolve in a makeshift or contingent manner in relation to changing ecological conditions. New organs do not suddenly appear that seem to have been specially created for some purpose. The appearance of a species results from numerous forces that combine at a certain epoch in a certain place. Had the conditions been different, the natural world would be different today: nothing is necessary, nothing is purposeful, and nothing therefore is beyond investigation. Evolutionary theory and rationalist explanation do not necessarily preclude the concept of God: some evolutionists may be agnostic or they may invoke God to explain the origins of the natural laws through which life evolves.

Judeo-Christian theology places humans outside and above nature. Accordingly, we were formed in the image of God and given dominion over nature. Darwinian evolutionary theory places humans within nature, as members of the animal "kingdom." Of course, humans have created culture, and it in turn has its own history. Yet, ever since Darwin, there have been biologists who advocate that culture is also, at least in part, biologically determined. The extent to which human social relations are determined by natural evolutionary processes remains a subject of heated debate (see chapter 20).

Scholars have probed various aspects of the genesis of evolutionary theory, especially the development of Darwin's theory of natural selection, its reception by the scientific community, and its impact on society. No one before Darwin had thought out and marshaled evidence for evolution in a manner that compares to that in *The Origin of Species*, published in 1859. But the seeds for his evolutionary theory had been planted earlier in the century, and there were evolutionists before him, including his grandfather Erasmus Darwin and Robert Chambers, editor of *Chamber's Encyclopedia* (see chapter 2). But the most prominent evolutionist before Darwin was Jean-Baptiste Lamarck (1744–1829), who, in 1802, coincidentally with Gottfried Reinhold Treviranus (1776–1837), coined the term "biology" for the study of the manifestations of life and the conditions and laws under which it occurs.[3] Yet, Lamarck's place in the history of biology is both controversial and ironic. Although he is the only other evolutionist of the nineteenth century who has followers today who carry on his name, he is also one of that century's most misunderstood, misrepresented and, in some ways, enigmatic figures.[4]

Revolution to Evolution

Lamarck was born into a poor noble family from Northern France. He studied under the Jesuits, but at seventeen, he joined the troops to fight in the Seven Years' War (1756–63). After the war, when he was nineteen, he went to Paris to study medicine, supporting himself by working in a banker's office. He became interested in meteorology, chemistry, and especially botany, writing prolifically in all three. Lamarck's recognition in the scientific community began in 1779 with the publication of his four-volume treatise *Flore française*. A significant contribution to the botanical naming of species, it was warmly embraced by Buffon, who that year engineered Lamarck's election to the Académie des Sciences. A few years later, Lamarck wrote the first two and a half of the eight volumes of the *Dictionnaire de Botanique*. In 1788, he was appointed botanist at the Jardin du Roi, which had become an important scientific center, headed by Buffon.

The great change in Lamarck's thinking about evolution had occurred during the French Revolution. Before the Revolution, French society was essentially a feudal system, a static hierarchical order, from serfs to landowners and king: everyone born in his or her own station in life, and upward mobility was scarcely possible. That order of things was recast in 1789 with the uprising of the peasants, artisans, and middle classes of France against the privileges of nobility and mismanagement of the country by kings. In 1799, Napoleon Bonaparte seized control and by 1804 had become emperor of France. Although the Revolution had come to an end, the ideas it inspired and the belief that everyone should be free and equal spread to many other countries. This change in social structure and thought certainly did not lead Lamarck to think in terms of a parallel sort of evolutionary change in the natural world. However, as we shall see, republicanism and liberal thinking influenced the reception of his ideas, and of evolution more generally.

The effect of the Revolution on Lamarck's thinking was indirect, mediated by a shift in career from botanist to zoologist. In 1793, the year in which Louis XVI and Marie Antoinette were beheaded and critics of the new government were imprisoned or executed in the Reign of Terror, the Jardin du Roi was reorganized into the Muséum National d'Histoire Naturelle, which became the European center for zoological research. Lamarck was given a new position as professor of the insects, worms, and microscopic animals. His new duties consisted of giving courses and classifying the large collection of these "animals without vertebrae," which he named invertebrates. He excelled at classifying them, and their study led him to explore fundamental questions about the causes of life processes and about evolution—from the most simple forms to the most complex.

Three convergent interests led Lamarck to evolution: his thinking on what constituted the essence of life in the simplest organisms (caloric heat and electricity), his view of the "natural" way to arrange taxa, and his geological thinking (i.e., of gradual change over long periods of time). Unlike his contemporaries, he believed the earth to be very old, almost incalculably so, involving thousands or even millions of centuries. He developed his evolutionary thinking in several books, beginning when he was in his mid-fifties: *System of Invertebrate Animals* (1801); *Studies on the Organization of Living Bodies* (1802); *Zoological Philosophy* (1809), a classic, translated into many languages; and *Natural History of the Invertebrate Animals* (1815).

Lamarckian Myths

Two other general myths about Lamarck still circulate. One is that of a romantic genius, isolated and ignored by his contemporaries, persecuted by state power (Napoleon) and the scientific establishment of his time, but rediscovered in the late nineteenth century. The other is of someone who tried, unsuccessfully, to tackle the problems of adaptation and the origin of species before Darwin. Although these views continue to be perpetuated in textbooks and popular writings, historians have presented a dramatically different image of Lamarck and his times.[5] They emphasize that Lamarck was not a precursor of Darwin. His questions (and answers) differed dramatically.

Lamarck is remembered by biologists today for having supposedly originated a mechanism of evolutionary change that contradicted Darwin's. It was based on the inheritance of acquired characteristics: that the characteristics you (or any organism) acquire in your lifetime may be passed on to subsequent generations—not unlike property and wealth. Some acquired characteristics may be a direct response to some external change in the environment—say, darkening of the skin with exposure to the sun. Others come from use or disuse of a part, or from the activity of the individual generally, such as music appreciation or athletics. The inheritance of such characteristics meant that they tended to be enhanced in the newborn child before it was ever exposed to sunlight or music. Today, biologists generally contend that the inheritance of acquired characteristics in this sense does not occur. But, as we shall see, their attribution of this belief system to Lamarck is fallacious.

Ascribing the belief in evolution by the inheritance of acquired characteristics to Lamarck and contrasting it to Darwin is misleading on three counts. First, the idea did not originate with Lamarck; it can be traced back to Hippocratic writers and was common in folklore and in the writings of philosophers and naturalists of many countries.[6] Before Lamarck, Darwin's grandfather Erasmus Darwin had used it as the basis of his own theory of evolution.[7] Second, although Lamarck is often the source of ridicule and error, present-day neo-Darwinians too often forget that Charles Darwin also maintained a belief in the inheritance of acquired characteristics—as did other evolutionists of his day. Third, the inheritance of acquired characteristics was only one aspect of Lamarck's theory of evolution.

Lamarck thought the environment might bring about heritable changes in several ways. For example, he argued that if seeds of a plant that had grown in a meadow for generations were blown onto a neighboring dry, barren, and stony hill, some plants might be able to survive. Malnourished by the drier soil of the hill, the offspring of these seeds would constitute a new race: "The individuals of this new race will have small and meagre parts; some of their organs will have developed more than others, and will then be of unusual proportions."[8] Animals, on the other hand, might develop new types of behavior due to environmental change, and the resulting effects of use and disuse of parts would be inherited. The giraffe, seeking to forage higher and higher on the leaves of the trees on which it feeds, stretches its neck. "From this habit long maintained in all its races," Lamarck wrote, "it has resulted that the animal's fore-legs have become longer than its hind legs, and that its neck is lengthened to such a degree that the giraffe, without standing up on its hind legs, attains a height of six metres."[9] He suspected that the horns of ruminants emerged as a result of their butting their heads together during combats. Flat-bodied fishes emerged from the habit of turning on one side in shallow water: the eye on the lower side moved toward the upper side as a result of its need to pay attention to anything above.

Lamarck's descriptions of how such behavioral changes could bring about new structures sometimes left him open to crude caricatures by his critics for advocating the ludicrous notion of evolution by desire on the part of animals. Part of the problem, historians argue, was in the less than lucid way he expressed himself. His depiction of how the wading bird got its long legs was often quoted by his detractors:

> one may perceive that the bird of the shore, which does not at all like to swim, and which however needs to draw near to the water to find its prey, will be continually exposed to sinking in the mud. Wishing to avoid immersing its body in the liquid, [it] acquires the habit of stretching and elongating its legs. The result of this for the generations of these birds that continue to live in this manner is that the individuals will find themselves elevated as on stilts, on long naked legs.[10]

It is easy to misinterpret this statement to imply evolution by volition. But what counted for Lamarck was not an animal's desires, but its habits, and the way it responded to its environment. Nonetheless, his ideas would be belittled and corrupted by both opponents of evolution and by some of its champions. Darwin went to lengths to distance his theory from that of Lamarck. As he wrote to his botanist friend Joseph Hooker in 1844, "Heaven forfend me from Lamarck's nonsense of a 'tendency to progression,' adaptations from the slow willing of animals etc.!"[11]

Simple to Complex

The inheritance of acquired characteristics was only one facet of Lamarck's theory of evolution, and not even the most important. To understand the whole we must first consider the way in which he arranged life-forms. He ordered the great classes of life in a linear, graded series toward "perfection," from simple to complex, like the great chain of being.

But he also recognized that there were genera and species that branched off and formed dead ends.[12] He offered two matching mechanisms. The branching or deviations from the progressive linear order would be due to the influence of particular environmental circumstances: the inheritance of acquired characteristics would account for the characters of organisms that distinguished genera and species, as well as their instincts and habits. But the general trend in evolution of ever-increasing complexity, of classes and families, from Infusoria to humans, was due to something else, an unknown inner force in nature which he referred to as "the power of life":

> It is quite clear that both animal and vegetable organisation have, as a result of *the power of life*, worked out their own advancing complexity, beginning from that which was the simplest and going on to that which presents the highest complexity, the greatest number of organs, and the most numerous faculties; it is also quite clear that every special organ and the faculty based on it, once obtained, must continue to exist in all living bodies which come after those which possess it in the natural order.[13]

Lamarck thus turned the great chain of being into an escalator. Organisms lower on the ladder of life were transformed from those higher up. All the great classes and families of animals could be arranged in a single series (plants in another). He adamantly opposed notions about the extinction of species, arguing that organisms resembling strange fossils may still be found somewhere on Earth.

To the modern reader, the "power of life" may have the ring of an unknowable supernatural vital force. However, Lamarck was not a vitalist. Life, he argued, was a phenomenon of organized beings—not an attribute of matter per se, but the way it was organized. Was there a preestablished plan directing change from simple to complex? Although some historians are uncertain about his beliefs, Lamarck himself seems to have been less equivocal. As he commented in his *Philosophie zoologique*, "everything is thus preserved in the established order . . . everywhere and always the will of the Sublime Author of nature and of everything that exists is invariably carried out."[14]

The inheritance of acquired characteristics by use and disuse of parts due to changing environmental circumstances would account for the "numerous anomalies or deviations" from a linear increase in complexity of classes. Those naturalists who believed in the fixity of species maintained that the structure of an animal is always in perfect adaptation to their functions, and that the structure of a part determined its function. But Lamarck reversed that causality: it was new functions and new habits brought about by needs that led to changed structures and the irregularities in the straight-line gradation from simple to complex.[15] In effect, these deviations from the great chain of being offered him evidence of evolution by the inheritance of acquired characteristics:

> It will in fact become clear that the state in which we find any animal, is, on the one hand, the result of the increasing complexity of organisation tending to form a regular gradation; and on the other hand, of the influence of a multitude of very various conditions ever tending to destroy the regularity in the gradation of the increasing complexity of organisation.[16]

Certainly, the idea of evolution was not novel in Lamarck's time. It had been well known in France since the Enlightenment and had been considered by leading zoologists, including George Louis Buffon. Buffon, who became director of the Jardin du Roi in 1739, was largely responsible for the great interest in France in natural history, and no other naturalist before him had done more for its study. The other great naturalist of the eighteenth century, Carl Linnaeus (1707–1778), in Sweden created the binomial system for naming organisms used today (for example, the willow oak, *Quercus phellos*, and the red oak, *Quercus rubra*). He classified enormous numbers of species sent to him from the four corners of the earth. Buffon, on the other hand, was noted for his study of living organisms and their characteristics in life. His forty-four-volume *Histoire naturelle* (1749–1804) dealt with many of the problems subsequently raised by evolutionists.[17] In fact, later in life Buffon adopted a theory of evolution according to which some few original types of animals developed, through hybridization and the direct effects of the environment, into all the species seen today. Linnaeus had held similar views.[18]

The works of Lamarck's predecessors were vital, but he himself was the only one at the time who presented a broad theory of organic change.[19] In the conclusion of his *Philosophie zoologique* he commented on an oft-repeated theme in regard to scientific discovery:

> Men who strive in their works to push back the limits of human knowledge know well that it is not enough to discover and prove a useful truth that was previously unknown, but that it is necessary also to propagate it and get it recognized; now both individual and public reason, when they find themselves exposed to any alteration, usually set up so great an obstacle to it, that it is often harder to secure the recognition of a truth than it is to discover it. I shall not dwell on this subject, because I know that my readers will see its implications sufficiently, if they have any experience in observation of the causes which determine the actions of mankind.[20]

To understand why Lamarck had largely failed to convince his contemporaries, one has to do more than measure the weight of religious opinion or state authority. Historians point to a number of issues, including his scientific style and his social standing among his peers as well as the evidence he had available to him.

By the time he announced his theory of evolution, Lamarck had acquired a reputation as someone who liked to speculate on the merest of facts. He was a synthesizer and a builder of grand speculative theoretical systems. He fancied himself as the universalists' naturalist-philosopher, one who set basic principles for all fields. His evolutionism was one aspect of a broad ambitious program of what he called terrestrial physics, which encompassed the study of the atmosphere, the changes in the surface of the earth, and the organization of living beings. He had written several books on physical and chemical problems, meteorological phenomena, and causes of atmospheric change, and in 1800, he advanced a theory on the origin of life.[21]

Alas, his peers were far from impressed by his ideas, nor by his "retrogressive," speculative approach to what was becoming ever more professionalized science in France, based on a more cultivated quality of careful observation.[22] In the late eighteenth century, new institutional developments provided a considerable number of positions for practicing and teaching science full time. Science was moving in the direction of a more specialized and disciplinary approach, narrowing the scope of problems an individual could competently address. Lamarck was clearly out of step.

Crucial evidence for evolution was lacking in the fossil record.[23] Lamarck could point out similarities between some living forms and some fossils, but the fossil record indicated that whole groups of species appeared suddenly. There was simply no evidence of a succession of transitional forms. All he could do was to point to his classification of the animals and claim that it represented the true order of the formation of living things. But Lamarck's order of things was vehemently attacked by his adversaries, who debunked the whole concept of a straight progression of life from the simple to the complex as depicted in the great chain of being. A linear order of the major taxa from lower to higher simply did not conform to evidence from an effectively new science of comparative anatomy.

Disconnecting the Unity of Life

Disconnecting the great chain of being was the task of Lamarck's greatest and formidable antagonist at the Muséum National d' Histoire Naturelle, Georges Cuvier (1769–1832).[24] Professor of natural history at the Collège de France and appointed to a new professorship of comparative anatomy at the Muséum in 1802, Cuvier built up a rival biological system based on a whole new approach to comparative anatomy. Comparative anatomy had been an adjunct of medicine; Cuvier made it a science, with close ties to natural history. While Lamarck was responsible for the "lower" invertebrates, Cuvier was in charge of "higher" organisms, though he too was an expert on the invertebrates and had made his reputation on studies of their comparative anatomy.

Their worldviews could not have been more different. Lamarck was an evolutionist; Cuvier, a fixist. Lamarck denied species extinction; Cuvier claimed there had been several mass extinctions. While Lamarck gained a reputation as one who speculated using scanty data, Cuvier established a reputation as a patient observer who stuck to the facts.[25] Unlike Lamarck, Cuvier stood especially high in Napoleon's favor. Indeed, they had similar characters. Cuvier was a fierce personality who shared the emperor's despotic disposition.[26] He was also politically astute. He became *inspecteur général* in the department of education at the time when the education system of France was thoroughly reformed and new universities founded. Even after Napoleon fell and the throne of France was restored, Cuvier managed to maintain his authority in science and education.

Cuvier's disdain for Lamarck's theories was matched only by his repulsion to the speculative way they were formulated.[27] The history of nature, for Cuvier, was not gradual and continuous from beginning to the present. There were no traces of a single line of descent in the animal kingdom, and no series of changes by which each species might have been gradually transformed from another, as Lamarck had supposed. In 1812, Cuvier declared that there were four distinct and completely unrelated divisions (*embranchements*) of animals, and moreover there was absolutely no evolutionary connection between them: vertebrates (fish, reptiles, birds, and mammals), mollusks (snails, squids, and octopuses), articulates (annelids and arthropods), and radiates (starfish, jellyfish, anemones, corals, and hydras).[28] The key to this reorganization of nature was in the techniques he used. He characterized these four basic types of the animal kingdom not on the external appearance or on readily visible morphology, but on the basis of an animal's internal anatomy, its inner structural organization, its anatomical plan, which he believed had been designed by the Creator to suit the animal's particular functional needs.

Cuvier established a great new tradition in comparative zoology and taxonomy. Taxonomists had a tendency to treat every characteristic as if it were independent of every other characteristic. However, Cuvier and his followers emphasized that an organism was not a jumble of characteristics: there was a correlation of parts. The various parts of an organism were so interdependent that if given the tooth of an extinct animal such as a dinosaur, the anatomist could make numerous deductions about the probable structure of other parts of the anatomy of that animal.

For Cuvier, the fossilized creatures buried deep in the earth were milestones of geological time, and the grouping of different fossils in specific layers into distinct regions of the earth's crust indicated that revolutions had suddenly over come the earth on several occasions. But he saw no connection between these extinct creatures and those alive today. Instead, he argued that the history of life on Earth was profoundly punctuated by major catastrophes and mass extinctions. The last cataclysm was the biblical flood. Thus, there were breaks in geological time, and complete breaks in life on Earth; the organisms that lived in earliest times had nothing in common with those alive today.

When catastrophes killed off animals in an area, Cuvier proposed, that area would later be invaded by new animals from another place, and some of these would become fossilized following a subsequent mass extinction. In other words, new animals and plants would not evolve; instead, preexisting species from neighboring unaffected areas would migrate to the area of the extinction. This would account for the changes in the fossil record in different regions and in different strata.

This interpretation had an obvious flaw: had there been a long succession of catastrophes, the number of species would have declined rather increased, as they apparently had. Therefore, some of Cuvier's followers adopted an alternative solution. To account for an actual increase in new species, they simply suggested that there was not one divine creation, but many—after each geological catastrophe.

Some historians have interpreted Cuvier's views about the fixity of species as the expression of a retrograde, conservative, and theological attitude that retarded the advent of evolutionary thinking.[29] Though well informed, industrious, clear, and methodical in his thinking, he was no intellectual revolutionary.[30] Yet others, including the historian and philosopher Michel Foucault and the Nobel Prize–winning biologist François Jacob have argued that, by undoing the great chain of being, Cuvier's system provided some of the crucial conceptual conditions for Darwinian evolutionary theory.[31] The contingent or makeshift nature of evolutionary change characteristic of Darwin's model, they observed, could not be entertained as long as species were set in a rigid framework, progressing single-file toward perfection. They suggested that because Lamarck's theory was based on a linear order of life from simple to complex, it really belongs to the previous natural order of things, whereas Cuvier's system of classification based on the internal organization of animals looked to the future.

Other scholars have noted, however, that by 1815 Lamarck had recognized that the single series of increasing complexity from "infusoria [microbes] to man" on which he had first constructed his evolutionary theory did not reflect nature. In fact, his final presentation of animal relationships did not differ in principle from the phylogenetic tree one would find in the literature at the end of the nineteenth century.[32] Lamarck's health began to fail in 1809, when he developed eye problems, and in 1818 he became completely blind. Indeed, his life was marked by tragedy and poverty. When he died in 1829, his family did not have enough money for his funeral and had to appeal to the Académie des Sciences for funds; his books, and scientific collections were sold at public auction. By that time, his evolutionary writings had been taken up by others in France and Britain.

The Cuvier-Geoffroy Debate

Scientists and scholars like to use the word "revolution" to describe great changes in science, but Lamarck's evolutionism was revolutionary in the original sociopolitical sense of the word as well. Evolutionary thinking was positioned in direct opposition to the privileges of nobility and the church and against the conservative and new professional social control of science in France. Toby Appel unearthed this aspect of Lamarckism when she reexamined one of the great debates of the early nineteenth century over comparative anatomy between Cuvier and another champion of evolutionary theory at the Muséum d'Histoire Naturelle: Etienne Geoffroy Saint-Hilaire (1772–1844), professor of zoology.[33] The antagonism between Geoffroy and Cuvier persisted through out the 1820s and finally ruptured into a famous public confrontation over a period of two months in 1830 at meetings of the Académie des Sciences, the supreme arbiter of science in France. These were weekly events attended by all the leading scientists of Paris.

At the heart of the Cuvier-Geoffroy debate were two opposing approaches to comparative anatomy. Cuvier adopted a functionalist approach: he viewed every part of an animal as having been designed by the Creator to contribute to the animal's functional integrity. Thus, for him, function—the animal's needs—sufficed to determine its structures.

Geoffroy and his followers challenged this functionalist understanding of organismic organization with one based on the primacy of structure over function. His new set of doctrines, called transcendental anatomy or philosophical anatomy, centered on the concept that all animals had a fundamental structural plan that preceded all other particular modifications to suit functional requirements. Transcendental anatomists insisted that animals had a constancy in the number and arrangement of parts that was independent of the form of the parts and the uses to which they were put, a unity of plan that transcended function. The way to find the plan was to ignore differences and search for resemblances in the relative positions of the parts

Cuvier and his followers had broken the unit of life with the four *embranchements* (the vertebrate, articulate, mollusk, and radiate plans); the transcendental anatomists, led by Geoffroy in the 1820s, sought to reestablish the unity of life. They argued that all animals, vertebrates and invertebrates alike, were built on the same basic plan. The structure of the invertebrates could be seen in stages in the embryonic development of vertebrates. Therefore, animal life could be strung into a more continuous, related series, rather than broken into discrete "divisions," as Cuvier had claimed. This series implied that the history of each organism, rising in complexity from starfish to humans, could be interpreted in an evolutionary manner.

Geoffroy was a follower of Lamarck and referred to his friend as a "genius." At the same time, however, he gave his own interpretations to Lamarck's evolutionary writings. As Geoffroy saw it, the effects of the environment and use and disuse altered some structures, but the basic plan of animals was always conserved. All extant vertebrates and all fossils of vertebrates exhibited a unity of plan and composition: "All of them, being composed of similar organs, are merely modifications of the same being, which we call the 'vertebrate animal.'"[34] Geoffroy's structuralism, in fact, became the basis, in the nineteenth century, for determining homological relationships (a bird's wing, a seal's flipper, and a monkey's arm, for example).

Historians judge both Cuvier and Geoffroy to have fared well in the debates. Succeeding generations hammered out reconciliations between their extreme views. In keeping with the structuralists' approach, comparative anatomists came to see that there was a unity of plan within each *embranchement*.[35] For example, the fundamental characteristic of radiata was radial versus bilateral symmetry. A jellyfish could be cut in half along many different lines: it was radially symmetrical. But humans and all other bilateral forms could be cut along just one line to yield halves that are roughly mirror images. Echoes of the Cuvier-Geoffroy dispute continue today in debates over structuralism and adaptationism. And the history of Geoffroy's structuralist approach has been reconsidered by both philosophers and developmental evolutionists.[36] Geoffroy's structuralism and the search for the unity of life was also instrumental in the development of Darwin's own theory of evolution.[37]

Like most great controversies, the Geoffroy-Cuvier was multifaceted, involving issues far removed from the anatomical approaches per se.[38] Geoffroy enlarged the debate beyond the purely anatomical basis to include the broader issue of evolution. Evolutionary thinking was tied to politics, and Geoffroy found support from various groups outside the scientific community: literary figures and scientific popularizers who saw science in France as increasingly professionalized, centralized, and confined to government institutions, staffed by scientists who no longer seemed to pose the interesting questions. They saw Geoffroy as a philosopher dedicated to unraveling the mysteries of nature for the common man, a scientist for the people. And they saw Cuvier as a political elitist, a fact collector, an upholder of biblical orthodoxy, a manipulator of patronage and suppresser of the great ideas of Lamarck and Geoffroy.

Appel has argued that the whole controversy of the late 1820s was entwined in political and religious tensions of the period.[39] After the removal of Napoleon by the allied forces, the Bourbon family was restored to the throne of France, much against the will of the French people. At first the royal family made some slight concessions to the spirit of democracy the French Revolution had aroused, but gradually, as the people continued to advance their claims for liberty, the royal family and court became more and more reactionary and opposed to democratic reform and freedom of the press. The result was the Paris Revolution of July 1830, after which the old aristocracy was reduced to a more minor role in national politics, though its members retained major influence in the church, military, and foreign service.[40] Geoffroy's view of nature, his evolutionism and transcendental anatomy, was associated with the liberal ideology of the times, and he carried a bloc of republican sympathizers with him, as well as a younger generation of comparative anatomists.

Adrian Desmond has described a similar pattern in England during the 1830s. There were very few evolutionists at Cambridge or Oxford, but one can find them among the working class, and the radical artisans, and the cosmopolitan medical schools.[41] Evolution imported from France had disturbing social and political associations for the privileged classes inside and outside the universities in England.[42] At the time of the Paris Revolution, the Anglican elite in Britain were staving off concerted attacks by radicals trying to secularize and democratize their society. The Parisian radicals included Lamarck in their rhetorical armory, and the British gentry remained suspicious of the republican masses across the channel, portraying them as the "national enemy." "And if France's periodic convulsions were fueled by poisonous, naturalistic, evolutionary philosophies," Desmond remarks, "then the conservatives were determined to keep them off English soil."[43]

The "scientific gentry" undertook a massive campaign to discredit evolutionary theory, but evolutionists could be found in the secular anatomy schools and radical nonconformist colleges. British advocates of Geoffroy's transcendental anatomy included a number of Edinburgh-educated anatomists who with political radicals opposed the Oxbridge aristocratic domination of the medical profession. Inside the medical schools, discussions and disagreements broke out over the relationship between anatomy and evolution. The mingling of reform and evolution was intimate, as Lamarck's ideas were mixed with demands by radical artisans for democracy and attacks on aristocracy and the clergy. Desmond suggests that one of the earliest uses of the term "evolved" (in 1826) to signify the transmutation of one species into another may have been by the Scottish comparative anatomist Robert Grant, a Lamarckian, who attacked corruption in medicine and society.[44]

It was during these times that Charles Darwin secretly wrote his theory of evolution by natural selection, which he did not publish until twenty years later, when mass unrest was past. Even then, he had to be pushed into it.[45]

2

The Origin

> Thus from the war of nature, from famine and death, the most exalted objects which we are capable of conceiving, namely, the production of higher animals, directly follows. There is grandeur in this view of life, with its several powers, having been originally breathed into a few forms or into one; and that, whilst this planet has gone cycling on according to the fixed law of gravity, from so simple a beginning endless forms most beautiful and most wonderful have been, and are being evolved.
>
> —Charles Darwin, 1859

NO ONE HAD DEVELOPED the idea of evolution into so consistent and well documented a theory as did Darwin in *The Origin of Species*. He did it by postulating that evolutionary change occurred by a struggle for existence, giving rise to a natural selection of the most fit. His theory was a triumph of synthesis. In contrast to today's science, in which one small aspect of nature is investigated by many individuals, Darwin investigated many aspects of nature. He wrote about coral reefs and coral islands in the South Pacific and theorized about how atolls are formed on the top of extinct volcanoes.[1] He wrote a four-volume treatise on barnacles that marked the beginning of the science of cirripedology.[2] He developed a theory of inheritance based on small genelike particles he called gemmules, and he carried out cross-breeding experiments.[3] He wrote about the pollination of orchids by insects,[4] the evolution of humans,[5] and the expression of the emotions.[6] No other individual in the history of biology has been given more attention by scholars: how he became an evolutionist, how he came to write *The Origin*, and the central tenets of his theory.

Paleontological and geological studies were crucial. But the new concepts Darwin developed relied on two other approaches: study of the geographical distribution of species; and study of the ecological processes involved in the formation of species. To do this, a different kind of naturalist was required. Unlike Lamarck and Cuvier, who worked in museums and zoological gardens, Darwin was a traveler, and so too was Alfred Russel Wallace (1823–1913), who developed the theory of natural selection independently of Darwin.[7] Naturalists such as Darwin and Wallace went from island to island and from continent to continent to study living organisms in their natural settings, and compare their habitats, form, and behavior. They collected an enormous number of data on relationships between species, their geographical distribution, and how they varied slightly from one geographical region to another.

Certainly, the theory of evolution was well known and widespread in England before Darwin's *The Origin*. Lamarck and Geoffroy's ideas were known, as were the Lamarckian views of Robert Grant (1793–1874).[8] A few other intellectuals had also spoken out in favor of transmutation of species. Darwin's grandfather Erasmus Darwin, a physician and naturalist, was one of the earliest and best known of the transmutationists. He had devoted a chapter of his famous his two-volume work *Zoonomia*, (1794–1796) to the idea.[9] Although his grandfather died before Charles was born, the book was a much-discussed part of family lore. But perhaps the most important pre-Darwinian evolutionary book in England was *Vestiges of the Natural History of Creation*, a best-seller first published in 1844 that went through ten editions in as many years. It was published anonymously to protect its author from scandal.[10] In 1884 the author was revealed to be Robert Chambers, the well-known Edinburgh publisher and editor of *Chambers' Encyclopaedia*.[11]

For Chambers, evolution, or development, as he called it, was a matter in which new species and the ascent of life were planned linear developments, controlled by natural law as preordained by God. His book was an immediate success among lay middle-class readers, but his arguments for evolution were so at a variance with conservative religious and social views that all the Oxbridge professional elite spoke out against it, including Charles Lyell, William Whewell, Adam Sedgwick, and other outraged critics. Chambers helped accustom readers to think about evolution. As Darwin later wrote to a colleague, "The publication of *Vestiges* has done excellent service in calling in this country attention to the subject and removing prejudices."[12] Another reason *Vestiges* would be of great value to Darwin was that the book's critics supplied him with a list of objections, which he took care to address in *The Origin*.

When Making Other Plans

Darwin (1809–1882) was born in the small medieval town of Shrewsbury, son of a very wealthy physician, Robert. His mother, Susannah, a daughter of Josiah Wedgwood, the well-known potter in Staffordshire, died in 1817 when Charles was only eight years old; he was raised by his elder sisters, Susan and Caroline.[13] He studied at Shrewsbury School, and when he was sixteen his father sent him to Edinburgh to study medicine. Medicine appalled him. Disgusted by the sight of operations performed without anaesthetic, he left Edinburgh before completing his medical degree.

In 1828, his father sent him to Christ's College, Cambridge, to take the common arts degree, with the idea that he should become a clergyman. Even in his youth Charles had expressed great interest in natural history, and at Edinburgh he learned of Lamarck's theory of evolution under the tutelage of Grant, who directed him in a study of marine invertebrates and who, as Darwin later recalled, "burst forth in high admiration of Lamarck and his views on evolution."[14] At Cambridge Darwin associated with several university faculty, including John Stephens Henslow, professor of botany; Adam Segwick, professor of geology; and William Whewell, professor of mineralogy. Darwin took his degree in 1831, tenth in the list of those who did not seek honors. He did not pursue a religious vocation. Instead, his career took an unforeseen turn.

That year, at Henslow's recommendation, Darwin obtained a position aboard the *H.M.S. Beagle*, which was about to set off on a five-year voyage surveying South American waters and circumnavigating the globe. The voyage of the *Beagle* was the real preparation for his life's work.[15] At first his job was simply to be a companion of Captain Robert Fitzroy with whom he debated questions of theology and politics. But after a few months both his zeal and his means for collecting enormous numbers of specimens were evident. Having come on board with a servant and a personal fortune, he was able to disembark at every port of call, hire a few willing natives, and go hunting for specimens. He soon took over the position of the ship's official naturalist. In regard to geological knowledge and the relationships between extant and extinct species, nothing was more important for him than Charles Lyell's *Principles of Geology* (1830–1831). Darwin dedicated his journal from the *Beagle* voyage to Lyell (1795–1875).[16]

Darwin's Bible

Lyell's contribution to Darwin's thinking was significant in two ways. First was his radical view that geological changes occurred gradually over a vast amount of time.[17] The subtitle to his *Principles of Geology* said it all: *Being an Attempt to explain the Former Changes of the Earth's Surface by Reference to Causes Now in Operation.* He saw no need for mysterious cataclysms such as Biblical floods. Opposing the catastrophism of Cuvier and others before him, he maintained that all the ancient changes produced in the earth's surface were brought about by factors similar in nature and intensity to those that operate today (uniformitarianism).[18] Certainly, violent events had shaped the earth's surface; whole mountain ranges had surged forth and then been submerged; valleys had opened up, been filled in, then opened up again; the seas had invaded the land, then withdrawn. But these changes were not the result of the supernatural catastrophes of the Bible's Book of Genesis.

Following the famed geologist James Hutton (1726–1897), Lyell postulated that the earth was billions of years old and that it had been sculptured by the action of the sea, rain, volcanoes, and earthquakes. Lyell's uniformitarian geological theory was vital to Darwin's evolutionism: it allowed enough time for evolution to occur, and it allowed evolution's mechanisms to be known by investigation. For if, by extrapolation, the living world was formed by causes still at work today, it should be possible to study those evolutionary processes in action. Uniformitarianism did not necessarily imply evolutionism, however. Lyell's own viewpoint is better understood in the context of other ways of interpreting the fossil record.[19]

During the first three decades of the nineteenth century, it became apparent that fossils followed a certain progressive order, from the simplest to the most complex forms. Some naturalists saw in this record actual confirmation of the biblical creation sequence from plants and lower animals to higher animals and finally to man. This linear order was modified a bit with the further observations that plants, invertebrates, and vertebrates all start very close together in some of the oldest known fossil deposits. Although invertebrates did precede vertebrates, progressive development was a matter not so much of a single-file temporal order of creation, a great chain of being, but of concurrent trends, with major subdivisions of organic life existing simultaneously. One then spoke of fossil progress separately for plants and for each of Cuvier's four groups: vertebrates, mollusks, articulates, and radiates. Nonetheless, the realization that, among vertebrates, mammals appeared late, and that human fossils were absent, strengthened the idea of a general progressive development.

The sequence of the fossils from the simple to complex could be interpreted in radically different ways. For some it indicated a real succession of increased morphological complexity produced by a transmutation of species, as Lamarck, Geoffroy, and Grant advocated. For others it suggested not a temporal or historic sequence but an ideal succession—progressive perfection driven by a spirit imposing its will and purpose on the overall structural plan or unity of nature.[20] This concept was the basis of *Naturphilosophie*, developed in Germany by Friedrich Schelling, Lorenz Oken, and others heavily influenced by the idealist philosophy of Plato, Immanuel Kant, and Friedrich Hegel.[21]

Other naturalists associated the progressive development they saw in the fossil record with geothermal theory. Accordingly, the earth had originated as an incandescent blob, and its subsequent history had involved a process of gradual cooling accompanied by other physical changes of climate, atmospheric conditions, and the distribution of the land and seas. The flora and fauna of each successive geological period had been designed by God to fit the physical conditions of the time. This view was developed by Cuvier's school in France and readily accepted by the geologists Adam Sedgwick and William Whewell at Cambridge in the 1830s and 1840s, and by Louis Agassiz at Harvard.

Lyell denied the reality of progressive development and explained away evidence in support of it as an artifact of sampling and the result of incomplete data.

He also opposed species transmutation, and in his *Principles of Geology* he severely criticized Lamarck's theory. Only later, when Darwin convinced him, did he accept evolution (with the provision that it did not apply to man). Nonetheless, his *Principles of Geology* played an important part in setting the framework for Darwin's theory not only by accustoming readers to the vast changes brought about by natural processes over unthinkable periods of time, but also by focusing attention on the question of the transformation of species. Lyell did not believe in species transmutation, but he did believe in their extinction. He adopted a steady-state theory to offset the loss of species, according to which the Creator replaced extinct species with new ones at a constant rate.

Thus, Lyell supposed that new species were introduced to replace the extinct, but he failed to provide any natural mechanism by which this might occur. Therein lay the problem that Darwin aimed to resolve. The question of transmutation was redirected away from progression and complexity to the origin and extinction of species.[22] *Principles of Geology* was Darwin's bible. It posed the right questions, but Darwin offered different solutions. As one reader of *The Origin* wrote to Darwin a month after its publication: "How could Sir C. Lyell, for instance, for thirty years, read, write, and think, on the subject of species and their succession, and yet constantly look down the wrong road!"[23] When Darwin first read Lyell's book, he still believed in special creation; he was on the *Beagle*, circumnavigating the globe, collecting data of his own about geology and the distribution of species extant and fossilized.

The *Beagle* Voyage

Darwin was naturalist for the *H.M.S. Beagle* from December 1831 to October 1836. After visiting the Cape de Verde and other Atlantic islands, the expedition surveyed on the South American coasts and adjacent islands, including the Galapágos, afterward visiting Tahiti, New Zealand, Australia, Tasmania, Keeling Island, Maldives, Mauritius, St. Helena, Ascension, and Brazil again. The ship went to de Verdes and Azores on the way home. Darwin's shift from believing in the fixity of species to evolutionary thinking occurred eight months after returning from *The Beagle* expedition. It was stimulated by three major observations from his voyage. Each of them, he argued, could be explained in no other way than by species transmutation.

One observation concerned the relationships between living animals and fossils of recently extinct animals in the same general location. Darwin postulated that the fossils from South America were related to the living organisms of South America, not to fossils of some past epoch and from another continent, as Cuvier and Lyell had understood them. For example, in rich fossil deposits in southern South America, he saw fossils of extinct armadillos that had characteristics in common with armadillos still living in the same area and yet different from those elsewhere. Why would there be living and fossil species in the same place, unless one had given rise to the other?

Darwin's second observation was that species manifested subtle differences as they migrated from one place to another. He had been much impressed by the manner in which closely associated animals replaced each other in proceeding southward in South America. He reasoned that animals of different climatic zones in South America were related to each other, rather than to animals of the same climatic zone on different continents.

His third observation was that the animals and plants of the Galapágos Islands resembled those of the nearby coast of South America. The species of birds in the Galapágos Islands existed nowhere else on Earth, yet there was a remarkable similarity between them and birds on the nearest continent. The differences of those Galapágos birds stood out against a background of similarities, as if these various species of birds were derived from a common ancestor and their species-specific characteristics were simply the result of their isolation in their geographical territories. As Darwin wrote:

> Here almost every product of the land and water bears the unmistakable stamps of the American continent. There are twenty-six land birds, and twenty-five of those are ranked by Mr. Gould as distinct species, supposed to have been created here; yet the close affinity of most of these birds to American species in every character, in their habits, gestures and tones of voice, was manifest So it is with other animals, and with nearly all the plants, as shown by Dr. Hooker in his admirable memoir on the Flora of this archipelago. . . . It is obvious that the Galapagos Islands would be likely to receive colonists, whether by occasional means of transport or by formerly continuous land, from America; and the Cape de Verde Islands from Africa; and that such colonists would be liable to modification; the principle of inheritance still betraying their original birthplace.[24]

When comparing the birds from separate islands of the Galapágos archipelago with one another and with those on the mainland, Darwin also noticed that species simply lacked the uniformity and clear-cut differences insisted on by those who believed in the fixity of species. All this shed a new light by the summer of 1837, as he recalled in his autobiography:

> During the voyage of the *Beagle* I had been deeply impressed by discovering in the Pampean formation great fossil animals covered with armour like that on the existing armadillos; secondly, by the manner in which closely allied animals replace one another in proceeding southward over the continent; and thirdly, by the South American character of most of the production of the Galapagos archipelago, and more especially by the manner in which they differ slightly on each island of the group; none of these islands appearing to be very ancient in a geological sense. It was evident that such facts as these, as well as many others, could be explained on the supposition that species gradually become modified; and the subject has haunted me.[25]

One entry in his journal is legendary:

> In July [1837] opened first note-book on Transmutation of Species. Had been greatly struck from about month of previous March on the character of South American

> fossils—and species on the Galapagos Archipelago. These facts (especially latter) origin of all my views.[26]

Natural Selection and Natural Theology

Darwin was reluctant to publish his arguments about evolution. He waited to do so for more than two decades, during which time he developed key principles to explain it.[27] From 1838 to 1841, he was secretary of the Geological Society, and he saw a great deal of Lyell. In January 1839, he married his first cousin, Emma Wedgwood (1808–1896). They lived in London until 1842, when they bought a country home in Down, about sixteen miles south of London, where Darwin resided for forty years. From 1846 to 1854, he was mainly engaged in four manuscripts on the recent and fossil barnacles. But from the time he opened his notebook on the transmutation of species in July 1837 he was compiling data and arguments from plant and animal breeders, who created new forms of the species they domesticated. Darwin read enormous amounts of literature and combed through abstracts of many natural history journals. He was also aided by his own observations at Down, where conservatories, fowl, pigeons, and gardens, the things of a country gentleman's life, all added grist to his evolutionist mill.

After collecting facts on the formation of breeds of domestic animals and plants, he saw "that selection was the keystone of man's success in making useful races of animals and plants. But how selection could be applied to organisms living in a state of nature remained for sometime a mystery to me."[28] He understood that breeders could select for desired traits and transform their stocks, but at first he failed to see how such a mechanism might operate in nature. Recognizing that more individuals would always be produced than could possibly survive, he found the conditions under which selection could operate naturally: "the struggle for existence." In other words, under conditions in which there is a struggle for existence, favorable variations would tend to be preserved and unfavorable ones destroyed. The result of this would be the formation of new species. This, in brief, was his theory of natural selection.

Darwin was certainly not the first to recognize "the struggle for existence" in nature. Lyell, whose *Principles of Geology* he had read so closely, had written about "the struggle for existence" and had noted, for example, that "the most fertile variety, would always in the end, prevail over the most sterile."[29] Lyell also recognized that there was a great deal of variation in the world and that these changes were transmitted to successive generations. But he did not believe that such deviations from the species type were endless. Species were fixed; they had limits of plasticity that they could not exceed.

In the 1830s, the British naturalist Edward Blyth also emphasized that the struggle for existence operated in nature and that it weeded out unadapted individuals.[30] However, the way in which Blyth understood the effects of the struggle for existence also differed dramatically from the way Darwin understood them.

For Blyth, all species were perfectly adapted by God to suit certain environments; each had a preestablished place in the economy of nature. The struggle for existence would weed out the sickly and the ill adapted; it was a conservative principle that actually maintained, not changed, the fixity of species. The slightest deviation in the coat color of an adapted cryptic species would lead to its discovery and destruction by predators. Blyth's arguments fit squarely within the natural theology of those who believed that the wonderful adaptations they saw in nature were testimony of God's divine providence.[31]

Natural theology had been expressed in the works of Linnaeus and his school, which were motivated by the search for an overriding purpose and agency in nature. "By the economy of nature," Linnaeus wrote in 1749, "we understand the all-wise disposition of the Creator in relation to natural things, by which they are fitted to produce general ends, and reciprocal uses."[32] Other leading works of natural theology included John Ray's *The Wisdom of God Manifested in the Works of Creation* (1691), William Derham's *Physico-Theology* (1713), and William Paley's *Natural Theology* (1802). In addition, there were the famous eight texts that comprised the *Bridgewater Treatises* (1833–36). These originated through the patronage of Reverend Francis Henry Egerton, eighth and last Earl of Bridgewater, who, upon his death in 1829, provided funds for selected individuals to write, print, and publish one thousand copies of a work titled "On the Power, Wisdom and Goodness of God . . . as manifested in the variety and formation of God's creatures, in the animal, vegetable and mineral kingdoms."[33]

Science and religion were very common subjects for many Victorian naturalists. Darwin himself was as well versed in the Bible as he was in natural theology, and there is no question that his own theory of natural selection had evolved, at least in part, from such arguments. In fact, the religious works of natural theology about adaptations were precisely the ones that he would address.[34] He problematized the adaptations of natural theologians, treating them not as states of nature, perfect products of divine wisdom, but as processes to be explained naturally.[35]

All those who wrote about the struggle for existence before Darwin did so in the conceptual world view of natural theology.[36] In Blyth's model, each species was bound to a certain geographical area; each had been assigned a place in the economy of nature by the Creator and kept there as the struggle for existence weeded out deviants from the norm. Because each species was exquisitely adapted to a particular mode of existence, when circumstances in a locality changed, Blyth thought, a species must perish with it. Darwin turned the effect of the struggle for existence on its head: natural selection would be a creative force in the production of new species, in effect selecting for variations favorable to an individual's reproduction in new environments. Crucial to this break away from Providence and special creation was his experience in new lands—his biogeographical observations. Species boundaries were less rigid and sharp than previously thought.[37]

Scholars have combed through Darwin's detailed notebooks to understand the thought processes that led to him to the theory of natural selection.[38] All agree that the catalyst was his reading, in September 1838, of Thomas Malthus's *Essay on the Principle of Population*.[39] Malthus argued that there would always be poverty, hunger, and war in the world because there was an inescapable imbalance between nature's supply of food and the human need for food and sex. More will be said about this book and its historical context in chapter 4. It is enough to know here that Malthus posited that, if there were no constraints on growth, human populations would increase geometrically or exponentially (i.e., 2, 4, 8, 16, 32, 64) while food supply could only increase at most arithmetically (i.e., 1, 2, 3, 4, 5, 6, 7). The idea of a struggle for existence in nature was common, but Malthus's essay indicated to Darwin the intensity of the struggle—that there would be a constant selection pressure on organisms engendered by an incessant war.[40]

Wallace's Manuscript

In June 1842, Darwin wrote out a sketch, which, in 1844, he expanded to an essay of 231 pages. As he wrote to his wife, Emma, in July 1844, "I have just finished my sketch of my species theory. If, as I believe, my theory in time be accepted even by one competent judge, it will be a considerable step in science. I therefore write this in case of my sudden death."[41] He had the manuscript copied and sent to his close friends, including Lyell and the botanist Joseph Hooker. On September 5, 1857, he explained his theories to the American botanist Asa Gray in a now famous letter. He had completed about half of another expanded treatise of ten chapters when, in June 1858, he received a manuscript from Alfred Russel Wallace, who was in the Moluccas Islands (part of Malay Archipelago, north of Timor). The manuscript was titled, "On the tendency of Varieties to Depart Indefinitely from the Original Type." Wallace requested that Darwin, if he found the arguments to be novel and interesting, to forward the manuscript to Lyell. In it, Wallace outlined the mechanisms thorough which transmutation of species might occur. He argued that the variations within species in the wild occurred by the same laws that produced domestic varieties. "The life of wild animals is a struggle for existence," he wrote.

> The full exertion of all their faculties and all their energies is required to preserve their own existence and provide for that of their infant offspring. The possibility of procuring food during the least favorable seasons, and of escaping the attacks of their most dangerous enemies, are the primary conditions which determine the existence of both of individuals and of entire species. These conditions will also determine the population of a species.[42]

Here, then, was the very same theory for the transmutation of species that Darwin had developed over the previous two decades.

Wallace knew of Darwin through his book *The Voyage of the Beagle*, and they had corresponded about some curious varieties that had interested them both. In

1855, Wallace published a paper stating that species came into existence coincident in time and space with preexisting closely allied species, but he had no idea Darwin was working on a theory of evolution based on similar assumptions.[43] When, three years later, Darwin received a manuscript with an argument that looked like his own, he sent it to Lyell with a letter:

> Your words have come true with a vengeance—that I should be forestalled. You said this, when I explained to you here very briefly my views of "Natural Selection" depending on the struggle for existence. I never saw a more striking coincidence; if Wallace had my MS. sketch written in 1842, he could not have made a better short abstract! Even his terms now stand as heads of my chapters . . . so all my originality, whatever it may amount to, will be smashed.[44]

Lyell and Hooker decided to stake a priority claim on Darwin's behalf. They sent Wallace's essay to the Linnaean society, together with an abstract of Darwin's essay of 1844, and the letter he had written to Asa Gray so as to confirm his priority. The joint essay, "On the tendency of Species to Form Varieties; and on the Perpetuation of Varieties and Species by Means of Natural Selection," was accompanied by an explanatory letter to the secretary.[45]

Unlike Darwin, who was a man of independent means, Wallace came from a poor family.[46] He was a professional collector who supported himself by selling specimens he had gathered in remote parts of the world. On his first expedition, from 1848 to 1852, he set out to explore the Amazon and Rio Negro Rivers, and he observed the same things that Darwin did about geographical distribution of species and ecology. Like Darwin, Wallace had taken Lyell's *Principles of Geology* with him and soon found himself questioning Lyell's assumptions about the fixity of species. On his way home, though, disaster struck on the high seas. His ship caught fire and had to be abandoned. Luckily the crew and passengers were rescued by a passing vessel, but Wallace lost his entire collection and most of his notes. After writing an account of his time in Brazil—his travels on the Amazon and Rio Negro—Wallace set sail in 1854 on a natural history collecting expedition in the far east, and traveled to the islands of the Malay archipelago (modern Indonesia). There he conceived of the idea of natural selection, as he recollected decades later; he was lying in bed "in the hot fit of intermittent fever, when the idea suddenly came to me. I thought it almost all out before the fit was over, and the moment I got up began to write it down, and I believe finished the first draft the next day."[47]

Wallace returned to England in 1862 an established natural scientist, geographer, and collector of more than 125,000 animal specimens. He married Annie Mitten (1848–1914), with whom he raised three children. Although he applied for several jobs, he was never to hold a permanent position. He lost the profits from his collections through bad investments and other financial misfortunes, and his income was limited to earnings from his writings, from grading school exams (which he did for some twenty-five years), and from a small inheritance from a relative. In 1881, thanks largely to the efforts of Darwin and Thomas Henry Huxley, he was added to the government Civil List and was granted an annual pension of 100 pounds for his services to science.

Wallace was a prolific writer. He published twenty-one books and more than seven hundred articles, essays, and letters in periodicals. His two-volume *Geographical Distribution of Animals* (1876) and *Island Life* (1880) became standard authorities in zoogeography and island biogeography. In them, he synthesized knowledge about the distribution and dispersal of living and extinct animals in an evolutionary conceptual framework. He received many honors and awards, including the Royal Society of London's Royal Medal (1868) and the Darwin Medal (1890) for his independent origination of the origin of species by natural selection.[48]

To regard Wallace as the unsung hero in the history of evolutionary biology, overshadowed by Darwin, would be simplistic. Darwin had developed the idea of natural selection many years before Wallace and had collected an enormous amount of data that converted many antievolutionists to evolution. There was nothing in Wallace's sketch that was not written out much fuller in Darwin's manuscript of 1844. Discovery entails more than just a good idea. As Darwin wrote to Lyell after commenting that all his "originality will be smashed": "though my book, if it will ever have any value, will not be deteriorated; as all the labour consists in the application of the theory."[49] Wallace expressed a similar opinion when he later remarked, "I can truly say *now*, as I said many years ago, that I am glad it was so; for I have not the love of *work, experiment* and *detail* that was so preeminent in Darwin, and without which anything I could have written would never have convinced the world."[50]

Concepts of *The Origin*

At the advice of Lyell and Hooker, on November 24, 1859, Darwin published *On the Origin of Species by means of Natural Selection, or the Preservation of Favoured Races in the Struggle for Life*. It was meant to be an abstract of a longer treatise that he had been working on for two decades. The book was literally an overnight sensation: all 1250 copies sold out to booksellers the next day. It went through six editions in Darwin's lifetime and has been analyzed and scrutinized by admirers and detractors ever since. A rich vein to mine indeed: one long argument with multiple complementary theories and concepts, some of which remain issues of debate among evolutionists to the present day.

Common Descent

Naturalists ever since Linnaeus, who had arranged species, genera and families, believed their "natural system revealed "the plan of the Creator"; but Darwin argued that "all true classification is genealogical; . . . community of descent is the hidden bond which naturalists have been unconsciously seeking, and not some unknown plan of creation, or the enunciation of general propositions, and the mere putting together and separating objects more or less alike."[51] All the innumerable species, genera, and families of organic beings descended, each within its own class or group, from common parents, and have all been modified in the course of descent. He inferred further from analogy that "all plants and animals have descended from one common prototype" and that "probably all organic beings which have ever lived on this earth have descended from some one primordial form, into which life was first breathed."[52]

Divergence

Darwin argued that the emergence of a new variety in a species enables it to better exploit the resources of its environment, owing to specialization, or a kind of "division of labor." More will be said about this concept in chapter 8. It was an ecological concept: a locality can support more life if it is occupied by diverse organisms partitioning the resources than if it is occupied by similar organisms. Divergence into specialized niches would be of adaptive advantage because organisms avoid competition that way. Ever-increasing specialization of niche within larger niche would create the hierarchical order of taxa within taxa.[53] Thus, in the course of evolutionary time, a small number of similar organisms could produce a large number of descendants which have diverged from the original type. In Darwin's words:

> This tendency in the large groups to go on increasing in size and diverging in character, together with the almost inevitable contingency of much extinction, explains the arrangement of all the forms of life, in groups subordinate to groups, all within a few great classes, which we now see everywhere around us, and which has prevailed throughout all time. This grand fact of the grouping of all organic beings seems to me utterly inexplicable on the theory of creation.[54]

Divergence specialization into new niches would thus generate a branching genealogy, not a linear, escalating order of living things. All organisms could be arranged on the branches of the same genealogical tree: species, genus, family, order, class, phylum, and kingdom:

> As buds give rise by growth to fresh buds, and these, if vigorous, branch out and overtop on all sides many a feebler branch, so by generation I believe it has been with the great Tree of Life, which fills with its dead and broken branches the crust of the earth, and covers the surface with its ever branching and beautiful ramifications.[55]

Gradualism

For Darwin, evolution was a matter of continuous gradual growth. The clear-cut divisions that naturalists observed among living taxa were merely illusions resulting from the extinction of intermediate forms. Birds were profoundly separated from other vertebrates only because a large number of species connecting their ancestors to those of other vertebrates were extinct. One could see species linking fish to amphibians because far fewer of those in-between species have disappeared, and the division is therefore less abrupt.

Natural selection acted "by the preservation and accumulation of infinitesimally small inherited modifications."[56] It was no more difficult to believe that the gradual accumulation of individual differences was the force shaping the diversity of life on Earth, Darwin argued, than it was to believe that the gradual wind and waves were responsible for shaping the earth's surface. There was no need to invoke the creation of new species by any great and sudden modification of their structure, any more than it was to suppose that a great valley was formed by a single diluvial wave.[57] There was no need to invoke causes of species formation in the past that are strikingly different today. All changes occurred gradually, without major jolts; varieties differentiated by divergence and separation. The sudden appearance of species in the fossil record was the argument used by those, including Agassiz, Sedgwick, and Lyell, who had opposed the transmutation of species. Darwin insisted that "as natural selection acts solely by accumulating slight, successive, favorable variations, it can produce no great or sudden modification; it can act only by very short and slow steps." This, he noted, both confirmed and made intelligible the old cannon of natural history *natura non facit saltum*.[58]

Natural Selection

Perhaps the most radical transformation in biological thought that Darwin introduced was his focus on species as dynamic populations, not as types. Individual differences in offspring arose from common parents. These slight differences between individuals were the sources of evolutionary change and would accumulate by natural selection to give rise to differences between species, genera, and all the great classes of organisms.[59] In this set of assumptions, Darwin rejected the notion that there were fundamental or important traits that never change within a species and that define that species.[60]

For Darwin, individual differences were real, and species and other types were not—they were abstractions. In taking this position, Ernst Mayr has argued, Darwin broke with one of the central concepts shared by philosophers since the ancient Greeks. The *eidos* (idea, type, or essence) had been part of philosophy since Plato. As applied to nature, it meant that species were real; they were natural types around which individual variations occurred. But Darwin was a nominalist: species were populations. Certainly, one species could be distinguished from another at any moment in time—fir tree and spruce, for example. But there was a great amount of variation within species as well. This was the premise behind the idea that gradual evolution occurred by natural selection. As Mayr put it,

> The assumptions of population thinking are diametrically opposed to those of the typologist. The populationist stresses the uniqueness of everything in the organic world. What is true for the human species—that no two individuals are alike—is equally true for all other species of animals and plants. . . . All organisms and organic phenomena are composed of unique features and can be described collectively

> only in statistical terms. Individuals, or any kind of organic entities, form populations of which we can determine the arithmetic mean and the statistics of variation. Averages are merely statistical abstractions, only the individuals of which the populations are composed have reality. The ultimate conclusions of the population thinker and of the typologist are precisely the opposite. For the typologist, the type (eidos) is real and the variation is an illusion, while for the populationist, the type (average) is an abstraction and only the variation is real. No two ways of looking at nature could be more different.[61]

Evolution, for Darwin, was a two-step process resulting from chance and necessity, from the production of variation, and subsequently from natural selection: the preservation of favorable variations and the rejection of injurious ones. Lamarck had assumed that the heritable variations that arose among individuals were directed by use and disuse, and by the environment. Darwin also considered that this might be true of some cases. But many variations in nature, he argued, appeared randomly—that is, some would be useful and some would not. He looked to artificial selection by breeders of domesticated plants and animals for a model. Breeders were able to select those varieties of particular interest to them; they did not act directly to produce the variability itself. Heritable modifications appeared occasionally; they appeared "randomly" and were selected purposefully.

The Struggle for Existence

Darwin linked evolution and diversity to the two most obvious characteristics of organisms and their environment: excessive reproduction on the one hand, and ecological checks on population growth on the other. While variations provided the fuel for evolution, overreproduction would create a struggle or competition, the motor for selection. Each organism, or each pair of organisms, had the ability to produce offspring in ever-increasing numbers from generation to generation. If only one species existed on Earth, without limits to its expansion, and nothing to prey on it, it would multiply in geometric progression. In the eighteenth century, Linnaeus had calculated that if an annual plant and each of its descendants produced only two seeds a year—and there is no plant so unproductive as this—then, in twenty years, there would be more than a million plants. Darwin made the same calculation for the slow-breeding elephant: assuming that it procreates beginning only when it is thirty years old, that it lives a hundred years, and that it brings forth six offspring in this interval, then the descendants of a single pair of elephants would total about 15 million in five hundred years. As for slow-breeding man, who doubles in number in twenty-five years, Darwin commented, "in a few thousand years, there would literally not be standing room for his progeny."[62]

But all this was far from earthly reality. There were checks on population growth. Individuals and groups always competed with each other for territory, food, and light in a struggle for existence. The effect of the environment would be to favor the multiplication of some species at the expense of others. Some are doomed to die out, others to expand. As soon as new varieties appear, they take part in the competition. They win or lose depending on whether or not the differences between them and their ancestors favor their multiplication. Darwin believed competition would be most severe between closely related individuals, who would share a common way of life and compete for the same food and space. Hence, selection driven by competition would favor divergence: the emergence of new specialized varieties and new species.

3

Darwin's Champions

> As for your doctrines I am prepared to go to the stake if requisite. . . . I trust you will not allow yourself to be in any way disgusted or annoyed by the considerable abuse and misrepresentation which, unless I greatly mistake, is in store for you. Depend upon it you have earned the lasting gratitude of all thoughtful men. And as to the curs which will bark and yelp, you must recollect that some of your friends, at any rate, are endowed with an amount of combativeness which (though you have often and justly rebuked it) may stand you in good stead.
>
> I am sharpening up my claws and beak in readiness.
>
> —Thomas Henry Huxley to Charles Darwin,
> November 23, 1859

THE CONCEPT OF EVOLUTION appealed particularly to nonscientists, and although the professional scientists who were best informed about biology believed in the fixity of species, Darwin was able to convert some of them. *The Origin* had profound implications. What came to be called Darwinism represented a new naturalistic methodology or scientific approach to life, our place in nature, our ethics, and our societies. These issues shaped the rhetoric over evolutionary theory in the nineteenth century as they do today. In the decades following the release of *The Origin*, those who saw themselves as Darwinists positioned their arguments against supernatural beliefs and Judeo-Christian theology. And Darwin had outstanding champions, none more prominent than Thomas Henry Huxley in England and Ernst Haeckel in Germany.

Man's Place in Nature

T. H. Huxley (1825–1895) was one of three people (along with Joseph Hooker and Charles Lyell) to whom Darwin sent parts of the manuscript of *The Origin* before its publication.[1] In mid-November 1859, after a jubilant Huxley had read the book in its entirety, he wrote to Darwin. "I think you have demonstrated a true cause for the production of species, and have thrown the *onus probandi*, that species did not arise in the way you suppose, on your adversaries."[2] Huxley knew that there would be an uproar of indignation and attacks by the antievolutionist philistines. Darwin had a reserved style; he deplored confrontations. Huxley, on the other hand, was a polemicist, an activist who aimed to change science education in England.

Huxley was born above a butcher shop in Ealing, a small village about twelve miles west of London, the seventh of eight children. "Not for him Darwin's silver spoon; he had no fortune to inherit, tradition to uphold."[3] At the time he was born, his father was a mathematics teacher at Ealing School, but the decline of the school in hard times left him penniless. Thomas had only two years of formal childhood education. Nonetheless, he read voraciously in science, history, and philosophy, and he taught himself German. When he was fifteen years old, he began a medical apprenticeship, and soon was awarded a scholarship to study at Charing Cross Hospital.

His turn to natural history occurred when at age twenty-one, he signed on as assistant surgeon on the *H.M.S. Rattlesnake*, a Royal Navy frigate assigned to chart the seas around Australia and New Guinea. On that voyage, he collected and studied marine invertebrates, in particular cnidarians (hydra and jellyfish) and tunicates (i.e., sea squirts, or ascidians, sedentary filter feeders with cylindrical bodies, usually found attached to rocks) as well as cephalopod ("head-foot") mollusks (octopus, squid, and nautilus). His studies of fossils and his contributions to comparative anatomy and embryology earned him acceptance into the highest ranks of the English community of naturalists. Professional positions as "scientists," as the Cambridge don William Whewell had named them in 1840, were rare. Most naturalists were affluent amateurs, but Huxley managed to support himself on a stipend from the navy and by writing popular science articles. After leaving the navy in 1854, he secured a lectureship as professor of natural history at the Royal School of Mines in London. His public lectures earned him great fame.

Thomas Henry Huxley was the beginning of a remarkable line of Huxley scientists and thinkers, a tradition of his own making.[4] His son Leonard was a noted biographer and man of letters. Leonard's sons were all distinguished intellectuals: Aldous, a novelist, screenwriter and essayist, was the famed author of the dystopian book *Brave New World*. Julian was one of the architects of the "evolutionary synthesis" of the 1940s, which brought Darwinism together with genetics, natural history, ecology, and population studies (see chapter 15). Andrew shared the 1963 Nobel Prize in Physiology or Medicine for his work on nerve impulse and muscle contraction.

Huxley made sure that Darwin's *Origin* got off to a fine start in the media, himself writing laudatory reviews in the *Times* and elsewhere. But he was critical of the idea that natural selection acting gradually on slight variations was the sole mechanism of evolution. He thought that evolution might proceed more quickly at times—by rapid jumps, or saltations. As he wrote to Darwin just before publication of *The Origin*, "you have loaded yourself with an unnecessary difficulty in adopting *Natura non facit saltum* so unreservedly."[5]

Huxley's most famous work was his *Man's Place in Nature* (1863).[6] This was the first comprehensive overview of what was known at the time about primate and human palaeontology and ethnology. It was also the first attempt to apply evolution explicitly to humans. Darwin avoided much mention of human evolution in *The Origin*, stating only in the conclusion that in research fields that would open up in the future, "Light will be thrown on the origin of man and his history."[7] He discussed human evolution in detail twelve years later in The *Descent of Man* (1871).[8] But Huxley had immediately begun to collect the evidence for human evolution, the links between *Homo sapiens* and his apelike ancestors. And he directly confronted theological beliefs that man arose through divine creation. "The question of questions for mankind—the problem which underlies all others, and is more deeply interesting than any other," he wrote,

> is the ascertainment of the place which man occupies in nature and of his relations to the universe of things. Whence our race has come; what are the limits of our power over nature and of nature's power over us; to what goal we are tending; [these] are the problems which present themselves anew and with undiminished interest to every man born into the world. Most of us, shrinking from the difficulties and dangers which beset the seeker after original answers to these riddles, are contented to ignore them altogether, or to smother the investigating spirit under the featherbed of respected and respectable tradition.[9]

These questions about human origins led him into direct confrontation with the most famous religious authorities of his day.

Natural Theology and Agnosticism

Remembered today as "Darwin's bulldog," Huxley wrote about theology and philosophy from the point of view of an "agnostic," a term he coined. Much of his polemics were directed at the anti-intellectualism of church dogma. He introduced the term "agnostic" as "suggestively antithetic to the Gnostic" of church history, who professed to know so very much. Agnosticism, for Huxley was not a creed, but an injunction about the way to approach knowledge: to follow reason as far it could go without consideration for where it might lead, and not to pretend to know things with certainty that had not been demonstrated or were not demonstrable.[10]

By introducing contingency in nature, the theory of natural selection replaced design and purposefulness as an explanation for adaptations. It displaced God but did not necessarily replace a First Cause. Darwin himself recognized that there might well be a "First Cause having an intelligent mind in some degree analogous to that of man." However, he considered such matters to be beyond the intellectual reach of man. As he wrote in 1876,

> But then arises the doubt—can the mind of man, which has, as I fully believe, been developed from a mind as low as that possessed by the lowest animals, be trusted when it draws such grand conclusions?
>
> I cannot pretend to throw the least light on such abstruse problems. The mystery of the beginning of all things is insoluble by us, and I for one must be content to remain an Agnostic.[11]

Darwin did not make his views on such matters public.

Huxley is remembered for his part in one of the most famous confrontations between religion and evolution: the debate on June 30, 1860, at the British Association for the Advancement of Science meeting at Oxford, with Archbishop Samuel Wilberforce. Wilberforce was a well-known public figure at the height of his fame; he had the ability to speak on platform and pulpit and in parliament and was the model of the English churchman. His nickname, "Soapy Sam," he earned from his slippery argumentative style. The British Association met for a week in a different town each year, and the public flocked in huge crowds. Huxley knew that Wilberforce was a "first-rate controversialist," and he had no intention of attending the meeting. However, he met up with the evolutionist Robert Chambers, the anonymous author of *Vestiges of Creation*. As Huxley recalled,

> On Friday I met Chambers on the Street, and in reply to some remark of his about the meeting, I said that I did not mean to attend it; did not see the good of giving up peace and quietness to be episcopally pounded. Chambers broke into vehement remonstrances and talked about my deserting the cause. So I said, "Oh! If you take it that way, I'll come and have my share of what is going on."[12]

Around a thousand people gathered at Oxford on that famed Saturday morning late in June with the expectation that the Darwin issue was going to be aired. They were not disappointed. One moment in the confrontation is legendary: when Wilberforce attacked Darwin and ridiculed evolution, he asked Huxley whether it was from his grandmother's side or his grandfather's that he descended from an ape. This snide remark was a strategic rhetorical mistake, and Huxley seized on Wilberforce's error. Accounts vary on the exact words, but according to Huxley's own version, he muttered to his colleague next to him, "The Lord hath delivered him into mine hands!" He then rose to give a brilliant defense, concluding with the rejoinder that he would rather have an ape for a grandfather than "a man highly endowed by nature and possessed of great means and influence, and yet who employs these faculties and that influence for the mere purpose of introducing ridicule into a grave scientific discussion."[13]

The legend of this confrontation has been often told as a victory of reason and intellectual progress over ignorance and conservative authority. But this is far too simplistic—first, because the confrontation continued, and evolution is still not taught in many schools;[14] and second, because Wilberforce was not as ignorant of science as has been supposed. He had studied with one of the greatest scientists of the nineteenth century, Richard Owen (1804–1892), president of the Royal Society. Owen had already crossed swords with Huxley over Huxley's statement that the brains of humans and apes were very similar. Owen was a prominent opponent of the Darwinians and one of the most distinguished anatomists and paleontologists of the nineteenth century, and his place in Victorian natural history has been the subject of intensive historical study.[15]

Archetype and Idealism

Owen named and described a vast number of living and fossil vertebrates. His experience with the anatomy of exotic animals began when he worked for the London Zoo as a prosector, dissecting and preserving animals that died in captivity. He rose to fame as "the British Cuvier" in 1839, when, presented with a bone fragment from New Zealand, he noted that it resembled an ostrich bone and suggested that giant flightless birds had once lived there (as was later confirmed). Owen named the giant bird Dinornis—the extinct moa. But his most famous taxonomic act resulted from his examination of reptile-like fossil bones of Iguanodon, Megalosaurus, and Hylaeosaurus found in southern England. In 1842, he named this taxon the Dinosauria, from the Greek *deinos* ("terrible") and *sauros* ("lizard"). He also described the anatomy of a newly discovered species of ape reported in 1847—the gorilla—and in 1863 he described the first specimen of an unusual Jurassic fossil from Germany, the famous bird *Archaeopteryx lithographica*.

Owen served on a series of government committees, took part in the London Crystal Palace Exhibition of 1851, and served as an advisor and expert witness to the government on all sorts of scientific matters. He taught natural history to Queen Victoria's children (astonishing the court with the news that tadpoles turned into frogs). Unfortunately, Owen was not easy to get along with. His vain, arrogant, envious, and vindictive personality seems to have inspired distaste in most of his colleagues.

Owen may have been called the British Cuvier, but by the mid-1800s, his and Cuvier's approaches were in fact poles apart. Recall from chapter 1 the debates between comparative anatomists over the teleological functionalism of Cuvier, who saw all organic structures as designed by God to suit particular functional needs, and the search for a structural unity of life by Geoffroy and the transcendental anatomists, whose approach was imported into England by anatomists who challenged the traditionalist Oxbridge dons. Some transcendental anatomists were evolutionists, some were not. Owen developed his own brand of transcendentalism.

Owen was heavily influenced by idealist thought of the early nineteenth century, and he was certainly driven to find the unity of nature and rationality of nature's plan. In his book *Archetype and Homologies of the Vertebrate Skeleton* (1848), he imagined a principle of transcendental unity existing at a deeper level of reality than the physical. This Platonist approach led him to formulate the concepts of homology and analogy still used today. Homology refers to the same organ in different animals under every variety of form and function. Structures as different as a bat's wing, a cat's paw, and a human hand nonetheless display a common plan of structure, with identical or very similar arrangements of bones and muscles. There is a structural relationship. Homologies are basic structures that various creatures use for different purposes. Analogies refer to cases in which parts that have no structural resemblance serve a similar purpose in different animals: the flippers of dolphins and penguins are analogous structures, but the underlying structures are different. Analogies would have no significance for classification. Owen reasoned that there must exist a common structural plan, a *bauplan* or "archetype" or for all vertebrates, the essence of the vertebrate form.[16]

Although historians and philosophers have traditionally placed Owen on the side of natural theology, conservativism, and staunch antievolutionism, this caricature has been recently challenged.[17] Owen had been on the side of the Oxbridge conservatives in the 1830s, when he was a follower of the functionalist teleological tradition of Cuvier, but by 1848 his aim was to form a reconciliation between Cuvierian functionalism and Geoffroyian structuralism. In the years following the publication of Darwin's *Origin of Species*, Owen portrayed himself as a misrepresented evolutionist who had been unfairly and erroneously cast by Darwin as an antievolutionist.[18] In reality, he sought a middle ground between natural theology and materialistic evolution. He maintained that while each species had its origin in natural causes, the course of evolution was directed in accordance with predetermined law (such as First Cause or the Creator).

Owen's idealism and his opposition to Darwinian transmutation had led him into battle with Huxley and to assert that gorillas and other apes lack a structure of the brain known as the hippocampus minor.[19] The structural similarity between humans and apes was a fearful problem for nonevolutionists. Owen searched for an anatomical difference in the brain that might explain man's uniqueness as a conscious decision maker. He insisted that one lobe, the hippocampus minor, was peculiar to the human brain, and that the cerebral hemispheres completely covered the cerebellum, whereas in apes, he alleged, the hippocampus was absent and the cerebellum was always exposed from above. Owen refused to retract even when Huxley and his disciples showed conclusively that Owen was mistaken. Apes do have a hippocampus, and Huxley successfully exploited it as a great triumph for Darwinian naturalism. The idealism that had informed Owen's research had been especially prominent in Germany early in the nineteenth century, where *Naturphilosophen* searched for the unity of nature according to divine plan.

Ontogeny and Phylogeny

If Huxley was Darwin's English bulldog, then Ernst Haeckel (1834–1919) was surely his German shepherd. Haeckel is celebrated as one of the outstanding theoreticians of nineteenth century biology. He coined the terms "ecology," "phylum," "ontology," and "phylogeny." He is well known today for his microscopic work during the 1860s on the Radiolaria (jellyfishes), Porifera (sponges), Annelids (segmented worms), and Protista. He is also remembered by evolutionists today for the first to attempt to construct a universal genealogical tree, stemming from the simplest one-celled organisms, which he called Monera. Some historians have suggested that Haeckel was even more influential than Darwin in convincing the world of evolution.[20] He especially drew out what he saw as evolution's implications for philosophy, religion and human social progress. His most popular book, *The Riddle of the Universe* (1899), was among the most spectacular successes in the history of printing. It sold 100,000 copies in its first year, had gone through ten editions by 1919, was translated into 25 languages, and had sold almost half a million copies in Germany alone by 1933.

Born in Potsdam, Prussia (now Germany), Haeckel studied medicine at Würzburg and at the University of Berlin, and he was Professor of Zoology at Jena from 1865 until his retirement in 1909. He was a Christian before he read *The Origin*, translated into German in 1860. He then became the champion of materialism and anti-idealistic thinking. While materialists were shaking traditional beliefs in England, German thinking had been strongly influenced by the idealist philosophy of Friedrich Hegel, Immanuel Kant, Friedrich von Schelling, and others who had turned away from the materialism of the enlightenment. German *Naturphilosophen* maintained that an unknowable spirit or creative organizing force in nature gave it purpose and accounted for progressive perfection of God's creatures. Haeckel aimed his polemics directly at their idealism and supernaturalism. Nothing caused nonevolutionists more trouble than organs and muscles that serve no purpose. Consider, for example, the muscles for human ear movement. Although most of us are incapable of moving our ears as we wish, individuals who deliberately exercise these muscles may succeed in moving their ears. Haeckel thought these muscles were on the road to complete disappearance. But he focused especially on the embryological evidence for evolution.

Darwin himself felt that embryology provided some of the strongest evidence for the transmutation of species. If each species had been created independently by divine inspiration, he argued, then one would expect that the route from egg to infant would be direct. But embryologists had found exactly the opposite: there were extraordinary detours.[21] Embryos of land-living vertebrates go through a stage of having gill slits. Embryonic baleen whales develop teeth, and higher vertebrates have a notochord, a flexible rod that develops along the back of the embryo, a characteristic of chordates such as tunicates and the very primitive filter-feeding fish known as lancets.

Comparative embryology provided the criteria for homology and so was the means for uncovering the relationships and ancestry.[22] One of the aims of this discipline was to see how far one could go in revealing ancestral forms common to all animals. Many who entered this field were inspired by the view, championed so forcefully by Haeckel, that the course of development, from embryo to infant, documented the path of evolution. In other words, that during the development of the embryo, key steps in evolution are repeated, or, in short, ontogeny recapitulates phylogeny.

The idea that there is some sort of parallel between the stages in the course of individual development and the natural order of life-forms from lowest to highest can be traced to the ancient Greeks.[23] Aristotle classified beings into those with a nutritive soul (plants), those with a nutritive and sensitive soul (animals), and those that also had a rational soul (human beings). He imagined that during the course of development these three kinds of souls came into operation in succession. Some transcendental anatomists of the early nineteenth century believed that the development of a higher animal actually passed through adult stages that lay below it. So, for example, birds and mammals pass through a fish stage, evidenced by gill slits in the embryo. This idea of a parallelism between the stages of ontogeny and the stages in a great chain of being became known as the Meckel-Serrés law. For the *Naturphilosophen*, it was evidence of the unity of nature and of God's divine plan.[24] Lorenz Oken, for example, argued that all the animal kingdom was actually a complete individual from protozoan to man, as illustrated in the mammal fetus, which passed through all stages, from polyp, snail, fish, and amphibian, to its own class.[25]

This view of a transcendental recapitulation of adult types in the course of development was modified by Karl Ernst von Baer (1792 –1876) into a recapitulation of embryonic types. In 1828 he advanced his biogenetic law, according to which development progressed from the general or primitive to the specific or advanced.[26] Von Baer noticed that during development, individual embryos successively added organs characteristic of their phylogenetic place in the natural order. The first features to appear were those of the phylum, and these were followed by those of the class, order, family, genus, and species. The human embryo was first a single cell, then a colony like a sponge, and when a liver was added it reached the level of organization of a mollusk. For von Baer, this was only a comparative argument, not an evolutionary one: developmental progression of the embryo simply paralleled the taxonomic order of life from simple to complex.

But in 1866, Haeckel transformed von Baer's law into a materialist evolutionary law of recapitulation. "The history of individual development, or Ontogeny is a short and quick recapitulation of palaeontological development, or Phylogeny,"[27] he argued, because phylogeny is the actual cause of the ontogenetic stages. Thus, all organisms are historical records because each preserves the forms of ancestors at key stages in the growth of the embryo. While von Baer emphasized the embryonic similarities, Haeckel's recapitulation theory proposed that descendants recapitulate the adult stages of their ancestors. His idea was that new adult modifications were piled onto the relatively unchanging embryo in the course of successive generations.

The embryological search for common ancestral forms was central to the development of evolutionary theory. But this approach greatly declined at the end of the nineteenth century as embryologists came to recognize that Haeckel's model did not hold: embryos pass through stages that often resemble ancestral *embryonic* types, not *adult* types. Haeckel has taken recapitulation to extremes with his notions of progressive evolution, with all of its social connotations (see chapter 4). Today, developmental evolutionists blame him for effectively driving a wedge between development and evolution with his "second-rate generalities,"[28] a wedge that "is only now beginning to be dislodged as the stigma of recapitulation is slowly removed from evolutionary developmental biology."[29]

Not surprisingly, given his status as the author of *The Origin*, the extent to which Darwin himself may have been committed to recapitulationism and to "progress" in evolution has become a subject of debate today. Scholars such as Ernst Mayr and Stephen Jay Gould have insisted that Darwin did not view evolution as progressive and that he rejected the idea that developing embryos of descendants are duplicates of their adult ancestors. They argue that Haeckel's erroneous views about recapitulation resulted from the fact that in reality he was a neo-Lamarckian who maintained a false belief in the inheritance of acquired characteristics.[30] However, Robert Richards has argued that Darwin was both a progressionist and a recapitulationist like Haeckel, and that Mayr and Gould are guilty of distorting history to protect Darwin from any taint.[31]

The link between embryology and transformationism begins with the word "evolution" (unfolding) itself. In the eighteenth century, the word was used by advocates of "preformationism" to describe the unfolding of the organism from the egg during development. Their views opposed those of "epigenists," who argued that an organ developed gradually from an amorphous homogeneous condition to an articulated heterogenous state. With the development of a transcendental form of recapitulation theory during the first decades of the nineteenth century, the term "evolution" came to refer to the sequence of events during development. As recapitulation came to imply phylogeny, Richards argues, the word "evolution" took on both meanings and came to signify embryological progression as well as species transmutation. Both Charles Lyell and Herbert Spencer gave currency to the word "evolution" to describe progressive change toward ever-increasing complexity.[32] But what about Darwin?

In *The Origin*, Darwin considered the evidence from comparative embryology for a common plan within groups. He recognized that there were fundamental embryological characters that were typical of the large classes of animals, and he thought that a common plan for all vertebrates might eventually be found in the fossil record.[33] He considered such embryological characteristics to be evolutionary relicts. They were conserved because evolution operated on inherited variations that occurred later in development. "This process, whilst it leaves the embryo almost unaltered, continually adds, in the course of successive generations, more and more differences to the adult." Thus, he reasoned, "the embryo comes to be left as a sort of picture, preserved by nature, of the ancient and less modified condition of each animal."[34]

In effect, for Darwin, at least in this statement, ontogeny recapitulates phylogeny. That Darwin believed in the progressive evolution from lower to higher forms is also doubtless (see chapter 4). Between those scholars who have divorced Darwin's thinking from recapitulationism and Richards, who sees it as central, are still others who argue that although Darwin certainly had a general theory of developmental change in evolution, he was not was not deeply committed to recapitulationism as such and changed his views in accordance with the fluctuating evidence for it.[35]

Recapitulationism declined in the late nineteenth and early twentieth centuries as it became clear to evolutionary embryologists that there was no simple linear progression from invertebrates to vertebrates and to humans.[36] Francis M. Balfour (1851–1882) at Cambridge and Walter Garstang (1868–1949) at Oxford argued that while embryos do pass through a series of stages that correspond to those of their ancestors, not all embryological features reflect ancestral patterns. The reason we and all mammals have gill slits in a "fish stage" of our ontogeny is still not fully understood. But the important point is that development also evolves. During the late nineteenth century, comparative embryology was overshadowed by experimental embryologists, who recognized that Haeckel had the cause and effect reversed (see chapters 9 and 10). Rather than phylogeny being the cause of ontogeny, changes in the course of ontogeny led to major evolutionary change.[37] As Garstang phrased it, "ontogeny does not recapitulate phylogeny, it creates it."[38]

Materialism for Mysticism

Haeckel reveled in the philosophical and religious implications of Darwinism. It provided the basis for a nonmiraculous, nonmystical history of creation and showed that man is an animal. "The dimming mirage of mythological fiction," he wrote, "can no longer exist in the clear sunlight of scientific knowledge."[39] Of course, this was not the first time science and the church had entered into conflict. The most notorious case had begun centuries earlier in 1513, when the Polish astronomer Nicolas Copernicus (1473–1543) proposed that the sun (not the earth) was at the center of the universe. Copernican theory gained support in the seventeenth century from Galileo Galilei (c. 1564–1642). Prima facie, Copernicanism contradicted the church, which supported the Ptolemic view of an earth-centered universe. In 1616, Galileo was secretly warned by the Inquisition not to defend Copernicanism. When in 1632 he published his *Dialogue on the Two Principal Systems of the World*, again supporting Copernican theory, he was summoned to Rome by the Inquisition and ordered, under the threat of torture, to deny his beliefs. He did so and was eventually condemned to house arrest, for life, at his villa at Arcetri (near Florence) and forbidden to publish anything else.

Copernicanism displaced the earth from the center of the universe, and Darwin's theory took the next step in displacing man from the center of the earth. Haeckel compared its importance to Isaac Newton's (1642–1727) theory of gravity/ Newton explained the motion of planets in term of the same mathematical laws of weight or attraction that applied to falling bodies on earth. Darwinism complemented and extended the mechanistic-materialistic outlook of Newtonian physics by bringing the living world into union with knowable physical laws.[40] One would seek a mechanico-causal explanation of the origin and structure of an organ in the same way as one would seek a mechanical explanation for any physical process—for example, earthquakes, the directions of the wind, or ocean currents.[41] Haeckel called this unity of explanation his "monistic" conception of nature, and he held it up against the church's belief in the miracles of a supernatural history of creation and belief in a Creator with human attributes.[42]

In Haeckel's view, nothing could be further apart socially and intellectually than scientific materialism and what he called the ethical materialism of the high priests and the privileged aristocracy of Europe. Scientific materialism affirms that everything in the world goes on naturally—that every effect has its cause, and every cause its effect. There are no supernatural processes.[43] Ethical materialism is based on the premise that only purely material enjoyment can give satisfaction; it proposes no other human aim than the most refined possible gratification of one's senses. Haeckel claimed that it was absent among naturalists and philosophers, "whose greatest happiness is the intellectual enjoyment of nature, and whose highest aim is the knowledge of her laws." But, he remarked,

> We find it in the palaces of ecclesiastical princes, and in those hypocrites who, under the outward mask of a pious worship of God, solely aim at hierarchical tyranny over, and material spoliation of, their fellow men. . . . They stigmatize all natural science, and the culture arising from it, as sinful "materialism," while really it is this which they themselves exhibit in a most shocking form. Satisfactory proofs of this are furnished, not only by the whole history of the Catholic Popes, with their long series of crimes, but also by the history of the morals of orthodoxy in every form of religion.[44]

The materialism of Darwinism, based largely on a view of the natural word as a place of persistent struggle, conflict, and war, confronted the optimistic view of natural theology of a harmonious natural world of give and take, of mutual cohabitation and a balance of nature in which each of God's creatures was created to play its part for the benefit of all (see chapter 5). The methodological naturalism of Darwinism lent weight to the belief that the immanent caring God of monotheism might not be anything more than a myth devised by our species to place itself above the rest of nature. As one commentator has put it, "Darwin made it possible to be an intellectually fulfilled atheist."[45] Of course, this did not mean that Darwinism required disbelief in God. Rather, by showing how natural order might arise without a designer, Darwinism robbed natural theologians' argument of its logical strength.

Confrontations between evolutionism and the Christian church continued into the twentieth century, one of its most famous events being the Scopes trial. When the Tennessee legislature passed the Butler Act in 1925 forbidding the teaching of evolution in the public schools of that state, John Scopes, a science teacher, consented to be the defendant in a court test of the law. He was arrested, stood trial, and was defended by the American Civil Liberties Union. Representing him was the famed trial lawyer Clarence Darrow. The case broadened to a debate over whether evolution and religion (especially fundamentalism) could coexist. In the latter decades of the twentieth century, the retrenchment called creationism and even a so-called scientific creationism gained strength in the United States among conservative groups concerned that the social order itself might dissolve without a belief in a God-given morality.[46] If nature lacked moral character, what did this mean about us, and how could we best construct a moral and prosperous society? These issues were paramount in larger discussions of evolution in the nineteenth century as well, as evolutionists extended their views about nature to human societies. As we shall see in the following chapter, they reached different conclusions.

4

Darwinism and Sociopolitical Thought

He had the luck to please everyone who had an axe to grind.
—George Bernard Shaw, 1921

DARWIN WAS ABLE TO CAPTURE a wide audience of scientists and nonscientists. *The Origin of Species* resounded with people for whom the Industrial Revolution epitomized the struggle for life. In effect, it dramatized what many people felt to be true about the way their world worked, and they used Darwinian natural law to explain their world. Many evolutionists treated organic evolution and human social relations as one subject, just as science and religion had been for natural theology. The use of natural law as the basis for a given view of society became commonplace in social, political, and economic theory of the nineteenth century.[1]

Darwin was, so to speak, all things to all people, as the Nobel Prize–winning Irish playwright George Bernard Shaw (1856–1950) noted. Darwinian theory was summoned to support all types of political and ideological positions, from the most reactionary to the most progressive, including racism, militarism, laissez-faire economics, unfettered capitalism, Marxism, and anarchism. Whatever their political inclinations, writers often drew on evolutionary theory and "the laws of nature" to bolster their views of progress. But did evolution imply progress? Today evolutionists would argue that words such as "progress" and "improvement" are inappropriate because their criteria remain ill defined. If one criterion of progress means adaptation to survive, then a bacterium appears just as well adapted as humans. But, for most biologists of the nineteenth century, evolution did imply progress in nature and society, and for most, this meant that "the struggle for existence," understood as conflict and competition, should be encouraged.

Laissez-faire

Laissez-faire economic theory was developed by the Scottish philosopher Adam Smith (1723–1790) in his celebrated book *The Wealth of Nations* (1776). Essentially, it was an argument against mercantilism, with its assumption that the wealth of nations is derived from government-regulated exports, gold and silver, and colonial expansion. Smith and his followers maintained that the wealth of nations was produced by labor within the nations of the Industrial Revolution with such improved technologies for producing goods as the cotton gin and, later, the steam engine. Government intervention in trade and industry was harmful. What was needed was unbridled competition among businesses, the desire for profits, and individual freedom to make decisions. Acting to pursue one's own economic interest would naturally be in the collective good of the community. Individual workers sold their labor to whoever paid the highest wages. Capitalists would compete among themselves for consumers and decide what to produce and how much according to consumer demand. Thus "natural laws" regulated economic processes; one obstructed these laws at one's peril.

The writings of Herbert Spencer (1820–1903) exemplify social Darwinism in this liberal tradition. Enormously popular as the prophet of progress, he used evolutionary theory to champion governmental nonintervention in "the natural laws" of the free market. He was the son of a schoolmaster in Derby, and as a young man he worked as an engineer in the construction of the London to Birmingham and Birmingham to Gloucester railways. He later gave up engineering and devoted himself to journalism and to writing books: *The Principles of Psychology* (1855); *Education: Intellectual, Moral, and Physical* (1861); *First Principles* (1862); *The Principles of Biology*, 2 vols. (1864–67); *The Study of Sociology* (1873); *The Principles of Sociology*, 3 vols. in 8 parts (1876–97); *The Principles of Ethics,* 2 vols. in 6 parts (1879–93); and *The Man versus the State* (1884).

Spencer's "synthetic philosophy" aimed to unite all knowledge under one explanatory concept: evolution. All of evolutionary progress in nature and in society, he argued, proceeds from the homogeneous to the heterogeneous, from the incoherent to the coherent. Progress through open competition the law of organic creation would push humanity onward to perfection. As long as competition and "that natural relation between merit and benefit" were maintained, all members of society would be increasingly integrated into a differentiated, mutually dependent, and efficient higher social organism in which violent competition (such as war) would be replaced by the peaceful competition of the free market.[2] This superorganismic view predicted that the evolutionary process in society would eventually lead it to an end state of equilibrium.

Spencer had developed his theory of evolution independently of Darwin, at least initially. He drew on ideas from his lower-middle-class background in England; from Thomas Malthus, Lamarck, and Lyell; from laissez-faire economics;

and from certain principles of physics.[3] In an essay of 1852, in which he coined the expression "the survival of the fittest," he came very close to offering a statement of the principle of natural selection. He argued that population pressures on resources lead to a struggle for existence and the survival of the fittest, with the more intelligent surviving. Population pressure was the motor of progress; it forced people to become more efficient, to become better adapted, and to be less fertile: "From the beginning, pressure of population has been the proximate cause of progress. . . . It forced men into the social state; made social organization inevitable; and has developed the social sentiments. . . . It is daily pressing us into closer contact and more mutually dependent relationships."[4]

Spencer's socioevolutionary views continue to provoke debate among historians. Many have emphasized that his writings stood in the way of important social reforms. Laissez-faire capitalism had led to great advances in trade and industry, and some people made huge fortunes. But when trade and manufacturing slumped in the nineteenth century, many workers suffered poverty and unemployment. In industrial England, with its sweatshops, child labor, homelessness, and poverty, many writers called for social reform and the need for state charity. Spencer, however, decried state intervention, whether it be education for the poor or privileges for the church. While some historians portray him as a brutal social Darwinist who "expounded the idea of struggle for survival into a doctrine of ruthless competition and class conflict,"[5] others have argued that this view is somewhat misleading.[6] For although the struggle for existence provided a plausible explanation for all the selfish behavior of which man is capable, Spencer also maintained that cooperation and altruism were widespread in nature, though they were secondary to and dependent upon egoism.[7]

Spencer opposed state charity and "state meddlings with the natural play of actions and reactions," but he insisted that "it does not follow that the struggle for life and the survival of the fittest must be left to work out their effects without mitigation."[8] He maintained that aid for the "inferior" should be supplied by the "superior" on a voluntary basis and thereby kept within moderate limits to "the benefit of both—relief to the one, moral culture to the other. And aid willingly given (little to the least worthy and more to the most worthy) will usually be so given as not to further the increase of the unworthy."[9]

Some biologists today feel justified in ignoring Spencer totally in a history of biological ideas because, they assert, his positive contributions were nil.[10] Yet biologists of the nineteenth century often referred to Spencer's biological writings, especially when defining the organism (see chapter 8).[11] He also had an important influence on anthropology, psychology, and the social sciences. Darwin adopted his expression "survival of the fittest" and used it in the new heading of chapter 4 (previously titled "Natural Selection") of the fifth edition of *The Origin*. Spencer's battles with religious authority and idealism earned him the friendship of T. H. Huxley. However, as much as Huxley applauded Spencer's attacks on church privileges, he cringed at his denunciation of state education.[12]

Social Darwinism Exported

Spencer was especially popular in the United States, where social Darwinism of the laissez-faire school received great support. To understand why, historians point to the American Constitution's concept of the inalienable right of the individual to personal freedom, and in particular to freedom from interference from government. Indeed, government's major function was thought to be the maintenance of individual freedoms. In academic circles, the leading social Darwinist was William Graham Sumner (1840–1910), professor of political economy and social science at Yale, who declared:

> Millionaires are a product of natural selection, acting on the whole body of men to pick out those who can meet the requirement of certain work to be done. . . . They get high wages and live in luxury, but the bargain is a good one for society. There is the intensest competition for their place and occupation. This assures us that all who are competent for this function will be employed in it, so that the cost of it will be reduced to the lowest terms.[13]

Sumner glorified the achievements of unfettered capitalism as reflecting the laws of nature, and he argued that the schemes of socialists ("social meddlers") were the greatest danger to society:

> We can take the rewards from those who have done better and give them to those who have done worse. We shall thus lessen the inequalities. We shall [thereby] favor the survival of the unfittest, and we shall accomplish this by destroying liberty. Let it be understood that we cannot go outside this alternative: liberty, inequality, survival of the fittest; not liberty, equality, survival of the unfittest. The former carries society forward and favors all its best members; the latter carries society downwards and favors all its worst members.[14]

Sumner believed that the only way to social progress lay through sobriety, industriousness, prudence, and wisdom. If people would cultivate these qualities, then perhaps poverty might be eliminated. However, unlike Spencer, he did not predict a future state of peace and equilibrium. Instead, he was worried that there would be a great conflict between the working class and the capitalist middle class.

Sumner's views typify social Darwinism in the American universities; those of Andrew Carnegie (1835–1919) characterize its expression in the business world. Carnegie's is a rags-to-riches story of a man who went from bobbin boy to founder of the United States Steel Corporation, a man who founded trusts that benefit people all over the world to this day. He emigrated from Scotland to the United States in 1848, when he worked in a cotton mill and a bobbin factory. Soon afterward he became a telegraph messenger and later a telegraph operator for the Pennsylvania Railroad. He bought shares in the company and then bought oil-rich land in the west of Pennsylvania. After the Civil War, he formed Carnegie Steel, the largest steel company in the world. He retired in 1901 with a fortune said to be $500 million. In his later life, Carnegie developed the "business of giving," asserting that "the man who dies . . . rich dies disgraced." He founded many libraries, art museums, natural history museums, and science centers. And in 1911, he started a foundation to award grants to promote "the advancement and diffusion of knowledge and understanding."

Carnegie is also remembered today as the author of *The Gospel of Wealth.* In it, he argued that individuals, private property, the "law" of accumulation of wealth, and the "law" of competition were the "highest result of human experience, the soil in which society, so far, has produced the best fruit":

> The price which society pays for the law of competition, like the price it pays for cheap comforts and luxuries, is also great; but the advantages of this law are also greater still than its cost—for it is to this law that we owe our wonderful material development, which brings improved conditions in its train. But, whether the law be benign or not, we must say of it, as we say of the change in the conditions of men to which we have referred: it is here; we cannot evade it; no substitutes for it have been found; and while the law may be sometimes hard for the individual, it is best for the race, because it ensures the survival of the fittest in every department.[15]

The best way to benefit society, Carnegie argued, was not tax schemes, but a system that allowed the accumulation of large quantities of money in private hands, which might be returned to the community in the form of generous philanthropies for purposes chosen by the donors. Of course, most businessmen did not see themselves as social Darwinists, and they attributed their success to their industry and virtue, rather than to any achievement in trampling on their less successful competitors.

War and Racism

Conservatives were generally reluctant to accept evolutionary theory, but there was one field in which Darwinism was vigorously applied by conservative politicians and ideologues: international relations. Taking the nation as a unit of struggle, British social Darwinists of the nineteenth century, for example, validated their empire building by claiming that uncivilized races were being taken over by a "superior" social order. Darwinism was used to justify war and struggles for social and/or racial supremacy. When the First World War began, British writers turned again and again to Darwinian analogies to stir up enthusiasm for it.

Evolutionary militarism reached its pinnacle with German racial theories and policies, some of which are traceable to the ideas of Haeckel and the Monist League he founded (see chapter 3). Haeckel's monist philosophy had an appeal across the political spectrum. His blasts against religion and established privilege interested the political Left. His prophecies that mechanistic evolutionary theory would bring humanity out of the dark ages of superstition endeared him to liberals. Although the Monist League included a wing of pacifists and leftists, it made an easy conversion to actively supporting Hitler. Unfortunately, Haeckel's greatest influence was on National Socialism.[16] An outspoken social Darwinist, he maintained that the overpowering laws of evolution ruled society and nature alike, and that natural law conferred on favored races the right to dominate others.

Perhaps the best-known nineteenth-century writer who proclaimed that the evolutionary progress of humanity could be furthered by interracial or international struggles was German historian Heinrich von Treitschke (1834–1896), who wrote in his two-volume work *Politics*: "Brave peoples alone have an existence, an evolution or a future; the weak and cowardly perish, and perish justly. The grandeur of history lies in the perpetual conflict of nations, and it is simply foolish to desire the suppression of their rivalry."[17] Von Treitschke spoke about "the moral majesty of war" and "the sublimity and grandeur of war," and he upheld the doctrine that might is right: "between civilized nations . . . war is the form of litigation by which States make their claims valid." Certainly these arguments are not derived solely from Darwinism. They show strong evidence of the influence of the theories of Hegel, and of the Prussian or Teutonic militaristic traditions. But von Treitschke drew on Darwinism to provide his ideas with a natural or scientific justification.

His diatribes were echoed in the writings of Adolf Hitler (1889–1945). For example, von Treitschke objected that instead of belonging to a single geographically defined *Volk*, Jews held allegiance both to the nation in which they lived and to Jewry as a whole. Hitler, in his infamous *Mein Kampf* (1925–26), drew analogies to the Jews from the world of animals. Each animal, he said, mates only with animals of its own species. He employed a long erroneous theory of blending inheritance, together with the survival of the fittest, to argue for the need for racial purity, the philosophical foundation of the Nazi movement.[18]

Darwinism on the Left

Darwinian theory was also embraced by the political left. Karl Marx (1816–1883) drew his ideas from several sources, but particularly from Adam Smith, from the dialectics of Hegel, and from his knowledge of economic conditions in the capitalist countries of Europe. He saw that the application of laissez-faire economics had given rise to a large class of wage laborers (the proletariats), who were working diligently in factories or fields, producing wealth that was not accruing to them. They were paid subsistence wages, and the surplus value of their labor was skimmed off by the capitalists or bourgeois class, who were gaining all the material and social benefit from the great wealth generated by the Industrial Revolution. The majority of wage laborers lacked the capital to break out of their miserable situations. There seemed to be no escape from their drudgery, poverty, ill health, and death within the industrial system that liberalism had produced. Such problems led to the formation of trade unions to seek improvements in workers' conditions, and to the growth of socialist ideas. The solution, according to Marx, was for the workers to rise in revolution against the ruling bourgeoisie and establish a new social system in which the state would at least temporarily take over all the means of production.

Marx argued that this class struggle and revolt was part of natural law. Both he and his colleague Friedrich Engels (1820–1895) were greatly interested in Darwin's work. While liberals applied it to competition between individuals, Marx and Engels applied it to conflict between groups, or class struggle. Marx wrote to a colleague on January 16, 1861, "Darwin's book is very important and it suits me well that it supports the class struggle in history from the point of view of natural science. One has, of course, to put up with the crude English method of discourse."[19]

Darwinism appealed to the political Left also because it offered a materialistic explanation in biology in place of vitalism and teleology. Marx also believed that Darwin provided a means of achieving a unified science encompassing both man and nature.[20] Engels placed Darwin and Marx on equal thrones in a speech delivered at Marx's grave side in Highgate in 1883: "Just as Darwin discovered the law of development of organic nature, so Marx discovered the law of development of human history."[21] A similar comparison was also made in Engels's preface to the English edition of the *Manifesto of the Communist Party* (1888).

Of course, those seeking causal laws of history also had to recognize the role of chance or accident in history. Marx addressed the issue only once, in a letter in which he asserted that it was not very important; it could accelerate or retard, but not radically alter the course of events.[22] Leon Trotsky used Darwinian theory to better account for the role of accidents: "The entire historical process is a refraction of historical law through the accidental. In the language of biology, we might say that the historical law is realized through the natural selection of accidents."[23]

Was Darwin a Social Darwinist?

What about Darwin himself? Was he a social Darwinist or in any way part of or responsible for social Darwinism? These questions have been debated for decades,[24] and many scholars and biologists desperately wanted his thoughts to be pure and untainted by such worldly social implications. Darwin was well aware of the debates around him. He also wrote in an anthropomorphic manner, making it easy for his arguments to resonate into the sociopolitical context of English life. However, consensus about his political views has been difficult to reach. Some writers insist that the young Darwin hardly fits the image of the individualistic social Darwinism,[25] and that there was nothing ideological in his view of nature.[26] Others, following the distinguished historian John C. Greene, emphasize that Darwin saw in natural selection a powerful means of interpreting human social evolution.[27] They point to several sources of evidence.

In his private correspondence, for example, Darwin expressed his belief that "man in the distant future will be a far more perfect creature than he now is," and that natural selection, driven by the struggle for existence between races, would continue to play a major role in human evolution.[28] In 1881, Darwin interpreted the Crusades of the Middle Ages in terms of such a struggle for existence:

> Lastly, I could show fight on natural selection having done and doing more for the progress of civilisation than you seem inclined to admit. Remember what risks nations of Europe ran, not so many centuries ago, of being overwhelmed by the Turks, and how ridiculous such an idea now is! The more civilized so-called Caucasian races have beaten the Turkish hollow in the struggle for existence. Looking to the world at no very distant date, what an endless number of the lower races will have been eliminated by the higher civilized races throughout the world.[29]

The best source for understanding Darwin's socioevolutionary views is *The Descent of Man* (1871). He had collected data for thirty years, and he embellished his work with writings of many social evolutionists of the 1860s.[30] Indeed, "social Darwinism" may itself be a misnomer because it presupposes that there was a Darwinism that was not. The term reinforces the notion that true science is objective and free from ideology. But one cannot draw a neat boundary between objective science on the one hand and ideology on the other, and thus distinguish Darwinism from its sociopolitical abuses. As James Moore put it, "The routine distinction made today between 'Darwinism' and 'Social Darwinism' would have been lost on the author of *The Descent of Man* and probably on most of his defenders until the 1890s."[31]

Social Theory in Evolution

Darwin was not a political pamphleteer. What is important is that Darwinian theory and social theory shared a common context. Indeed, one should not be left with the impression that the relationship between Darwin's theory and social theory was only a matter of applying naturalist theories to human history in order to justify certain forms of social order. The relationship was much more entangled. When formulating his theory, Darwin drew on socioeconomic concepts of the times.

Although sociopolitical thought is usually omitted when the theory of natural selection is taught, cut and dried, to biology students, it was crucial to the theory's formulation. In fact, some scholars have gone so far as to assert that "Darwinism was an extension of laissez-faire economic theory from social science to biology."[32] Two fundamental social concepts are at the heart of Darwin's theory: the concept of division of labor, which he used in his theory of divergence, and Malthusian population theory, which he used as the driving force behind natural selection.

The Division of Labor

Adam Smith used the expression "the division of labor" to refer to the separation of work into a number of tasks, each performed by a separate person or group of persons. For Smith, this specialization was key to economic progress because, he argued, it provided a cheap and efficient means of producing economic goods. Darwin used the concept in his theory for the origin of species through divergence.

His argument began with the ecological premise that a locality can support more life if it is occupied by diverse organisms partitioning the resources. Divergence and specialization would be adaptive advantages because organisms avoid competition and elimination in the struggle for existence if they separate into specialized niches in the economy of nature. The origin of varieties—"incipient species" for Darwin—would occur by vigorous selection for such specialization, and over evolutionary time, this process would generate a branching genealogy comprising all taxa from species to phylum.[33]

Darwin did not draw the concept of the division of labor directly from Smith, however. He adapted it from the concepts of the French zoologist Henri Milne Edwards. Whereas Darwin used the idea as a way of understanding the relations between varieties and between species, "an ecological division of labor," Milne Edwards had used it to describe a "physiological division of labor" when explaining the increased complexity of organization in animals who have diverse organs with specialized functions.[34] Darwin saw the cause of both to be one and the same: "The advantage of diversification of structure in the inhabitants of the same region is, in fact, the same as that of the physiological division of labour in the organs of the same body."[35]

Milne Edwards understood the difference between simple and "higher organisms" on the basis of political economists' views about the benefits of the division of labor in manufacturing. He taught in an institution that was particularly opportune for developing conceptual associations between natural history and political economy: the Ecole Centrale des Arts et Manufactures. Milne Edwards wrote in 1827:

> The bodies of these [simple] animals may be compared to a workshop where all the workers are employed at similar tasks and where, therefore, their number effects the amount, but not the nature of the result. In higher organisms, where life manifests more complex phenomena, organs become specialized. There is a location of functions and the life of the individual is more than the sum of the workings of elements of the same nature: it becomes the result of a set of essentially different actions produced by distinct organs. The diverse parts of the animal economy cooperate towards the same objective, but each one in its own manner; and the more numerous and developed the faculties of a given being, the more elaborate are the diversity of structure and the division of labor. . . . In a word, it is always according to the principle of the division of labor that nature perfects the results it wants.[36]

For Milne Edwards, the division of labor reflected the divine order of life from the simple to the complex. Darwin explained it in ecological and evolutionary terms. He argued that the more diversified organisms become through variation, the easier it would be for them to occupy specialized niches in the economy of nature and to escape elimination through competition to survive and reproduce. In other words, a physiological division of labor reflected an ecological division of labor.[37] I will return to the principle of the division of labor in later chapters when I examine cell theory, concepts of symbiosis, and the growth of biology itself.

Darwin and Malthus

Some of the most widely discussed evidence for a direct link between Darwin's theory of natural selection and sociopolitical issues of the period comes directly from his own notes. Eight months after returning from the *Beagle* voyage, in the spring of 1837, Darwin abandoned his belief in the fixity of species. But it took some time before he developed the concept of natural selection. He understood that breeders could select for desired traits and transform their stocks, but at first he failed to see how selection might operate in nature. Scholars agree that September 1838 was decisive. It was then that the twenty-nine-year-old naturalist read Thomas Malthus's *Essay on the Principle of Population*. Realizing that more individuals would always be produced than could possibly survive, Darwin found the conditions under which selection would operate in nature: the struggle for existence.

Thomas Malthus (1766–1834) was educated at Jesus College, Cambridge; he took holy orders, although he was not particularly interested in ecclesiastical matters. In 1805, he became professor of history and political economy at Haileybury College, an institution established by the East India Company for the training of men going into its service. He was elected Fellow of the Royal Society in 1819 and was a founding member of the Political Economy Club. Although a prolific writer, he is remembered today exclusively for the *Essay on the Principle of Population*, which was first published anonymously in 1798. The second, modified edition appeared in 1803 with the author's name.

The book addressed two issues. In the first place, it confronted the optimistic ideals of the Enlightenment and the utopian views of such writers as the mathematician and philosopher Jean Marie Condorcet (1743–1794), who saw in the French Revolution great hope for a progressive reconstruction of society, and William Godwin (1756–1836), one of the founders of philosophical anarchism. Both men contemplated indefinite progress in the perfectibility of humanity toward the complete absence of struggle among men—no illness, no sexual urges, no cares. Malthus insisted that it would never be possible to realize such ideals, for they took no account of an absolutely fundamental issue: an inescapable imbalance between nature's supply of food and the human need for food and sex. He postulated that populations, if left unchecked, would increase geometrically (exponentially) (2, 4, 8, 16, 32, 64), while the means of existence—that is, the food supply—could only increase at most arithmetically (1, 2, 3, 4, 5, 6, 7). Therefore, he reasoned, humankind must always be subject to famine, poverty, disease, and war.

Malthus's *Essay* was also aimed against restructuring the Poor Laws in England.[38] Since the seventeenth century, every parish had been required to levy rates for the relief of the poor. Relief was provided by the parishes in terms of doles, family allowances, and "aid-in-wages." The money was to be spent to train children whose parents could not provide for them, to provide work for the able-bodied unemployed, and to assist those who could not work owing to sickness, old age, or other some infirmity. Prime Minister William Pitt put forward a controversial bill for extending poor relief to larger families. About 8 million pounds were already collected annually for the poor, and as Malthus saw it, it did not and could not do much good.[39] On the contrary, he argued, such state charity would only encourage them to breed more and would inhibit individual industry: if there were more charity, the demand would rise to exhaust it, and everyone would be dragged down together. Thus, he wrote:

> Mr Pitt's Poor Bill has the appearance of being framed with benevolent intentions, and the clamour raised against it was in many respects ill directed, and unreasonable. But it must be confessed that it possesses in a high degree the great and radical defect of all systems of the kind, that of tending to increase population without increasing the means for its support, and thus to depress the condition of those that are not supported by parishes, and, consequently, to create more poor.[40]

In overcrowded industrialized Britain, Malthus's views about state charity earned him the hatred of all social reformers. But it held considerable sway. Pitt dropped his bill. In 1834, the year of Malthus's death, the New Poor Law was introduced, in which the Victorian workhouse was instituted. Conditions in the workhouses were deliberately made harsh to dissuade people from coming into them, and once in them men and women were segregated in order to discourage the production of children.

One might think that Malthus's notion that mankind would always increase beyond the greatest possible increase in food might be inconsistent with his own theological belief in the harmony of the divinely created world. But this paradox is only apparent. He insisted that his population principle did not impeach "the goodness of the Deity," and was not "inconsistent with the letter and spirit of the Scriptures." He argued that a way to improve conditions was to exercise "moral restraint," by delaying the average age of matrimony and by sexual abstinence or restraint within marriage. He emphasized that most cultures had some means for limiting population growth artificially—by having a large priestly class of celibates, for example, or by late marriage. The letter and spirit of Revelation implied difficulties to be overcome and temptations to be resisted. Therefore, "the proper government of the passions" was not only consistent with Revelation, "but afforded strong additional proofs of its truth."[41]

Malthusian population concerns continue to the present day as the human population of the earth exceeds 6 billion. Certainly, the size and rate of growth of the population today are major problems for the improvement of living conditions for more than half the world's population. However, critics argue, Malthus did not foresee, nor would he have approved of, the contraceptive practices by which Europe reduced its birth rates drastically during the nineteenth and twentieth centuries. They also emphasize that he was wrong in his assumption that only suffering could be relied upon to induce restraint among the masses. In reality, poverty induces more of the same, and not prudence. It was actually the secure upper and middle classes of Europe's society, not the poor, that started the trend toward reduced reproduction. In other words, improved living conditions and rising aspirations, not poverty and disease, have motivated the trend toward birth regulation.

As one commentator, concerned with population control in the twentieth century, put it: "Malthus is important today as the father of the most regressive social doctrine of our time, and as an expositor of some fundamental principles of population. His truth lives but under the cloud of his errors."[42]

The influence of Malthus on Darwin is a pivotal issue in the historiography of Darwinism.[43] Malthus impressed upon him the intensity of the "struggle for existence" between individuals.[44] As Darwin recalled in his autobiography, the key insight came to him in a flash after reading Malthus. Everything seemed to fall into place. Overreproduction would lead to a struggle for existence, which in turn would provide the conditions for natural selection:

> In October 1838 [actually September 28], that is, 15 months after I had begun my systematic enquiry, I happened to read for amusement "Malthus on Population" and being well-prepared to appreciate the struggle for existence which everywhere goes on from long-continued observation of habits of animals and plants, it at once struck me that under these circumstances favourable variations would tend to be preserved and unfavourable ones to be destroyed. The result would be the formation of a new species. Here then, I had at last got a theory by which to work.[45]

Alfred Russel Wallace also acknowledged that "perhaps the most important book I read was Malthus's *Principle of Population*."[46] As he reflected in 1887, "The most interesting coincidence in the matter, I think, is, that I *as well as Darwin*, was led to the theory itself through Malthus."[47]

In *The Origin*, Darwin wrote that the struggle for existence "is the doctrine of Malthus applied with manifold force to the whole of the animal and vegetable kingdoms; for in this case there can be no artificial increase of food, and no prudential restraint from marriage. Although some species may be now increasing, more or less rapidly, in numbers, all cannot do so, for the world would not hold them."[48] The struggle for existence in this strict sense also applied to relations between species and groups as well as between individuals. "But, the struggle almost invariably will be most severe between individuals of the same species, for they frequent the same districts, require the same food, and are exposed to the same dangers."[49] Competition led to suffering, death, and extinction, because there were always too many mouths for the world to feed. But this suffering, by producing "fitter" individuals through natural selection, would ultimately produce better organisms and lead to evolutionary progress. Competition was both inevitable and desirable. Sociopolitical thought was embedded within Darwinian theory.

The "struggle for existence" became the revolutionary slogan of enlightened materialists who raged against supernaturalism and mysticism, and against any moral character in nature. No one said it better than Darwin himself. He concluded in *The Origin*: "Thus, from the war of nature, from famine and death, the most exalted object which we are capable of conceiving, namely, the production of the higher animals, directly follows."[50] But even among evolutionists there were those who cautioned against overemphasizing Malthusian population pressure, and conflict and competition, as motors of "progressive" evolutionary change.

5

Mutualism

> Don't compete!—competition is always injurious to the species, and you have plenty of resources to avoid it! That is the *tendency* of nature, not always realised in full, but always present. That is the watchword which comes to us from the bush, the forest, the river, the ocean. Therefore combine—practice mutual aid! That is the surest means of giving to each and to all the greatest safety, the best guarantee of existence and progress, bodily, intellectual, and moral.
>
> —Peter Kropotkin, 1923

DURING THE LATE NINETEENTH CENTURY, when "natural law" was used as the basis for a particular given view of society and when Darwin's champions emphasized conflict and competition, others pointed to the importance of cooperation or mutual aid in evolutionary progress. The best-known works on mutualism emerged from two sometimes overlapping philosophies. One was anarchism, the other natural theology.

Anarchism

Anarchism grew from a profound concern with the ills of society and from an optimistic view of nature and of human society. In his book *Enquiry Concerning Political Justice* (1793), William Godwin argued that humankind could transcend its physical nature, that reason was supreme, that society could approach perfect harmony, and this could occur without central government, which was more of an oppressor than a protector. People were, by nature, good and imbued with natural sympathies toward one another. War, anarchists argued, is created by nations, not by the masses, and if there were no national governments there would be no such thing as international war.

Anarchists positioned their arguments in direct opposition to those expounded by the seventeenth-century philosopher Thomas Hobbes (1588–1679). In his famous book *Leviathan* (1651), Hobbes argued that humans were, by nature, selfish, brutal, and warlike, and if there were no state, there would be "a war of everyman against every man." He reasoned that people live together in civil society because, despite all the objectionable aspects of such a rule-bound civil structure, it was better than life in a society lacking central government. Even a wholly despotic regime was better, provided the ruler was powerful enough to keep the peace. Without a powerful government—an almighty Leviathan—to restrain them, humans would live as animals without virtue and morality, without agriculture, arts, and letters. In brief, life would be "nasty, brutish, and short."

Hobbesian views were echoed in the writings of Darwin's champions, who portrayed nature as ceaseless conflict and competition in opposition to the image of a cooperative, peaceful natural world designed by a benevolent creator, as portrayed by natural theology and *Naturphilosophie*. Ernst Haeckel expressed these Hobbesian views succinctly:

> If we contemplate the common life and the mutual relations between plants and animals (man included), we shall find everywhere, and at all times, the very opposite of that kindly and peaceful social life which the goodness of the Creator ought to have prepared for his creatures—we shall rather find everywhere a pitiless, most embittered *Struggle of all against all*. Nowhere in nature, no matter where we turn our eyes, does that idyllic peace, celebrated by the poets, exist; we find everywhere a struggle and a striving to annihilate neighbours and competitors. Passion and selfishness—conscious or unconscious—is everywhere the motive force of life.[1]

Such statements were not left unchallenged.

Peter Kropotkin (1842–1921) was one of the best-known anarchists who argued against those he called "the champions of egoism." His book *Mutual Aid: A Factor of Evolution*, first published in 1902, was an attempt to put the case for cooperation on a scientific basis and to embed anarchism in evolutionary theory. Born a prince of the old nobility of Moscow, Kropotkin grew up in the midst of the revolutionary movement against the Russian czars during the years of intense struggle for the abolition of serfdom and for a constitutional government. He studied mathematics and geography at the University of St. Petersburg for five years. While there, was drawn to the cause of peasants and adopted an anarchist political philosophy.

The discovery that Kropotkin was engaged in revolutionary activities caused a sensation. He was arrested and held in prison without trial. After a year's confinement he escaped and found refuge in England. He remained in exile for forty-two years, engaged in scientific research and anarchist propaganda.[2] In England, he was recognized for his work in geography. He was elected to the British Royal Geographical Society, an honor he rejected because of his hostility to any association with a "royal" affiliation. He was offered the chair of geography at Cambridge but declined since it was plain that the university would expect him to cease his anarchist activities while in its service. Instead, he earned a living through his writings.[3]

Between Individuals

Mutual Aid was based on a series of magazine articles he wrote, beginning in 1890, in response to views expressed by T. H. Huxley. In 1888, Huxley published "Struggle for Existence and Its Bearing upon Man," in which he argued, like Hobbes, that human morality has no basis in nature. From the point of view of the moralist, Huxley wrote, "the animal world is on about the same level as the gladiator's show. The creatures are fairly well treated, and set to fight—whereby the strongest, the swiftest, and the cunningest live to fight another day. The spectator has no need to turn his thumbs down, as no quarter is given." Huxley continued that as among animals, so among men: "the weakest and stupidest went to the wall, while the toughest and shrewdest, those who were best fitted to cope with their circumstances, survived. Life was a free fight, and beyond the limited and temporary relations of the family, the Hobbesian war of each against all was the normal state of existence."[4]

In a lecture titled "Evolution and Ethics," Huxley insisted there was no trace of moral purpose in nature; no ethics were required, merely survival, and the fittest to survive in the struggle for existence may be, and often are, the ethically worst.[5] For humans to progress socially, then, required a checking of the natural process.

Kropotkin insisted that humans were naturally cooperative and that cooperative behavior and altruistic feelings were perhaps the most important progressive elements in organic evolution. He supported his claims with observations he had made during natural history expeditions to eastern Siberia and northern Manchuria. There, he failed to find that bitter struggle for the means of existence among animals of the same species that was considered by Darwinists to be the dominant characteristic of the struggle for life and the main factor in evolution. Instead, he saw "mutual aid and mutual support" carried on to an extent that made him "suspect in it a feature of the greatest importance for the maintenance of life, the preservation of each species, and its further evolution."[6]

Kropotkin's views on mutual aid were not exclusively those of an anarchist.[7] They were representative of many evolutionists in Russia who argued against the all-exclusive Malthusian principles underlying Darwinian evolution and who saw the overemphasis on struggle between individuals as merely a reflection of the dog-eat-dog character of British industrial competition.[8] Indeed, it was not just anarchists who emphasized mutualisms. For example, the famed Russian botanist Constantin Merezhkowsky, who advocated the fundamental importance of "symbiogenesis," was just the opposite: a czarist reactionary (see chapter 19).[9]

In *Mutual Aid*, Kropotkin asserted that, much to the regret of Darwin himself, both the meaning and extent of the struggle for existence in evolution had been exaggerated, while the importance of sociability and social instincts in animals for the well-being of the species and community had been underrated. In *The Origin*, Darwin premised that he used the expression "'struggle for existence' in a large and metaphorical sense, including dependence of one being on another, and including (which is more important) not only the life of the individual, but success in leaving progeny."[10] He discussed the mutual benefits of flowers and the insects that pollinate them when collecting nectar, and he mentioned the dispersal of mistletoe seeds by birds.[11] Nonetheless, the fundamental ecological and evolutionary thinking of Darwin and those who followed him was that adaptation and speciation could be explained by conflict and competition, especially between individuals who shared the same niche. But the concept of struggle had become narrower still, Kroptokin remarked, when followers of Darwin "raised the 'pitiless' struggle for personal advantage to the height of a biological principle which man must submit to as well."[12]

Whereas Darwin had argued that conflict and competition would be most intense between individuals of a species, Kropotkin saw the inverse: a tendency toward cooperation between individuals of species and toward conflict between species. In fact, he juxtaposed cooperation among individuals of a species against "the immense amount of warfare and extermination going on amidst various species."[13] He emphasized the herding instincts of deer against a common foe and the colonies built by termites and ants. He argued that those animals which practiced mutual aid were much more "fit," intelligent, and highly developed than those that were constantly at war with each other: "The ants and the termites have renounced the 'Hobbesian war' and they are the better for it."[14]

Kropotkin's discussions of mutualism, like those of many other biologists in Russia, were concerned primarily with cooperation within animal species. He made only the briefest mention of microbial cooperation—that is, among "the lowest animals"—when he commented that "we must be prepared to learn someday, from the students of microscopic pond life, facts of unconscious mutual support."[15] I will explore symbiosis involving microorganisms later when I examine the role of symbiosis in evolution (see chapter 19). Studies of mutualism between species also grew in Western Europe in opposition to the notion of all-exclusive conflict and competition.

Between Species

The term "mutualism" was introduced into biology in 1872 by the Belgian zoologist Pierre-Joseph van Beneden (1809–1888) at the Catholic University of Louvain.[16] He used the term in a communication to the Royal Academy of Belgium which he elaborated into a very popular book, *Les commensaux et les parasites*, translated into English as *Animal Parasites and Messmates* (1876).[17] In it, he argued that the kinds of social relations in animal societies were as varied as those found in human societies, and he classified them in terms of parasitism, commensalism, and mutualism. The word "parasite" had been used since the mid-sixteenth century to derogatorily describe someone who ate at the table or expense of another. As van Beneden wrote, "The parasite is he whose profession it is to live at the expense of his neighbour, and whose only employment consists in taking advantage of him, but prudently, so as not to endanger his life."[18] The commensal or "messmate" "is he who is received at the table of his neighbour to partake with him of the produce of his day's fishing. . . . The messmate does not live at the expense of his host; all that he desires is a home or his friend's superfluities."[19]

Van Beneden pointed to various species of fish that swim alongside larger individuals (such as pilot fish swimming alongside sharks) from which he believed they receive aid and protection. He found many examples among crustaceans: tiny crabs (Pinnotheres) lived inside shells and seemed to exchange food for lodging. Still other "messmates" become completely dependent on their host and "lose their former appearance: not only do they throw aside their oars and their pincers, but they cease sometimes to keep up any communication with the outer world, and even give up the most precious organs of animal life, not even excepting those of the senses; they are installed for life, and their fate is bound up with the host which gives them shelter."[20] He referred to barnacles that lodge on the back of a whale or the fin of a shark and subsequently become transformed in such a way that they no longer "have a mouth by which to feed, and are reduced to a mere case which shelters their progeny."[21] In his view, there existed "almost insensible gradations of differences between parasite, messmate and free animals."[22] The mutualists, he wrote,

> are animals which live on each other, without being either parasites or messmates; many of them are towed along by others; some render each other mutual services, others again take advantage of some assistance which their companions can give them; some afford each other an asylum, and some are found which have sympathetic bonds which always draw them together.[23]

Van Beneden pointed to insects that shelter in the fur of mammals or the down of birds and thereby care for the toilet of their host by feeding on epidermal debris, excretions, infusoria (protists, such as those living in the rectum of frogs), and the like. Other mutualists rendered services he compared to medical attendance. The Egyptian plover "keeps the teeth of the crocodile clean," and, he noted, there was a certain worm that lives in a lobster, and eats only the dead eggs and the embryos, the decomposition of which might be fatal to the host lobster and its progeny.[24]

Van Beneden's text also had a political context in anarchism (though it was much less explicit than that of Kroptokin), as well as a religious component in natural theology. While competition and progress through individual life struggle were dominant themes of both natural and social science in the nineteenth century, political and intellectual opposition developed in concert. In Britain, various associations such as trade unions and the "Friendly Societies" flourished in the nineteenth century to help members from falling into debt because of illness, old age, or a death in the family. The analogous organizations in France were the mutual aid associations. Beginning in the period of the French Revolution, the different *mutualité* societies were a hotbed of socialist ideas.

French *mutuellisme*'s most celebrated advocate was Pierre-Joseph Proudhon (1809–1865), recognized today as one of the founders of socialist and anarchist movements. He, along with Karl Marx and many others, repudiated Malthus for trying to lay at the door of the poor the responsibility for their own plight—which, they contended, was really the result of the selfishness of the rich.[25] Proudhon's mutualism was an antiauthoritarian ideology based on the abolition of governments and the reconstruction of society as an overarching federation of worker's cooperatives. He denounced political revolution as unnecessary and even dangerous to liberty. He believed that the path to socialism could instead be blazed through the development of a system of mutual credit in which workers could borrow the funds to amass capital and create cooperatives, which would eventually replace capitalism. He articulated a program of mutual financial cooperation among workers that he believed would transfer to them control of economic relations.

Proudhon was imprisoned from 1849 to 1852 for publishing articles criticizing Louis Napoleon Bonaparte. From 1858 to 1862, he lived in exile in Belgium. Pardoned by the emperor in 1862, he returned to Paris and continued to gain influence among the Paris workers with his mutualist ideas. In 1871, the Paris Commune took control of the city, and in the brief period available to it for reorganizing the economy before being bloodily repressed, it manifested a clear Proudhonian viewpoint. The Metal-Workers and Mechanics Unions of Paris expressed their aims as "the abolition of the exploitation of man by man, last vestiges of slavery; the organization of labor in mutual associations with collective and inalienable capital."[26] The Association of Women, led by the revolutionary socialist Elizabeth Dimitrieff, proposed organizations that would help provide work for women and would "instil into them a strong consciousness of mutualism."[27]

This was the sociopolitical setting in which van Beneden first introduced the term "mutualism" in his address in Belgium in 1872. He drew analogies from industry, human social relations, and morality to describe the social relations he saw in nature. He compared industrialists leading the life of noblemen to parasites:

> In the ancient as well as the new world, more than one animal resembles somewhat the sharper leading the life of a great nobleman; and it is not rare to find, by the side of the humble pickpocket, the audacious brigand of the high road, who lives solely on blood and carnage. A great proportion of these creatures always escape, either by cunning, by audacity, or by superior villainy, from social retribution.[28]

The word "sharper" is used as the English translation for the French term *chevalier d'industrie*. The implication in both is that the industrialist is a thief, and the ironic comparison of parasites and "knights of industry" is remarkably close to the anarchist perspective. Moreover, the term "mutualists" itself would surely have evoked thoughts of Proudhon and the mutualists of the Paris Commune.[29]

Roots in Natural Theology

Mutualisms were frequently used by ancient writers as examples of nature's balance: those tendencies that prevented any species from becoming either too abundant or extinct were due to divine providence. Herodotus told the story about a mutually beneficial relationship between Nile crocodiles and a species of plover. The plover ate leeches from the crocodiles' mouth, and the crocodile never hurts the bird.[30] Aristotle liked that story and mentioned it in three different books. He also reported that a mutual relationship existed between certain mussels (*Pinna)* and little crabs (*Pinnotheres*).[31] Similar descriptions were given by Cicero and Aelian, who drew the moral that humans should learn friendship from nature.[32] Pliny also remarked that "friendships occur between peacocks and pigeons, turtle-doves and parrots, blackbirds and turtle-doves, the crow and the little heron in a joint enmity against the fox kind, and the goshawk and kite against the buzzard."[33]

Mutual interactions were favorite examples used to illustrate divine providence in the natural theology of the seventeenth and eighteenth centuries, which found its ultimate expression in the works of Linneaus. In his "economy of nature," all natural things are fitted together by the divine wisdom of the Creator to produce reciprocal uses; all living beings were so connected, so chained together, that they all aimed at the same goal. The search for "general ends," for an overriding purpose and agency in nature, was the crucial impetus to the Linnean school of natural history. Besides Linneaus's pivotal essay of 1749, "The Oeconomy of Nature," the leading works in this field included John Ray's *The Wisdom of God Manifested in the Works of Creation* (1691), William Derham's *Physico-Theology* (1713), William Paley's *Natural Theology* (1802), and the famous eight books that comprised the *Bridgewater Treatises* (1833–36). All were united by their common repudiation of doctrinal schisms and of claims of private mystical revelations, and instead looked to reason and the testimony of nature to establish their faith on a firm, universally acceptable ground.[34]

Followers of natural theology implicitly accepted the assumptions underlying the Hobbesian view, but they argued that nature did in fact have laws that prevented such disorder and disharmony.[35] The Creator, they maintained, had established a vast system of subordination to assure peace in the natural world. Each species had been assigned a fixed place in a social hierarchy or scale of being, which was also a system of economic interdependence and mutual assistance. Even the most exalted creatures depended upon those lower on the scale for their very existence; man and worm alike lived to preserve each other's life.

All of these views are explicit in van Beneden's *Animal Parasites and Messmates*. "The assistance rendered by animals to each other," he argued, "is as varied as that which is found among men. Some receive merely an abode, others nourishment, others again food and shelter; we find a perfect system of board and lodging combined with philozoic institutions arranged in the most perfect manner."[36] Within this system of mutual aid, some organisms performed medical attendance, others carried out the role of menagerie keepers, some cleansed the animals themselves, and others kept their cages clean and removed the dung and filth. Paupers helped other paupers as well as the "higher classes." Some members of the higher classes helped the lower, as the little crabs helped the mussels in which they took shelter by dropping crumbs of food from their pincers. They were not all like the "rich man who installs himself in the dwelling of the poor, and causes him to participate in all the advantages of his position." They were good lodgers. But the point was equally true that they actually required assistance from the lower classes. As van Beneden wrote, the "noble crab" relied on lodging from the lower "blind and legless mussel."[37]

Van Beneden was a Catholic with deep religious convictions. He believed that evolution was designed by the will of God, and that mutualisms were examples of perfect adaptations created by the divine wisdom:

> All these mutual adaptations are pre-arranged, and as far as we are concerned, we cannot divest ourselves of the idea that the earth has been prepared successively for plants, animals, and man. When God first elaborated matter, He had evidently that being in view who was intended at some future day to raise his thoughts to Him, and do Him homage.[38]

Van Beneden remained a lifelong opponent of the view that evolution resulted from a "struggle for life" and natural selection. If this were true, he wondered, how is it that those beings that were powerfully armed for struggle, the giants of the animal world of diverse classes, were precisely those that had succumbed in the struggle for existence?[39]

Although zoologists and botanists adopted van Beneden's terminology, the meaning they placed on it generally differed. For most, mutualistic and parasitic associations involved no finality. They were not prearranged or laid down by divine providence. Nonetheless, from the nineteenth century onward, mutualisms were often discussed in opposition to a Hobbesian war of each against all and a belief in evolutionary and social progress resulting from a pitiless struggle for individual advantage.[40]

Understanding how mutualisms could have evolve by natural selection was only one problem left to the future (see chapter 20). Evolutionists of Darwin's day pointed to many other difficulties with the theory of natural selection, and as we shall review in the following chapter, they proposed additional mechanisms that they considered more important.

6

Dissent from Darwin

> Very different judgements have been passed on Darwin. Even on his first appearance he was either extolled as one of the greatest geniuses in the world or abused as an ignorant and unreliable dilettante. . . . Nor have subsequent generations been any more unanimous; especially since the theory of selection has been condemned—at least in its original form.
>
> —Eric Nordenskiöld, 1927

REVOLUTIONS ARE RARELY COMPLETE, and the Darwinian revolution was no exception. Darwin convinced many biologists of evolution, but he did not convince them that natural selection accounted for the origin of species. Between 1860 and 1930, his theory was eclipsed as evolutionists raised numerous problems and concerns and theorized about alternative mechanisms to account for various facets of life's diversity.[1] Historians have studied the reception of Darwinian theory in many countries, examining various religious, philosophical, social and political contexts[2] and discovering several global issues.

Above all was the sweeping concern that natural selection acting on chance variations was not enough to account for direction in evolution—for "evolutionary progress" or increased complexity. Evolution according to selection was a tinkering process: life, in all of its manifestations, *is* because it can be, with no design or purpose. Darwin's metaphor of natural selection was no more than that, and, as critics noted, it was somewhat misleading, for there is no known agent in nature that deliberately selects. Natural selection is not an active force or promoter of evolution, but rather a passive pruning device to weed out the unfit; it acts as a sifting mechanism, permitting some species to thrive while others perish. Evolution would be a matter of chance and necessity. Chance would provide the raw materials for evolution, but natural selection (necessity) or the "survival of the fittest" alone would direct evolution. And many evolutionists refused to leave the history of life to such a haphazard, apparently undirected process.

There were numerous other specific biological issues. Were all traits really adaptive, as the theory of natural selection presupposed? Why did the fossil record not show evidence of gradual transitions between species? Did species occur suddenly? Did speciation require geographic isolation? How could one account for the evolution of cooperative relationships among social animals, such as birds and deer? How could altruistic behaviors be explained in terms of individual competition for life? How could one account for the evolution of complex organs by natural selection acting on minute adaptive changes?

Criticisms of selection theory did not emanate from one particular biological specialty, nor from any specific country, although individuals naturally emphasized problems in their own specialties, and certain groups favored some theories over others. In order to appreciate the strengths and weaknesses of alternative theories, we must first understand some customary assumptions about the nature of the earth and about heredity and species that were not congruent with Darwin's model.

Is the Earth Old Enough?

Following Lyell's uniformitarian geology, Darwin had assumed that the earth was several billion years old, which he thought would allow ample time for evolution to occur by a slow, gradual process by natural selection, in which variations accumulated within populations. However, in 1868, the physicist William Thomson (later Lord Kelvin) challenged this notion of a very old earth.[3] Based on estimates of the internal temperature and the rate of cooling from its beginnings as a mass of molten rock, he calculated that the earth was no more than about 100 million years old, most likely about 24 million.

Kelvin's calculations mustered considerable authority for decades. But he was unaware of a long-lasting source of energy whose presence would have slowed cooling considerably: radioactivity. In 1903, Pierre Currie reported that the radioactive decay of such elements as radium liberated a slow, steady supply of energy as heat. In 1906, Lord Rayleigh calculated that the heat produced by radioactivity deep in the earth would balance the cooling effect postulated by Kelvin. Allowing for radiation, physicists calculated the earth to be about 4.5 billion years old, just about what Darwin had suspected.

Blending Inheritance

Darwin proposed that the differences between individuals that we see all around us provide the fuel for evolution. But the concepts of heredity of his day made it difficult to see how these differences could ever lead to new species. The problem was that most variations seemed to be blends of characteristics inherited from parents, and it was assumed that the actual germinal material from each parent was mixed together like two cans of paint. Certainly, there were cases in which characters were inherited on an all-or-nothing basis, but these seemed to be rare. If traits were blended, then any new characteristic that arose spontaneously would be passed on in ever-diminishing force to succeeding generations. Therefore, all new single heritable changes would soon be washed out of a population. Natural selection, as the engineer Fleeming Jenkin pointed out in 1867, would then have little that was stable enough to work on.[4] This difficulty would not be solved until after 1900 with the rise of Mendelian genetics (see chapters 11, 12, and 13).

What Is a Species?

In the eighteenth and early nineteenth centuries, the word "species" by definition implied its immutability. Georges Buffon offered the traditional definition, according to which "we should regard two animals as belonging to the same species if, by means of copulation, they can perpetuate themselves and preserve the likeness of the species."[5] A common view even in the nineteenth century was that diversity within a species was constrained, and the accumulation of individual variations could not step outside the species type. The individual differences on which Darwin believed natural selection acted seemed to be relatively trivial, not the fundamental characteristics that distinguished species and higher taxonomic groups. Yet, for Darwin there were no species-specific characteristics as such. All species arose from accumulations of small individual differences. Ernst Mayr has characterized these conflicting views of species as belonging to "typologists" and "nominalists." While typologists held that each species had its own essential type, nominalists maintained that species were not real entities, but rather arbitrary units of classification: individuals and the differences between them were the only things that were real.[6]

It seemed illogical to argue that species are not real while at the same time claiming that new ones gradually evolve from within them, as Darwin did. If there was no such thing as species, critics objected, how could they evolve? The American naturalist Louis Agassiz commented in 1860,

> It seems to me that there is much confusion of ideas in the general statement of the variability of species so often repeated of late. If species do not exist at all, as the supporters of the transmutation theory maintain, how can they vary? And if individuals alone may exist, how can the differences which may be observed between them prove the variability of species.[7]

Darwin was not impressed with such criticisms. In his response, he emphasized that species did have a reality in time, that is, a temporary constancy. As he wrote to Asa Gray, "I am surprised that Agassiz did not succeed in writing something better. How absurd that logical quibble—'if species do not exist, how can they vary?' As if anyone doubted their temporary existence."[8]

Scholars today find Darwin's views somewhat confusing—as did his own contemporaries, who insisted that a new definition was required.[9] Debates over the definition of species continue to the present day.[10]

Speciation and Isolation

In *The Origin*, Darwin argued that speciation was a process of divergence. The struggle for existence would be most intense between individuals occupying the same niche. Therefore, selection would favor those traits that would enable an individual to avoid competition. In short, the struggle for existence was sufficient to account for divergences of type and the branches of the evolutionary tree. Others insisted that geographical isolation was crucial. According to this view, some geographical barrier such as a river, a mountain range, or an island formation would isolate populations, which would then evolve separately, adapt to new circumstances, and eventually accumulate differences that would make them different species. The German naturalist Moritz Wagner was the first to challenge Darwin's account of divergence, in 1868, and there was a range of opinions about the importance of geographic isolation for speciation.[11]

Is Everything Adapted?

If natural selection was the sole mechanism of evolution, then every evolved character must, at some time or other, be useful or advantageous in the struggle for existence. They would all be adaptations. Indeed, Darwin offered a completely utilitarian view of life: if a characteristic was not useful, it would be lost in a population. Of course there were exceptions.[12] In *The Origin* he added the concept of sexual selection to account for such features as eye color, the bright plumage of a male bird, or the antlers of a deer. Such traits, though perhaps not useful in the struggle for existence, would be useful in attracting mates.[13] But naturalists pointed to various other aspects of organisms that they believed would not have contributed to the reproductive success of the individuals that possessed them. During the 1860s, Alfred Russel Wallace, who had been a strong selectionist, turned to spiritualism to explain some psychological traits. He pointed to a number of mental qualities that he suspected would have been useless for early man.[14] What, for instance, was the usefulness or purpose of the musical sense, or the ability to perform abstract mathematical calculations?

Similar problems were raised with animal evolution generally. What would be the selective value of rudimentary stages of complex organs? If all characteristics were useful, how could such complex organs as an eye or the wing of a bird or insect have evolved by selection? How can natural selection favor an eye, or a wing, before it is fully formed and functionally operational? This objection was emphasized by one of Darwin's most persistent critics in Britain, St. George Jackson Mivart, in his well-known book *Genesis of Species* (1871).[15] Mivart, a Catholic deeply concerned over the implications of human evolution, was prepared to accept the idea of evolution if it were supernaturally guided in certain directions. However, the objection that complex organs would have had earlier stages of no real value was certainly not exclusive to those who believed in supernatural forces.[16]

Many evolutionists pointed to the difficulty of explaining the emergence of mimicry by natural selection.[17] Stick insects use stealth so as not to be eaten by birds; leaf insects look like leaves. Many edible species of butterfly are protected by their resemblance to noxious or poisonous species. Would there not have been stages in the evolution of such complex characteristics that would have had no selective value? Debates over the limits of adaptationism continue today (see chapters 18 and 20).

Holes in the Record

The fossil record was the only real evidence of the course of the history of life on Earth. Yet, the gaps between species in that record was, then and now, something of an issue in regard to continuous and gradual evolutionary change as postulated by Darwin. New types or classes seemed to appear fully formed, with no signs of an evolutionary trend by which they could have emerged slowly from an ancestor. Darwin considered this lack of intermediate forms the greatest stumbling block to his theory, and he devoted a chapter of *The Origin* to explaining this absence. He insisted that the fossils brought to light by paleontologists represented only a tiny fraction of the species that actually lived. Many species, and many episodes in evolution, would leave no trace at all because they occurred in areas where conditions were not suitable for fossilization. Thus, the sudden leaps in the history of life are illusions, an artifact of an imperfect fossil record. As Darwin remarked, "He who rejects these views on the nature of the geological record will rightly reject my whole theory."[18]

Subsequent discoveries in paleontology filled in some larger missing links. The bridge between reptiles and birds attracted attention when T. H. Huxley suggested that the legs and feet of certain dinosaurs were almost indistinguishable from those of birds.[19] He claimed this to be the first sign of a "missing link" between two major classes. Later, *Archaeopteryx* was discovered: a creature with feathers like a bird, but a body with many reptilian characteristics, and a mouth with teeth rather than a beak. In 1880, the famed Yale paleontologist Othniel C. Marsh (1831–1899) discovered more toothed birds in the United States and named them the *Odontornithes*.[20] And other mammal-like reptiles showed some possible links between other classes.[21] All of these discoveries contradicted nonevolutionists' beliefs that different classes of organisms had completely separate origins.

A number of other illustrations of continuous evolutionary changes were constructed, but none attracted more attention than that of the horse, with its broad hoof derived from a single toe. Evolutionists argued that it must have evolved from an ancestor that had five toes like any other mammal. Accordingly, in the 1880s, Marsh arranged a whole series of fossils so as to trace the horse back to a small four-toed ancestor, which he named *Eohippus*, the dawn horse from the Eocene (54–38 million years ago).[22] Marsh's construction of horse genealogy was not only used as hard evidence of evolution; it also supported the view that the gaps in the fossil record may be a result of its imperfection. The fossils of the earliest known hominid species were first described by the Australian-born paleontologist Raymond Dart (1893–1988) in South Africa in 1925 as belonging to the genus *Australopithecus* (southern apes), which lived between 4 and 2 million years ago.[23]

Evidence of evolution was abundant, but Darwin's natural selection acting on random changes as the driving force of evolution faced three commonly accepted alternatives: neo-Lamarckism (the inheritance of acquired characteristics), orthogenesis (directed evolution of a characteristic along a certain path or straight line), and saltationism (evolution by sudden leaps).

Neo-Lamarckism

The inheritance of acquired characteristics offered a mechanism that could theoretically account for adaptations as readily as natural selection acting on random changes. Many biologists believed that adaptive variations due to environmental effects or to changes of behavior leading to the use and disuse of organs in an individual's lifetime could affect later generations. Darwin himself, especially in later editions of *The Origin*, also emphasized hereditary adaptative changes possibly arising through the effect of use and disuse, or direct action of the environment.

Neo-Darwinism emerged in the early 1880s, shortly after Darwin's death. The movement was led by the Freiburg zoologist August Weismann, who denied that the inheritance of acquired characteristics ever occurred. Fierce debates emerged between strict selectionists and those who maintained that the inheritance of acquired characteristics played the crucial role in progressive evolution. In the United States, the paleontologists Edward Drinker Cope (1840–1897) and Alpheus Hyatt (1838–1902) led a neo-Lamarckian school.[24] In France, a major tradition of neo-Lamarckism emerged in the 1880s and persisted throughout most of the twentieth century (see chapter 17).[25] Neo-Lamarckism was also strong in Germany, during the late nineteenth and early twentieth centuries; Ernst Haeckel himself was one of its chief defenders. Debates over the inheritance of acquired characteristics shaped the growth of evolutionary biology and led to proposals in the 1890s to set up institutions devoted to the study of evolutionary mechanisms.[26]

The inheritance of acquired characteristics would allow nature to be more progressive and directed than would natural selection alone. It offered a softer view of life, a more hopeful vision of nature, without the relentless individual life struggle that natural selection based on Malthusian tenets demanded. Neo-Lamarckism often carried teleological and spiritual overtones, especially among French biologists steeped in Catholicism. It also left room for consciousness as a directing force in nature: instead of being designed by a creator, species designed themselves as consciousness evolved.[27] Animals would control their own evolution because they consciously chose how to respond to their environment. Thus, neo-Lamarckism generated a sense that life controls its own destiny.[28]

Importantly, the inheritance of acquired characteristics also could account for the progressive evolution of morality.[29] Herbert Spencer, for example, argued that if the inheritance of acquired characteristics did not occur, then one could not hope for permanent improvement through the conscious exertions of individuals, groups, and nations. As he wrote about Weismann's views and his own in 1887: "I will add only that, considering the width and depth of the effects which acceptance of one or other of these hypotheses must have on our views of Life, Mind, Morals, and Politics, the question—Which of them is true? demands, beyond all other questions whatever, the attention of scientific men."[30]

Darwin also suspected that natural selection alone could not account for morality and altruism and included the inheritance of habit to explain them. Where did the benevolence of parents and human bravery come from, and how could they be perpetuated? How could natural selection operate on traits that would be for the good of the group but that would have an adverse effect on their bearer? Darwin thought that these traits would have been acquired and maintained in humans through a combination of the inheritance of acquired characteristics (over many generations, habits of mutual aid in early humans would have become hereditary) and the struggle for existence between groups.[31] The evolution of altruistic traits remains a subject of controversy among evolutionists (see chapter 20).

For neo-Lamarckians and other non-Darwinians, natural selection was about the survival of the fittest, not the arrival of the fittest. They admitted that natural selection might be able to explain the long neck of the giraffe if one assumed that the necks of the giraffe's ancestors were of different lengths, and the selection of the longest produced the long necks we see today. But what about the horns of ruminants? No one had ever seen the appearance of rudimentary horns on any other mammal as an occasional variation. It was not hard to imagine that butting their foreheads would produce them.[32] Neo-Lamarckians argued that the struggle for existence played the role of the gardener pruning the tree of life: it took away the twigs and branches to produce an arrangement with clearly distinguishable parts, and it made the phylogenetic differences stand out. But it no more made the twigs than did the gardener. Some nineteenth-century evolutionists espoused several mechanisms, and some included an unknown force through which evolutionary changes would be directed along a straight line.

Orthogenesis

Orthogenesis was often closely associated with neo-Lamarckism. The word was popularized by the German zoologist Theodor Eimer (1843–1898) in his book *Organic Evolution as the Result of the Inheritance of Acquired Characteristics According to the Laws of Organic Growth* (translated into English in 1890). He used the term to describe the regular trends mentioned by American paleontologists such as Hyatt and Cope, who argued, for example, that the evolution of the modern horse had proceeded in a regular single direction that was difficult to explain by random variation.

Such trends were often nonadaptive, and in some cases orthogenesis could lead the species to extinction. The most famous case was that of the Irish elk, whose relatively recent extinction was thought to be due to its antlers becoming too large. The giant deer had evolved from smaller deer with smaller antlers. While the antlers may have been useful at first, supporters of orthogenesis argued, the trend toward large antlers acquired a momentum of its own and went too far. Stooped by the weight of their ninety-pound antlers, or with the antlers entangled in trees or mired in ponds, the Irish elk were eventually wiped out. This theory of extinction came to be entertained widely by non-Darwinian paleontologists during the early twentieth century.[33]

Henry Fairfield Osborn (1857–1935), a former student of Cope, became one of the chief exponents of orthogenesis.[34] Osborn, a dinosaur expert, became the first curator of the American Museum of Natural History in 1891. He recognized that evolution of a class from a common ancestor represented a branching tree (he coined the expression "adaptive radiation" to describe the diversification of a class into many orders). But once an order emerged, its own evolution would be a linear process without the postulated small branchings of Darwin's bushier phylogenetic tree. What caused this direction in evolution was unknown.[35]

Some maintained that these straight evolutionary lines were due to some kind of mysterious force internal to the organism. The German physiologist Carl Ernst von Nägeli in 1884 proposed that some unknown "internal causes" of a "mechanical nature" effect transformations "towards greater perfection."[36] "Once the motion of evolution is started," he asserted, " it cannot cease, but must persist in its original direction."[37] By "greater perfection," he meant more complex structure and greater division of labor. Thus, neither the struggle for existence nor the inheritance of acquired characteristics had any effect on the formation of species.

Eimer, on the other hand, found Nägeli's interpretations of orthogenesis to be absurd. If evolution constantly worked in a specific direction toward increased complexity, "lower organisms," which existed millions of years ago, would have changed into "higher," and there should be no lower organisms around today.[38] Nägeli himself accounted for this by assuming that abiogenesis (spontaneous generation), the origin of very primitive life-forms, still takes place to the present day to refuel the orthogenetic process. In other words, the "lower" or "simplest" creatures we see would not be very old at all; on the contrary, they would be the most recently evolved. As Eimer saw it, Nägeli's argument about evolution by "internal causes" amounted to nothing more than vitalism and mysticism dressed in materialist clothes.[39] Nägeli had also gone too far in removing adaptation, "the Darwinian principle of utility," from evolution.[40] Eimer sought an explanation for orthogenesis in terms of both physiological and external causes, and he argued that not all evolution changes were directional.

Eimer developed an eclectic view of evolution based on five different causes: the direct influence of external conditions; the functional activity of organisms in relation to the external world (use and disuse); the struggle for existence; "sexual mixing, which may, without any influence of adaptation, lead to the formation of quite new material combinations, that is, to the production of new forms"; and "the sudden appearance of new formations through correlation (evolution *per saltum*)."[41] This last concept was very popular between 1860 and 1920.

Saltationism

The sudden appearance of species without the aid of selection could in principle resolve several outstanding issues. It could explain the absence of intermediate forms in the fossil record. It explained evolution in the face of blending inheritance because individual differences would not be the ones that fuel evolution. It opened the possibility for more rapid evolution and thus provided an answer to Lord Kelvin's critique of the insufficient amount of time necessary for gradual evolution. And it resolved difficulties about how complex organs such as the eye could have evolved, because one need not postulate small gradual steps of supposedly adaptive value for selection.

Evolution by large leaps could also account for correlative characteristics of the organisms.[42] For example, naturalists noted that hair, horns, and hooves were connected correlatively with one another, and simultaneous variations in these three organ systems lead to fundamental changes in the appearance of an animal.[43] Theodor Eimer drew an analogy with the kaleidoscope when imagining how such large changes in the correlative parts of the organism as a whole would occur:

> As soon as something or other in the original state, in the original arrangement of the parts of the organism, is changed, other parts also are set in motion, all arranges itself into a new whole, becomes—or forms—a new species,—just as in a kaleidoscope, as soon as on turning it one particle falls, the others also are disturbed, and arrange themselves in a new figure—as it were recrystallise.[44]

St. George Jackson Mivart combined saltationism with teleological principles to account for progressive evolution.[45] Similarly, Richard Owen speculated that there might be two kinds of variations with two different causes. Those changes within a species might be the result of accidental causes, and those discontinuous variations would be directed along a predetermined path.[46] Thomas Henry Huxley himself appealed to saltationism, without a directed plan, against Darwin's own insistence that *natura non facit saltum*.[47]

This enthusiasm for saltationism provided the setting for one of the great moments in the history of science: the rediscovery of Mendel's laws at the turn of the twentieth century (see chapter 11). The idea that new species were formed by sudden leaps without selection was advanced by William Bateson in England, who coined the term "genetics"; by Wilhelm Johannsen in Denmark, who issued the terms "genotype," "phenotype," and the "gene"; by Hugo de Vries in Holland, one of the codiscoverers of Mendel's laws; and by Thomas Hunt Morgan in the

United States, founder of the *Drosophila* school that developed the Mendelian chromosome theory.[48]

On the basis of a merger between Mendelian theory and microscopic studies of cells, geneticists were able to link the inheritance of variations to actual mechanisms operating within the germ cells of individuals. Studies of fertilization, the internal organization of cells, and the manner in which they reproduce developed in parallel to evolutionary theory. And by the late nineteenth century, it had become amply clear to a new generation of experimental biologists that the general problems of heredity and evolution could be fully understood only in the light of studies of the cell.

II

THE CELL IN DEVELOPMENT AND HEREDITY

7

The Myth of the Cell Theory

> It is one of the amazing facts of scientific history that in many biological textbooks Schleiden is called the founder of the cell theory, as if he had first discovered that all tissues of plants are composed of cells, or that the cell is the universal unit of organic function as well as of structure.
>
> —E. G. Conklin, 1940

BIOLOGISTS OF THE SECOND HALF of the nineteenth century who investigated the cell often proclaimed that it held the key to all biological problems. All distinctive vital processes—metabolism, growth and reproduction, sexual phenomena and heredity—could be reduced ultimately to activities taking place in cells. The cell theory was comprised of three tenets: that all plants and animals are made of cells, that all cells arise from division of preexisting cells, and that cells possess all the attributes of life (assimilation, growth, and reproduction). With the exception of the theory of evolution itself, it was argued, no other biological generalization brought so many diverse phenomena under one point of view or did more for the unification of biological knowledge. For medical research, the cell theory ushered in a new era in physiology and pathology based on the premise that all the various functions of the body in health and disease were but outward expressions of cell activities. For evolutionary biology, studies of the cell underscored the unity of life. A fundamental common plan of organization underlay the diversity of life—because every organism is, or at some time had been, a cell.

An Historical Paradox

Stories about the origins of the cell theory are puzzling. Generations of students have been taught that Matthias Jacob Schleiden (1804–1881) and Theodor Schwann

(1810–1882) were its founders. As the American cytologist Edmund Beecher Wilson wrote in 1925, "Among the milestones of modern scientific progress, the cell theory of Schleiden and Schwann, enunciated in 1838–1839, stands forth as one of the commanding landmarks of the nineteenth century."[1] Historians have also credited them as the founders of the cell theory.[2] Yet, as biologists decades ago emphasized, much of what has been attributed to them had actually been discovered by others, and what they themselves thought they discovered directly contradicted fundamental principles of the cell theory.[3] Herein, then, lies an historical paradox.

Schleiden and Schwann did maintain the first tenet—that all organisms are made up of cells—but this idea certainly did not originate with them. One can point to several predecessors.[4] In 1808 and 1809, Brisseau de Mirbel (1776–1854), professor of botany at the Musée d'Histoire Naturelle in Paris recognized the universal presence of cells in plants.[5] Jean-Baptiste Lamarck might also deserve to rank among the founders of the cell theory: he devoted an entire chapter of the second volume of *Philosophie Zoologique* to cellular tissues.[6] Although he recognized that organisms were composed of cells, neither he nor Mirbel made the leap to the second tenet: that the cell is an independent unit of life.

It has often been supposed that Schleiden was the first to propose the second principle: that the cell possesses the complete characteristics of life, and that plants and animals are the sums of these units. At the beginning of his famous paper of 1838, "Contributions to Phytogenesis," Schleiden wrote:

> Each cell leads a double life: an independent one pertaining to its own development alone, and an other incidental in so far as it has become an integral part of a plant. It is, however, apparent that the vital process of the individual cell must form the very first, absolutely indispensable basis of vegetable physiology and comparative physiology.[7]

This was indeed a revolutionary idea, but it also predated Schleiden's announcement. Henri Dutrochet wrote in 1824 that "this astounding organ [the cell] is truly the fundamental element of organization; everything, indeed, in the organic tissues of plants, is evidently derived from the cell, and observation has just proved to us that it is the same with animals."[8] In 1826, J. P .F. Turpin (1775–1840) published a mémoir in which he emphasized that not only are plant tissues composed of cells, but these cells are distinct individualities that form the composite individuality of plants.[9]

If Schleiden and Schwann were not the first to announce that all tissues of plants are composed of cells or that the cell is the universal unit of organic function and structure, perhaps they established the third tenet: that cells arise only by division of preexisting cells. But here we are met only with a more striking historical irony, for they actually opposed that very principle. In fact, Schleiden believed he had discovered that cells arose not by cell division, but by reproduction of new cells inside mother cells by a slimy substance that first made the nucleus, then the cell. In 1833, the Scottish botanist Robert Brown (1773–1858) emphasized that this round body, which he named the areola or nucleus, was a general feature of a number of plant cells.[10] Schleidin renamed it the cytoblast and argued that it was formed from the accumulation of granulated mucus, and that after it creates the cell, it is dispensable.[11] Schwann adopted this model and applied it to animal cells; he named the cell-generating slime cytoblastem. In his famous article of 1839, he asserted that "the process of crystallization in inorganic nature . . . is the nearest analogue to the formation of cells."[12]

Cells from Cells

That cells arose from undifferentiated chemical substances called cytoblastem was known as "free cell formation." It emerged in direct conflict with the theory that new cells always arise by division of existing cells.[13] Cell division had been had been reported in 1832 by Barthélemy Charles Dumortier (1797–1878), whose paper had been presented to the Académie des Sciences in Paris by Georges Cuvier. In Germany, the botanist Hugo Mohl (1805–1872) reported cell division in a celebrated paper of 1837.[14] Mohl debunked Schleiden's theory,[15] as did the Austrian Franz Unger, who confronted the theory in a series of papers in 1844.[16] Others, such as Carl von Nägeli, adopted an intermediate view that some cells arose solely by division and others arose in the manner described by Schleiden.[17] The theory of free cell formation was especially criticized by those who studied with Schwann's own mentor, Johannes Müller (1801–1858), at the University of Berlin.

Müller was well known for his diverse studies in physiology and comparative anatomy. He studied the passage of nerve impulses going to and from the brain and spinal cord, thus elucidating the concept of reflex action. He traced the development of the genitalia, discovering what is now known as the "Müllerian duct," which forms the female internal sexual organs. He also contributed to knowledge of the composition of the blood, the process of coagulation, the formation of images on the retina of the eye, and the propagation of sound in the middle ear. Muller also concentrated on the cellular structure of tumors. His work of 1838, *On the Nature and Structural Characteristics of Cancer, and of Those Morbid Growths Which May Be Confounded with It*, helped to establish pathological histology. Müller's students became leaders of cell research, and two of them, Robert Remak (1815–1865) and Rudof Virchow (1812–1902), helped to establish the theory that cells arise only from cells.

Remak considered the theory of free cell formation to be as unlikely as the spontaneous generation of organisms. To disconfirm one was to disconfirm the other. By 1852 he had been studying problems of cell generation for a decade, first as an unpaid assistant in Müller's laboratory, supporting himself by his medical practice. As a Jew, Remak was barred from university teaching by Prussian law. In 1843, he petitioned directly to Friedrich Wilhelm IV for a teaching position but was refused.[18] Not permitted to obtain a salaried university position, he focused on his clinical practice, where he specialized in neurology and introduced the use of electric therapy for nervous disorders. In 1847, after he obtained considerable fame, he was given a lectureship (*Privatdozent*) at the University of Berlin, the first Jew to teach there. He was promoted to assistant professor in 1859, but he never held a post as full professor (*Ordinarius*).[19]

No one did more to contest the view of free cell formation than Virchow. After completing his studies with Müller, he worked for a time at the Charité Hospital, where he acquired a reputation from his writings on pathology and from his journal *Archive für Pathologische Anatomie und Physiologie* (which he founded in 1847 and edited until his death). Although he was recognized as one of the main architects of cell pathology, Virchow was also known for his political activities as a liberal during the revolutionary year of 1848. Republican revolts against European monarchies began in Sicily and spread to France, Germany, Italy, and the Austrian Empire. They all ended in failure and repression and were followed by widespread disillusionment among liberals. But Virchow continued to support the cause of liberalism as a member of the German Parliament and founder of the liberal Progressive Party. He was also effective in reforming the sanitary system in Berlin and in organizing the medical corps during the war of 1870.[20]

Virchow is best remembered in the annals of biology for his now-famous dictum *omnis cellula e cellula* (every cell from a preexisting cell), an aphorism that first appeared in his paper "Cellular Pathology" in 1855.[21] He subsequently developed a series of lectures under this title and was appointed to the chair of pathological anatomy at the University of Berlin.[22] His lectures were published in his book *Cellularpathologie* in 1858, in which he expounded on the role of cells, cell division, and the origin of tumors and other growths. Although his views about cell division echoed those of Remak, Virchow failed to even mention him—to the dismay of Remak himself. *Cellularpathologie* was an immediate success: it was translated into several languages and went through three editions between 1858 and 1862.

By that time, both Schleiden and Schwann had left the field. Schleiden refused to give up his own model of free cell formation and modified it until it became virtually incoherent.[23] But in 1850, after twelve years as professor of botany at Jena, he resigned. Withdrawing from botany, he turned to anthropology for a brief period and then wrote popular articles on the three kingdoms of nature (minerals, plants, and animals) and on materialism in German philosophy. Schwann had moved to the Catholic University of Louvain in 1839 and then to Liège in 1852, where he spoke little about cell theory and free cell formation. In fact, of his more than fifty years of research, he was concerned with cell generation only during his five years in Muller's laboratory in Berlin.

More than Meets the Eye

Let us return to the historical riddle with which we began. If much of what is credited to Schleiden and Schwann had been stated by others before them, and if their assertions about the generation of cells was simply wrong, why are they credited as the founders of cell theory? In pondering this, we have to consider the personalities of the men themselves and their ability to champion the field of microanatomy. Scientific discovery is more than what meets the eye. Schleiden is one of the most colorful figures in the history of botany. He first studied jurisprudence at Heidelberg and became a doctor of law. He had so little success as a lawyer that in a fit of despondency, he shot himself in the head. It did not have the desired effect, and when he recovered, he resolved to study natural science. He earned a doctorate of philosophy and medicine at Göttingen.[24]

If Schleiden's contributions were measured only in terms of his observations and conclusions about cell genesis, he would perhaps rate less than the average scientist. But Schleiden was a great promoter for a cellular outlook on life when an advocate was needed.[25] Not only did he induce many scientists into the field, but he also encouraged the young Carl Zeiss to start his successful and important business, which improved optical instruments, particularly microscopes.[26] Schwann's book of 1839, *Microscopical Researches into the Accordance in the Structure and Growth of Animals and Plants*, also had an enormous impact. By examining embryonic tissues and following their subsequent development, he succeeded in demonstrating a cellular origin even for tissues such as bone, which, when full grown, show no trace of cellular origin.

Although Schleiden and Schwann's announcements about the importance of cells marked a turning point in the advance of biology, we can also appreciate the views of those who emphasized long ago that it was absurd to speak of them as the founders of the cell theory.[27] Promoting a field and popularizing concepts is an important aspect of science, but this story illustrates another lesson. In a world of heightened individualism, we often search for heroic achievements made single-handedly, and scientists often attribute major discoveries to individuals. Yet science is in reality a collective activity.

We also need to appreciate the technological basis of scientific discovery. Research on cells in the nineteenth century received a great boost from technical improvements in microscope construction with the rise of the precision optical industry and the availability and use in cytology of aniline dyes.[28] Since the seventeenth century, microscopy had been plagued by two main problems. At first, early lenses had the defect of surrounding objects in the field of view with distracting fringes of color, known as chromatic aberration, that made high magnification difficult. This problem was overcome by Dutch designers in the late eighteenth century who combined a convex lens made of crown glass with a concave lens made of flint glass. They also increased stability and focus precision by substituting all-brass construction for the wood and cardboard that had been used earlier. But the curvature in the lens glass introduced a second defect, spherical aberration, which often completely negated the benefits of the achromatic lens. This problem was resolved in the 1830s when Joseph Jackson Lister found that if two separate achromatic lenses were combined as a single lens, the outcome was complete freedom from both chromatic aberration and spherical aberration. This opened the way to the construction of high-power microscopes, which vastly facilitated research on the nature of cells.

The development of microscopy was also stimulated by concern on the part of the state for the health of its citizens and its military, and by medical research, not just on the cells of animals, but also on the world of microbes, or germs, as the cause of infectious disease. This research was led by Louis Pasteur (1822–1895) in France and by Robert Koch (1843–1910) in Germany. In 1887 the Pasteur Institute was founded in Paris; it later expanded, with institutes throughout France and internationally. In 1891, Robert Koch's Institute for Infectious Diseases opened in Berlin. In 1903, it added a special division for protozoology. A few years later, the Imperial Ministry of Health established two new specialized institutes: the Institute for Protozoology in Berlin and the Hamburg Institute for Naval and Tropical Diseases. In 1906 the Georg Speyr House for chemotherapeutic research in Frankfurt opened. Headed by Paul Ehrlich, it specialized in biochemical studies on the agent of syphilis and other protistan diseases.[29]

Vitalism, Materialism, and Spontaneous Generation

A driving philosophical debate also underlay cell genesis and the origins of germs: the ongoing contest between materialism and vitalism. That microbes did not arise spontaneously from bad air or from putrefied matter was an issue of profound importance for medical practice and epidemiology. Attempts to prove or disprove it led to innovative experimental designs and important sterilization techniques in the nineteenth century that became the basis of modern microbiology and the germ theory of disease.

The belief in spontaneous generation rose to the center of a heated debate in France in 1858, when Félix Pouchet, director of Rouen's Natural History Museum, reported that microbes were spontaneously generated after sterile air was passed through mercury and introduced into a flask containing hay infusions.[30] Pasteur entered the fray. He made long-necked swan-shaped flasks containing infusions of various organic substances. Unboiled flasks became infected with microbes, but flasks he boiled for a few minutes remained sterile even when air was free to pass through it. Germ-carrying dust particles passing down the long necks, he argued, would adhere to the sides before reaching the water.

Like all genesis problems, spontaneous generation debates intermingled with philosophical and religious issues. Spontaneous generation was debated in Victorian England by T. H. Huxley, the physicist John Tyndall, and others in the context of materialism versus belief in a divine primordial creation.[31] In his famous paper of 1869, "The Physical Basis of Life," Huxley lent support to the notion that microbes could arise from organic precursors ("heterogenesis"). By the mid-1870s, he had abandoned heterogenesis and restricted the question of the origin of life to what he called abiogenesis, predicated on the idea that microbes had originated from inorganic molecules at the dawn of life on Earth. Pasteur saw both issues in the same manner. Spontaneous generation now or in the remote past was an argument for materialism and atheism. Thus, he allied his arguments against spontaneous generation with conservative religious views to attack materialism and the heresy of evolutionists. As he exclaimed in 1864:

> What a victory for materialism if it could be affirmed that it rests on the established fact that matter organizes itself, takes on life itself; matter which has in it already all known forces! . . . Ah! If we could add to it this other force which is called life . . . what would be more natural than to deify such matter? Of what good would it then be to have recourse to the idea of a primordial creation, before which mystery it is necessary to bow? . . . Thus admit the doctrine of spontaneous generation and the history of creation and the origin of the organic world is no more than this. Take a drop of sea water containing some nitrogenous material . . . and in the midst of it the first beings of creation take birth spontaneously. Little by little they transform themselves . . . for example, to insects after 10,000 years and to monkeys and man after 1000 years. Do you now understand the link which exist between the question of spontaneous generation and those great problems I listed at the outset?[32]

The same issues about materialism and vitalism underlay the debates over cell generation in Germany. Schleiden and Schwann had placed their theory about cell formation from disorganized organic materials in direct opposition to the vitalism, idealism, and teleology of *Naturphilosophie*.[33] The "apparent gap between inorganic and organic form is not unbridgeable," declared Schwann: "an organized body is not produced by a fundamental power which is guided in its operation by a definite idea, but is developed, according to blind laws of necessity, by powers which, like those of inorganic nature, are established by the very existence of matter."[34] Cell research was a part of a materialist trend toward a way of knowing life through experimentation and closing the gap between the forces underlying life and nonlife. As discussed in the next chapter, the cell theory also led to a profound change in the concept of the individual.

8

The Body Politic

> On the same grounds that the sociologist affirms that a society is an organism, the biologist declares that an organism is a society. . . . A society is an organized whole, the unity of which consists in, and is measured by, the mutual dependence of its members. The living body is an organization of individual cells with the same bond of unity. The principle of organization in both cases is the division of labor or function.
>
> —C. O. Whitman, 1891

CELL THEORY REVOLUTIONIZED CONCEPTIONS of the individual. A plant or an animal was no longer conceived of as a singular entity constructed out of cellular tissue. What was important was not that cells were found in all tissues, nor that all organisms consisted of cells; it was that the cell was alive—it possessed all the attributes of life. Cells were "elementary organisms," as Ernst von Brücke (1819–1892) called them in 1861.[1] That all cells arise by division of preexisting cells, that the egg is a single cell, and that in sexual reproduction each parent contributes one cell from its own body to the formation of an offspring all magnified the importance of the cell as a universal unit of structure and function. As Virchow phrased it, "Every animal appears as a sum of vital units, each of which bears in itself the complete characteristics of life."[2] This revolutionary slogan took hold of biology and framed the way in which many biologists conceived of organisms.

The Cell State

The cells of the bodies of plants and animals were understood to be mutually interdependent and, with the exception of the mature germ cells, could not easily maintain their existence apart from their fellows. In other words, the only natural environment suitable for their continued existence was the complex body or "cell commonwealth" of which they formed an integrated part. This was no mere analogy. The only real difference between the cell of an animal and the single protist, whether amoeba or ciliate, was that in the latter case the whole body of the living individual could, for whatever reason, reach no higher degree of complexity than the single cell.

The common ancestor of all the cells of a plant or animal was traceable back to the fertilized egg. The first step in development consists in the division of the egg into two parts. The two divide to form four, eight, sixteen, and so on, until step by step the egg has split up into the multitude of cells that build the body of the embryo, and finally of the adult. The different cells of a plant or animal are of different types and perform different functions in different tissues. Each type carries out a particular mission for the whole community. In other words, there is a distribution of tasks, a division of labor.

Many biologists, though not all (see chapter 9), understood the development of a complex cellular organism in terms of Haeckel's biogenetic law: that the process of ontogeny recapitulates the process of evolution. One could imagine that sometime in the remote past, primitive protist colonies evolved into complex integrated plants and animals. As a first step, a simple colony of identical cells would have evolved into a commonwealth of differentiated and mutually dependent cells. The struggle for existence between individuals would lead to a division of labor, which in turn would lead to increased interdependence. The individual protists of the colony would then become more and more intimately associated. Thus, a multitude of independent individuals, adopting mutual service as the best economy, would find themselves in the end so firmly bound together in interdependence that they constituted a complex individual. The struggle for existence was supposed to extend to cells from the dividing egg to adult.

"Each cell leads a double life," wrote Schleiden, "one for its own development and the other as an integrated part of the plant." A plant or an animal represented a colony of these smaller individuals; it was a "'cellular state," a collective in which, Schwann said, "each cell is a citizen." All higher organization was supposed to evolve through the principle of physiological division of labor and to reach its fullest expression in the mutuality of the constituents. This was held to be as true for the development of complex organisms as it was for human societies. Early leaders in ecology—Arthur Tansley in England and Frederick Clements in the United States—following Herbert Spencer, extended the model to ecology as well, asserting that competition among plants resulted in a highly developed division of labor in some communities, thereby producing a stable superorganism.[3] The usual conception of this division of labor was, as Spencer stated it in 1893, "an *exchange of services*—an arrangement under which, while one part devotes itself to one kind of action and yields benefits to all the rest, all the rest, jointly and severally performing their special actions, yield benefits to it in exchange. Otherwise described, it is a system of *mutual* dependence."[4] Thus, biologists conceived of organisms as cooperative assemblages with parts integrated into organs that live for and by one another. The evolution of simple to complex organisms could be compared to that of underdeveloped and developed societies.[5]

The cell theory of the organism was indeed a social theory, as expressed in the "cell state" or "cell republic" and expounded by Schlieden, Schwann, Virchow, his former student Haeckel, and Haeckel's former student Oscar Hertwig (1849–1922), professor of anatomy at Berlin, who argued that because biology dealt with the organization of life, it was more akin to the social sciences than to physics and chemistry.[6] The cell theory held sociopolitical messages. Like the theory of evolution, it was sometimes portrayed as a description or prescription for our social world for how to best define the individual and organize the state.[7]

Liberals such as Theodor Schwann argued against what he called the "autocracy of the organism" when he asserted that "the cause of nutrition and growth resides not in the organism as a whole, but in the separate elementary parts—the cells."[8] In 1859, Virchow defined the organism as "a society of living cells, a tiny well-ordered state" consisting of living members of a common origin. Haeckel took a more hierarchical view when he compared cells to law-abiding citizens in an orderly state. In plants, cells formed republics, but in animals there was a cellular monarchy. The organs formed from tissues were like state departments, and rule by a central government was comparable to the power of the brain.[9]

Biologists of the nineteenth century applied the concept of the division of labor to understanding the state of biology as well. Intellectual progress relied on a division of labor no less than did social and economic progress.[10] Biology had become ever increasingly partitioned into diverse specialties: paleontology, anatomy, physiology, zoology, botany, embryology, cytology, bacteriology, and protistology. The days were over in which an individual could pretend to be an expert on all aspects of the life sciences. The idea that specialization was inevitable, progressive, and natural was promoted by many leaders in cell and developmental biology, including Charles Otis Whitman (1842–1910), founding director in 1888 of the Marine Biological Laboratory in Woods Hole, Massachusetts.[11] In 1891, he insisted that the metaphor of the living body as "a commonwealth of cells" was not based upon "superficial or fanciful resemblances," but upon "analogies that lie at the very foundation of organic and social existence." The division of labor, Whitman argued, had "taken possession of the biological sciences, and presides over their onward march, just as it determines and directs social and industrial progress."[12]

Although Whitman and many others later rejected the conception that animals "were only the sum of vital units" (cells) (see chapter 9), the concept of the cell state remained the central explanatory framework for cytology, embryology, and physiology.[13] As Edmund B. Wilson (1856–1939) at Columbia University commented in his famed book *The Cell in Development and Heredity*,

> The more complex life of the higher plant or animal arises through the specialization of the cells, this way or that, for the better performance of particular functions; hence that "physiological division of labor" which, as in organized human society, leads to higher functional efficiency. On such considerations was based the famous comparison of the multicellular body to a "cell-state," due especially to Virchow

> though foreshadowed by Schwann and other early writers, and later elaborated by Milne Edwards, Haeckel and many others. This conception of the multicellular organism brought about a revolution in the prevailing views of vital action, and gave an impetus to physiology and pathology as to morphology. . . . The conviction of its essential truth has survived all criticism, and as measured by its continued fruitfulness, it still stands among the most important generalizations of modern biology.[14]

The conception of the organism as a colony of mutually interdependent individuals was indeed progressive in terms of the biological problems it addressed. It also gained great support from comparative studies of free-living "unicellular" organisms.

The Dawn of Protistology

In the late nineteenth century, microbes were studied primarily from a medical perspective because some of them cause diseases.[15] But from an evolutionary perspective, they were defined as "unicellular organisms"[16] The first generation of microbial genealogists investigated them with the hope of gleaning some understanding of early life on Earth. They soon found that the age-old dichotomy of plants and animals did not hold true for the simplest organisms. Some possessed locomotion like animals, yet they had modes of living that were more like plants than animals. In some species, an individual could nourish itself at one time as a plant and at another time as an animal by eating, according to its circumstances. So it was supposed that here in the realm of the very small and relatively simple, one might be observing creatures that were the not-quite-animal and not-quite-plant ancestors of all living things. Several researchers proposed a new kingdom. In 1859, Richard Owen called it the Protozoa; in 1860, John Hogg called it the "Primigenum"; and in 1866, Haeckel designated it as the Protista, the plural superlative form of the Greek word *protos* ("first"). Protists were the first living creatures.[17]

In his great work of 1866, *Generelle Morphologie der Organismen*, Haeckel constructed the first phylogenetic tree ever published. The kingdom Protista included the Monera (bacteria), Protoplasta (amoebae), Flagellata (unicellular algae such as Euglena and Volvox, and some dinoflagellates), Diatomeae, Myxomycetes (slime molds), Myxocystoda, Rhizopoda (including Radiolaria), and the Spongiae. Twelve years later, Haeckel included the ciliates in Protista, and he placed sponges among the Animalia.[18] He further divided the Protista into several subgroups: he used the term Protozoa to refer to those microbes that were ancestral to animals, and the term Protophyta to refer to the ancestors of the plants. Those that were not ancestral to either group he called Protista Neutralia. But biologists argued that it was often difficult to place forms in the groups proposed by Haeckel, and therefore it was best to distribute microbes between the plant and animal kingdoms as best one could (see chapter 18).

Microbial phylogeny was difficult and speculative. Microscopists emphasized instead the intracellular organization of protists, whose structure, they argued, could be readily compared to the cells of multicellular organisms. This was the message of Richard Hertwig (1850–1937) at the University of Munich in 1902, in the lead article of the first issue of *Archiv für Protistenkunde*: protists were "single-celled" organisms, in every way comparable to the cells of an animal.[19] There was a deep unity of the living world. As the British protistologist Edward Minchin proclaimed in 1915, "We find in the Protista every possible condition of structural differentiation and elaboration, from cells as highly organized as those of Metazoa or even, in some cases, much more so, back to types of structure to which the term cell can only be applied by stretching its meaning to the breaking-point."[20]

Abundant experimental evidence also supported the theory of the cell state. Not only could tissue cells removed from an animal live and multiply in artificial culture media, but some cells of all animals permanently retained their complete independence of movement and action. This was especially apparent in sponges, jellyfish, and hydroids. In the course of ontogeny, entire groups of cells could alter their relative positions in the body as the result of migrations performed by individual cells. Furthermore, if the adult sponge or hydroid was broken up completely into its constituent cells, those cells could build up and regenerate the body of the organism. In 1882, the Russian biologist Elie Metchnikoff observed mobile cells of starfish larvae surround and engulf invading germs. He called this process phagocytosis and subsequently argued that protective white blood cells in animals were the basis of immunity, and that vaccination somehow strengthened them against invading germs.[21] These were only some of reasons many biologists regarded the body cells of animals as individuals complete in themselves, primitively as independent as the individual protozoan, and in every way comparable to it.

A Cell Is Not a Cell

The term "cell" is a misnomer. By the last decades of the nineteenth century, all biologists had come to recognize that whatever the cell was, it was not a hollow chamber surrounded by solid walls. The word was derived from Robert Hooke's (1635–1703) microscopic observations of cork. Hooke was one of the greatest experimentalists and inventors of the seventeenth century. As one of the founding members of the Royal Society, chartered in 1662, his duties were to conduct two or three experiments for the society's weekly meetings, which he did for forty years, and he himself made many of the instruments for experimentation. Among other accomplishments, he investigated celestial mechanics, invented the vacuum pump used by Robert Boyle, and devised the forerunner of the balance spring of the modern watch.

In 1665, Hooke published *Micrographia or Some Physiological Descriptions of Minute Bodies made by Magnifying Glasses*. It was a large collection of essays on such diverse observations as the structure of cloth, the intimate morphology of the gnat, and Hooke's own studies of thin sections of cork. His aim was to understand "the lightness and yielding quality of the Cork." He suspected it might be porous. With the compound microscope he made for himself, he was able to see that "the substance of Cork is altogether fill'd with Air, and that that Air is perfectly enclosed in little Boxes or Cells distinct from one another. It seems very plain . . . why the pieces of Cork become so good floats for Nets and stopples for Viols, or other close Vessels."[22] Thus, he used the word "cells" (Latin *cella*, "small room") to describe the chambers of the dead cells of cork, and he calculated that there were more than a million of them per square inch. But the word was not often used in the seventeenth and eighteenth centuries.

Of the others who naturally turned to the microscope during the seventeenth and eighteenth centuries, none was more prominent than Anthony Leeuwenhoek (1632–1723), who studied numerous microbes as well as the microanatomy of plants and animals. He discovered spermatozoa, and with his simple lens he thought he saw in the human spermatozoon the homunculus, or little man, long postulated by preformationists. During the eighteenth century, other words, including "utricles," "vesicles," and "globules," were used for the constituent parts of plants and animals.

The term "cells" came into general use in biology at the beginning of the nineteenth century, when they were understood to be structural elements only, and literally conceived of as a chamber as the name correctly implied. The cell wall seemed to be the important part; the cell content was thought to be an unorganized fluid, or a homogeneous glutinous material without a trace of organization, fibers, or membranes. This conception changed dramatically with advancements in microscope lenses by mid-century that resolved problems of chromatic and circular aberrations and thus allowed for higher magnification (see chapter 7). In fact, microanatomists concluded that a cell membrane or wall was not even a necessary part of a cell. After all, animal cells had no such wall. Focus shifted to the cell contents: the organization and behavior of protoplasm.

What's in a Word

In 1840, the Czech Johannes Evangelista Purkinje (1787–1869) used the word "protoplasm" to designate the true living substance in the interior of the cell. Theological writers had long used the word "protoplast" for Adam, "the first formed."[23] The word "protoplasm" came into prominence after it was used by the champion of agnosticism, T. H. Huxley, in his lecture of 1868, titled "The Physical Basis of Life." Though it was still a vague concept, he presented it as a victory for mechanistic materialism over vitalistic conceptions of life. "All vital action," he commented, "may be said to be the result of the molecular forces of the protoplasm which display it."[24]

Microscopists soon recognized that far from being bags or boxes of formative material, all cells had a basic structural organization that was essentially the same in all plants and animals and protozoa. The most conspicuous parts were the spherical body that Robert Brown dubbed the "nucleus" in 1831 and its surrounding material, named cytoplasm by Rudolf Kölliker (1817–1905) in 1862. With the possible exception of some of the "lowest forms" of life, such as bacteria (or the fission fungi, as they were often called), all cells contained a nucleus. Thus, by the 1880s, the cell had come to be defined morphologically as a mass of protoplasm containing a nucleus.

When it became clear that the cell was not an isolated chamber, there were repeated attempts to rid biology of the misleading metaphor. Lionel Beale (1870) proposed the term "bioblast," Johannes von Hannstein (1880) proposed the name "protoplast," and Julius Sachs (1892) suggested the term "energid."[25] Although some cytologists welcomed a name change, the old misleading word was difficult to dislodge.[26] Some argued that scientific names should not change every time some new or contradictory qualities are found in the object. Making name changes to suit new conceptions or information would only cause confusion.[27] The United State's premier cell biologist, E. B. Wilson, lamented in 1896, "Nothing could be less appropriate than to call such a body a 'cell'; yet the word has become so firmly established that every effort to replace it by a better has failed, and it probably must be accepted as part of the established nomenclature of science."[28]

The unprecedented work of microscopists of the late nineteenth century revealed a complicated organization of both nucleus and cytoplasm that would have greatly astonished the previous generation. The nucleus and cytoplasm held significant chemical and structural differences. The nucleus had an abundance of a substance rich in phosphorus, which, in 1871, the Swiss physiologist and chemist Friedrich Meischer (1844–1895) called nuclein. In 1889, his student Richard Altmann (1852–1901) showed that nuclein was made up of a protein base, rich in nitrogen, and a complex organic acid containing phosphorus to which he gave the name nucleic acid.[29] Microscopists observed a number of structures (ribbons, bands, and threads) that appeared in the nucleus during cell division. Since these structures could be stained, the Austrian Walther Flemming (1843–1905) called them chromatin (*chroma*, "colored"; *tin*, "thread"). Chromatin was shown to contain a high percentage of nucleic acid. The cytoplasm (sometimes referred to as the "cell body") was believed to contain no true nuclein or chromatin, but it was rich in protein.

The main focus of fin-de-siècle cytology was not on cell chemistry, but on the mechanism for cell reproduction. In a series of stunning papers published between 1878 and 1884, Flemming followed various stages of nuclear division in the epithelial cells of the salamander. Many of the terms introduced by him are in use today, including "mitosis" to describe the indirect division of the nucleus. In 1884, he, along with Edouard Strasburger (1844–1912) and Karl Rabl (1853–1917) in Germany, and Edouard van Beneden (1846–1910) in Belgium, reported how mitosis occurred. At first, the chromatin appears long and slender and coiled together, but then they grow shorter, thicker, and straighter to form "chromosomes" ("colored bodies"), so dubbed by Wilhelm Waldeyer (1837–1921) in 1888 because they were easily seen when stained by appropriate dyes. At a subsequent stage, the nuclear membrane disappears, the chromosomes move to the center of the cell, and each splits lengthwise into two "daughter" chromosomes that move apart toward the two poles of the cell, where they in turn form two daughter nuclei. The cell body then divides by constriction to form two daughter cells. Mitosis, then, conserves the number of chromosomes per nucleus. Each species had a fixed number of chromosomes—from two to two hundred, but typically between ten and thirty.

By the 1880s and 1890s, in view of such an elaborate mechanism for transferring chromosomes from one cell generation to the next, the nucleus was often regarded as a primary factor in growth, development, and the transmission of hereditary qualities from cell to cell, and so from one generation to another.[30] Mitosis was indeed an extraordinarily accurate mechanism for transmitting chromosomes from one cell generation to the next. But there was a problem. If chromosomes were transferred by egg and sperm during sexual reproduction, why did the number of chromosomes not double in every generation? Some had surmised that the number of chromosomes must be reduced somehow during the formation of germ cells to prevent this doubling, and in 1883, van Beneden showed that this was in fact the case.[31] The germ cells (eggs and sperm) of the parasitic worm *Ascaris* contained half the number of chromosomes characteristic of body cells. This halving of the chromosome number when forming germ cells was subsequently confirmed for many organisms.

The process of chromosome reduction was later (in 1905) called meiosis (Greek for "lessening"). However, a consensus about how meiosis occurred was difficult to achieve, for two main reasons. The process seemed to be different in different species, and a crucial idea was lacking: cytologists had not yet generally agreed on chromosomal individuality—the notion that different chromosomes had different properties.[32] In 1883 Wilhelm Roux (1850–1924) at the University of Breslau postulated that each chromosome carried different hereditary determinants.[33] But the experimental evidence for chromosomal individuality was not reported until 1902, when Theodor Boveri (1862–1915) at the University of Würzburg announced that each of the thirty-six chromosomes of the sea urchin were necessary for normal development. That year Walter Sutton (1877–1916) at Columbia University reported his studies showing that each of the eleven pairs of chromosomes in the grasshoppers he studied was different, as was the smaller accessory chromosome.[34]

Reduction of the number of chromosomes—meiosis—occurred when eggs and sperm were formed. Chromosomes derived from each of the animal's parents unite into pairs: big ones with big ones, little ones with little ones, and those of a peculiar shape with others of a similar shape. Subsequently, the pairs of chromosomes separate along the line of the junction, one member of each pair going to one pole of the cell, and the other to the other pole. The sperm or egg cell divides, and each of the daughter cells has only half the number of chromosomes of the body cells, one of each kind. It rarely happens that all the mother's chromosomes go to one pole and all the father's to the other. So meiosis produces a scrambling of chromosomes, and though each egg and sperm cell has a complete set of chromosomes, some of them come from the mother and some from the father. When the egg and sperm unite in sexual reproduction, the original number is restored.

Both Boveri and Sutton suspected that the pairing of maternal and paternal chromosomes and their subsequent separation during meiosis would be the physical basis of the newly rediscovered Mendelian laws of heredity (see chapters 11 and 12). The history of nineteenth-century cytological work usually ends here, and historians turn to the development of the Mendelian chromosome theory of inheritance developed during the second decade of the twentieth century. But there was more to nineteenth-century studies of the cell than this. There were also fundamental studies of intracellular bodies in the cell cytoplasm.

Organisms within Organisms

With the introduction of the oil immersion lens in 1870, the development of the microtome technique for making thin sections of tissues, and the use of new fixing methods and dyes, microscopists soon recognized that the cell cytoplasm was not a homogenous fluid any more than was the nucleus. It contained an abundance of various rods, threads, membranes, vacuoles, pigment bodies, and other granules. Among these, three bodies captured the limelight: mitochondria, chloroplasts, and centrosomes.[35]

Richard Altmann at Leipzig is usually credited for discovering mitochondria by means of a special staining technique. In his treatise *The Elementary Organisms and Their Relationship to the Cell*, he suggested that visible intracellular bodies he called bioblasts arose by division from preexisting bodies, just as did the chromosomes and indeed the cell itself. They were "elementary organisms," which, he argued, secreted various cell substances, including fat, glycogen, and pigments; he also argued that they could be transformed into or produce various rods and fibers.[36] In 1898, Carl Benda (1857–1933) developed another technique to fix and stain the granules with greater certainty and brilliancy. He renamed them mitochondria, from the Greek *mitos*" ("thread") and *chondros* ("granule"), since they seemed to exist as both threads and granules. During the first decades of the twentieth century, these bodies were intensely investigated, especially in Germany and France, under various aliases, including chondriosomes, ergastidions (little workers), eclectosomes, vacuolides, and plastidules.[37]

By the mid-1880s, many cytologists argued that the chlorophyll bodies in plants, which the famed botanist Andreas Schimper (1856–1901) named chloroplasts in 1883, arose only by division of preexisting bodies of the same kind. He compared them to independent organisms, like symbionts, living within a host (see chapter 19).[38] In the 1890s, no cell structure aroused wider interest than the central body, called the centrosome: dark-staining corpuscles just outside the nucleus that seemed to play a directive role in mitosis. Boveri and van Beneden are credited with independently discovering centrosomes in 1887. Later, in 1895, Boveri noted a minute granule inside the centrosomes of *Ascaris* that he called the centriole, but sometimes he could not clearly distinguish it from the centrosome. In subsequent years, the terms centrioles and centrosomes were often confused and used indistinguishably.[39]

In animal cells, centrosomes appeared to play an extraordinary role in separating daughter chromosomes during cell division.[40] At the onset of cell division, the centrosome seemed to divide in two; the pair then separated, and each one moved to opposite sides of the nucleus. When the nuclear membrane dissolved, starlike structures ("asters") formed around each centrosome, and rays of threads ran through the nuclear area so as to constitute a "spindle." One could observe chromosomes split into two and watch each daughter chromosome attach to a spindle and move to opposite poles where the centrosomes lay. Two daughter nuclei were then formed. Thus, it seemed clear that at least in animals, centrosomes, asters, and spindle constituted an apparatus for the accurately separating the daughter chromosomes and for the division of the cell body. There was still more. In 1898, Félix Henneguy in Paris and Mihaly Lenhossék in Budapest reported that centrioles could move from the cell center to the cell membrane, where they were converted to "kinetosomes" (often referred to as "basal bodies"), which lay at the base of cilia. They multiplied and functioned as organs of motility.[41]

Thus by the turn of the century, it was clear to cytologists that cell reproduction and organization were complex processes. Indeed, it was naive to believe that a cell could be generated *de novo* from disorganized material, as nineteenth-century publicists of materialism had hoped. As E. B. Wilson observed in the second edition of *The Cell*, in 1900, cytology had seemed "to widen rather than narrow the enormous gap that separates even the lowest forms of life from the inorganic world. I am well aware that to many such a conclusion may appear reactionary or even to involve a renunciation of what has been regarded as the ultimate aim of biology."[42]

The evidence that various organelles seemed to grow and divide led to the suggestion that cells themselves might have first arisen from a symbiosis of several dissimilar beings that came together in the remote past. Such speculations remained on the margins of biology until the late twentieth century, when new techniques became available for investigating cell origins (see chapter 19).

Weismannism

The evolution of the cell and the genesis of intracellular bodies during cell division or in the remote past were not at the center of cell studies of the late nineteenth century (see chapter 17). Theorists who aimed to unite cell theory and evolutionary theory focused on problems of herditary variation and embryonic development. They wrote of hypothetical "bearers of heredity" that might come together during sexual reproduction. They would be responsible for the development of an adult from an egg and also be the basis of the inherited differences between individuals that might fuel evolution. In regard to these problems, many leading cytologists focused on the cell nucleus and its chromosomes. It was often assumed that the nucleus was the governing organ of the cell, the basis of inheritance, and that all other cell structures and activities resulted from its activity.[43] This nucleocentric view formed the basis of an elaborate theory formulated by the Freiburg zoologist August Weismann (1834–1914).

Of the many speculations about heredity in the nineteenth century, none was as well developed and as influential as Weismann's.[44] No other cell theorist advanced more conceptual distinctions and drove scientists into the laboratory to test them than he did. His theorizing on the role of the nucleus in heredity and development stemmed from one of the most hotly debated questions in evolutionary biology: the inheritance of acquired characteristics (see chapter 6).[45] Although many evolutionists had taken it for granted that the characteristics one acquired in the course of one's life might be passed on to subsequent generations, very few had made an attempt to show how it was theoretically possible. Darwin was an exception. In his book *Variations of Plants and Animals under Domestication* (1868), in a chapter titled "Provisional Hypothesis of Pangenesis," he speculated about invisible hereditary particles he called gemmules.[46] These gemmules were supposed to multiply by fission and circulate throughout all parts of the body; they would then be collected in egg and sperm cells and subsequently be passed on to the next generation.

Weismann denied that such a transfer of hereditary units from body cells to the germ cells ever occurred. His theorizing about heredity culminated with his celebrated book *The Germ-Plasm: A Theory of Heredity* (1893). Its ambitious aim was to develop the outlines of a single model that would account not only for the inheritance of the variations that fuel evolution, but also for the underlying mechanism by which an adult organism arises from a fertilized egg.[47] His germ-plasm theory was based on three fundamental premises.

The first premise was that the primary constituents of the individual were as yet unknown ultramicroscopic vital units, each of which would grow and multiply by fission.[48] The idea that the body was somehow built up by a large number of submicroscopic elementary vital units or intracellular "elementary organisms" that grow and divide was pervasive since the 1860s. While Darwin had called them gemmules, Herbert Spencer called them physiological units, and in 1884 Carl von Nägeli coined the word "idioplasm" to refer to a vast number of fundamental hereditary determinants in the protoplasm.[49] Life (processes of assimilation, growth, and reproduction) was not something to be explained within this paradigm. While making a theory of life was impossible at the time, Weismann argued that it was possible to explain heredity if one simply took for granted the essential phenomena of life.[50] Thus, he bestowed the properties of life on the hereditary substance.

The second premise concerned the relationship of the germ cells to the body of an organism. Germ cells, he argued, descended from germ cells, not from the body, and inheritance did not take place from the body of the parent to the child. This was his theory of the continuity of the germ plasm. There could be no inheritance of acquired characteristics unless those characteristics were acquired directly by the germ cells. The germ cells were potentially immortal, whereas the body cells, or somatic cells, were transient. In effect, the body was merely the carrier of the germ cells, which are "held in trust" for coming generations.[51] Thus, Weismann proposed a fundamental duality of immortal germ cells giving rise to transitory bodies. As Samuel Butler (1835–1902) phrased it, "A chicken was just an egg's way of making another egg."

The third premise concerned the location of the hereditary substance within germ cells. Weismann situated it exclusively within the chromosomes in the nucleus. Wilhelm Roux had speculated in 1883 that hereditary units were arranged along chromosomes and that the nucleus would undergo differentiation as development proceeds.[52] In his book *Intracelluläre Pangenesis* (1889), Hugo de Vries postulated that "pangenes" migrated from the nucleus into the cytoplasm step by step during ontogeny, thus determining cell structure and the successive stages of development.[53] Weismann adopted both ideas.

The two main reasons Weismann gave for the nucleus as the sole bearer of hereditary qualities were echoed by leading geneticists throughout the twentieth century. The egg is many times larger than the sperm cell, which is little more than a motile nucleus. If the cytoplasm also contributed to heredity, the female would contribute more to heredity than the male. Yet, Weismann asserted that "we know that the father's capacity for transmission is as great as the mother's."[54] The nucleus provided a place for equal contributions of hereditary substance from both parents. Then there was that dazzling dance of the chromosomes at mitosis, the existence of a complex apparatus for assuring the exact distribution of the chromosomes.[55]

Weismann also appealed to the "economy of Nature" to argue for the exclusive chromosomal basis of hereditary material: "this substance can hardly be stored up in two different places, seeing that a very complicated apparatus is required for its distribution: a double apparatus would certainly not have been formed by nature if a single one suffices for the purpose."[56] Thus, he postulated that vital units must pass through the nuclear membrane into the body of the cell and there form its parts and structures.[57] He did not specify exactly when the first vital units constructed the cell, nor was he concerned with the actual processes involved. Only one issue was certain: "the nature of the cell is really decided by elements of the nucleus."[58]

Weismann's vital entities would have to do more than assimilate, reproduce, and migrate to the cytoplasm. In the course of development, cells become biochemically and morphologically different in an orderly way. To account for this, he supposed that the germ plasm was arranged in terms of a series of units, each with special properties. Myriad entities he called biophores would be grouped together to form a higher unit, the determinant, which would specify a certain type of cell or group of cells. Determinants were grouped into a still larger unit, the id;

each id contained all the kinds of determinants and so stood for the sum of all the characteristics of the individual organism. The ids were arranged in a linear series along chromosomes, or idants.

Weismann thus explained development as a mechanical marching parade of many kinds of determinants, which would be assigned to the different parts of the developing body through a progressive disintegration of the id as development proceeded. As he wrote, "the changes in the id of the germ-plasm during ontogeny consists merely in the gradual disintegration of the determinants into smaller and smaller groups, until finally only one kind of determinant is contained in the cell *viz.* that which has to determine it."[59] Weismann compared it to military organization:

> The development of the nucleo-plasm during ontogeny may be to some extent compared to an army composed of corps, which are made up of divisions, and these of brigades, and so on. The whole army may be taken to represent the nucleoplasm of the germ-cell: the earliest cell-division . . . may be represented by the separation of the two corps, similarly formed but with different duties; and the following cell-divisions by the successive detachment of divisions, brigades, regiments, battalions, companies, etc.; and as the groups become simpler so does their sphere of action become limited.[60]

Generations of researchers have asserted that Weismann had gotten much right, and much wrong. He had speculated that the hereditary units were arranged in a serial order on chromosomes, much like the beads-on-a-string model of classical geneticists. He had predicted that during the formation of all germ cells, there would be a reduction in the number of chromosomes to avoid an excessive build-up of germ plasm. Such a reduction (meiosis) did occur, but not precisely in the manner Weismann envisaged. His theory of the continuity of the germ plasm continues to be taught as a core concept of the organism. Nonetheless, many biologists then and now have argued that the distinction between the germ cells and somatic cells is less sharp than he had proposed (see chapters 15 and 19).

Sorting among Weismann's ideas, one can pick out concepts showing how close nineteenth-century biologists had come to mid-twentieth-century canonic genetic views about heredity. All this theorizing and cytological work on chromosomes of the late nineteenth century might be ordered in a progressive series leading nicely to a grand synthesis, especially with the Mendelian theory of inheritance on the horizon of the twentieth century. There was one enormous problem. Weismann could not have been more wrong about development as the unfolding or sorting out of a hierarchy of vital units in chromosomes. As we shall see in the subsequent chapters, the observations of a new generation of experimental embryologists seemed to be irreconcilable with Weismannian theory and with a concept of organic individuality in terms of a hierarchy of vital entities.

9

Evolving Embryology

> To many, the cell is always an independent living being, which sometimes exists for itself alone, and sometimes becomes joined with others—millions of its like in order to form a cell-colony, or, as Haeckel has named it for the plant particularly, a cell republic. To others again, to whom the author of this book also belongs, cell-formation is a phenomenon very general, it is true, in organic life, but still only of *secondary significance.*
>
> —Julius Sachs, 1887

EXPERIMENTAL EMBRYOLOGY BEGAN in the 1880s. Its aims were to discover the actual physicochemical processes by which the adult developed from the egg. Referred to by its founders as "developmental mechanics," it originated from the premise that organisms could be understood in the same way as machines. Its findings however, caused biologists to stand back somewhat as their studies revealed a complexity of development that confronted the general aim of understanding the organism in terms of its parts. Many leaders of the new generation of experimental embryologists opposed the conception of the organisms as a cell state.

Since the middle of the nineteenth century, biology had been teeming with theories of the organism that postulated smaller corpuscular determinants or "elementary organisms." In effect, the internal organization of an organism was regarded as a series of Russian dolls. An animal or plant was regarded as a colony of cells; the cell was regarded as a colony of simpler units—nucleus, centrosome, mitochondria, and chloroplasts; the nucleus was regarded as a colony of chromosomes; the chromosomes, according to Weismann's theory, were colonies of ids; the id was a colony of determinants; the determinant was a colony of biophores; and the biophore was a colony of molecules.

The first generation of experimental embryologists rejected such conceptions of elementary organisms in a cell state; they adopted "the organism as a whole" as a counterrevolutionary slogan while they redefined biological aims and explanations. They had three main objections to the previous concepts and approaches. First, models of the individual in terms of hypothetical "elementary organisms" took as their starting point the very thing that many young biologists wanted to explain: life. The previous generation of theoreticians had simply bestowed the properties of life on a multitude of hypothetical entities, capable of assimilation, growth, and reproduction. Second, the concept of elementary organisms, whether cells or intracellular determinants, suffered from reductionism, the fallacy of reducing the properties of the whole to the parts of which it is composed. Instead of emphasizing "determinants," experimental embryologists focused on forces, flows, structures, and interactions. Third, the postulated elementary organisms of Weismann, Darwin, or Spencer were purely speculative. The new generation of experimental embryologists aimed to discover the actual causes of development: they conducted experiments on the relationships among cells of the dividing egg, they traced cell lineages back to the fertilized egg and examined its internal organization, and they experimented on the roles of the nucleus and the cytoplasm in heredity and development.

Technical Virtuosity

New technologies as much as new theories have opened up new worlds for scientific exploration.[1] Studies of the cell in development and heredity relied on technical skill in the fixing and staining of cells, improvements in methods for cutting through tissues with the microtome, and adequate theories whose predictions could be tested. The emergence of experimental embryology also relied on domesticating appropriate organisms—model organisms—for the design, development, and execution of experiments.

Some organisms lent themselves particularly well to observations on the nucleus; others, on the cytoplasm, germ cells, or embryonic development. For examining the process by which the nucleus is duplicated during cell division, the ideal organism was *Ascaris*, a parasitic worm of the horse. Its qualities were emphasized by Edouard van Beneden, and Theodor Boveri extolled its virtues:

> *Ascaris* forms an unsurpassable material. The eggs can be stored for some months, dry, in the cold, without alteration. When one has time for work on them, this can be done at room temperature, where they continue to develop slowly. If one wishes to accelerate development temporarily, one brings the egg into an incubator. If one must interrupt work, one puts them back in the cold, and on returning, one finds them in the same condition in which they are left.[2]

A useful "material" indeed. It had other attributes. The nucleus of *Ascaris* is particularly simple: the number of chromosomes is small, generally four, and even two in a certain type, and it was easy to study their shape and behavior. During cell division one could watch chromosomes split into two and arrange themselves along a spindle that attracted them to two opposite poles where the centrosomes lay.

But if one wished to study fertilization and development, then the organism of choice was the frog or the sea urchin. Some sea urchin eggs are transparent and easy to observe; and sea urchin spermatozoa are small, each with a dense readily visible nucleus. If an egg and some sperm are placed in a dish of seawater, one can watch the spermatozoa adhere to the egg and watch the first one to touch the egg enter it. It was even possible to follow the path of the male nucleus as it fused with the female nucleus, and to watch the successive divisions that occurred in the egg in strict temporal and spatial order. Sea urchin eggs were also noted for the ease with which they could be kept under experimental control. And one could even fertilize the egg of one species with the spermatozoa of a foreign species or genus.

Experimental embryology was born in Germany. Wilhelm Roux had called it *Entwicklungsmechanik*; others called it physiological embryology, and during the 1880s and 1890s it grew up in marine biological laboratories on the Swedish coast and in the British Isles, France, Italy, Japan, and the United States.[3] Little wonder experimental embryology emerged by the sea, given the importance of aquatic animals. At the Marine Biological Laboratory at Woods Hole, Massachusetts, Ethel Brown Harvey (1885–1965) praised the many attributes of sea urchins:

> The Arbacia egg is an ideal cell. It is spherical. Thus, rendering changes in size easy to determine. It is fairly simple in comparison with most cells. It is quite hardy and can be subjected, without damage, to moderate changes in the sea water, produced by the addition of water, or salts, or anaesthetics, or other chemicals, and to changes in temperature, pressure, light, and other physical factors. Harmful effects and recovery can be readily detected by fertilizing the egg and watching its development. The granules in the egg can be moved by centrifugal force, and the egg can be broken into halves and quarters containing different kinds of materials in definite amounts. The experimental work on sea urchin eggs has included every line of approach, cytology, embryology, physiology, and biochemistry, and has been concerned with the solution of many fundamental problems.[4]

The technical virtuosity of embryologists was extraordinary. They were able to change the behavior of the germ cells and of the developing egg. By treating eggs with certain chemical compounds, Boveri succeeded in fertilizing each of them with several spermatozoa. And when he shook these eggs, he observed abnormal distribution of chromosomes in the dividing cells. By increasing the salt concentration in seawater or exposing eggs to various chemical or physical treatments, Jacques Loeb (1859–1924) induced development without the need for sperm, a process called artificial parthenogenesis.[4] Wilhelm Roux damaged one of the first segmenting cells (blastomeres) of a frog's egg to see what its neighbor would do. Would a whole organism develop or only a half? One of embryologists' main tasks was to experimentally demonstrate the roles played by its two chief components, nucleus and cytoplasm. They could remove the nucleus of an egg and replace it with another. The effects were measured by the stage the embryo reached in its development and by the kind of monsters created. Thus, the actual formation of the embryo was opened up to experimental analysis.

The Organism as a Whole

The methods used by embryologists consisted in amputating or incompletely separating parts of the embryo, or rearranging the parts by compressing them, transplanting them, or centrifuging them. These methods, of course, said nothing about the actual chemical changes underlying development, but from the large mass of work done with these techniques, one important generalization emerged: in practically all cases, a part the embryo had the power to give rise to more than it would if left in its normal surroundings; and in many cases, a part of the early embryo could give rise to a complete adult. Development was not due solely to the inherent properties of different cells themselves.

The fact that each of the cells of a dividing egg was capable of developing into a complex organism, and yet did not do so when left in its natural state, indicated to embryologists that the organism as a whole controls the formative processes going on in each part. The organism had supracellular properties. Cells, they argued, were the instruments, not the agents, of morphogenesis. Development was not the result of a colony or republic of cell individuals bound together by the division of labor and mutual dependence. Organisms made cells, not the inverse. Cells were the result, not the cause, of development.

That plants and animals were only collections of cells had been a simmering issue in Europe since the mid-nineteenth century. As early as 1853, T. H. Huxley argued against the "erroneous conception of the organism as a beehive": "They (the cells) are no more the producers of the vital phenomena than the shells scattered along the sea-beach are the instruments by which the gravitative force of the moon acts upon the ocean. Like these, the cells mark only where the vital tides have been, and how they have acted."[6] The German botanist Anton de Bary (1831–1888) wrote in 1862, "The plant builds cells, the cell does not build plants."[7] The plant physiologist Julius Sachs (1832–1897) made similar remarks in 1887. At Cambridge in 1895, Adam Sedgwick declared that the cell theory "blinds men's eyes to the true relations of cell organization and ontogeny."[8] Criticisms of the cell theory of development continued throughout the twentieth century.[9] Clifford Dobell (1886–1949), one of Sedgwick's former students, argued that it was simply anthropocentric to conceive of plants and animals as cell communities and to conceive of protista as primitive unicellular animals, as "proto-zoa." "The great importance of the Protista," he argued, in 1911 "lies in the fact that they are a group of living beings which are organized upon quite a *different principle* from that of other organisms."[10] They were noncellular organisms.

There were no unicellular "protozoan" precursors of animals, in Dobell's view. The protozoa-to-man hypothesis rested on the biogenetic law that ontogeny was the recapitulation of phylogeny. As applied to the cell theory, recapitulation theory assumed that when the egg undergoes segmentation in ontogeny, it repeats the process that occurred in phylogeny when animals arose from "unicellular" ancestors.[11] But Dobell denied any real analogy between an egg dividing into two blastomeres and a protist dividing into two protists. He maintained that multicellular organisms had evolved not by an aggregation of many individuals in the form of primitive protist colonies, but rather by the growth, differentiation, and division of one individual: the egg.[12]

Debates over the cellular or acellular nature of protists continued throughout the twentieth century.[13] Calling protists "unicells" or "protozoa" was indeed zoocentric. Certainly, when protists were studied without reference to animals, few biologists would disagree that they may be termed noncellular in the sense that they are not composed of cells. But, when their organization was compared with the cells of multicellular organisms, Dobell's antagonists insisted that the term "unicellular" was perfectly applicable.[14] Whether plants or animals were to be understood as colonies of cells or protists was a different matter.

Certainly, the evolution of the embryo could have arisen from a division of labor among protistan colonies in the remote past. After all, intermediate forms between a single cell and embryo-forming animals were known. Many protists do not separate their progeny, so the adult develops a colonial or multicellular form. There are a large number of such colonial protists, and many of them show differentiation and division of labor amongst the colony.[15] But there was also empirical evidence of the alternative view—that plants and animals could have originated by means of internal subdivision of a single protozoan cell. Protists were known that met those requirements.[16] For example, some ciliated protists possess many nuclei; to account for the genesis of animals, one could suppose that some of these protists developed membranes separating the nuclei. In this case, the primary unit is the animal, which later becomes multicellular. How multicellular plants and animals arose by selection remains largely unsettled (see chapter 20).[17] Those who protested against the cell state in the nineteenth century argued that whether or not animals had first evolved hundreds of millions of years ago by a struggle for existence and division of labor among cells, it was enough to know that embryonic development today had properties that were not reducible to parts.

The revolt against the cell state was especially strong in the United States. At the University of Chicago, C. O. Whitman rejected the notion that the development of animals could be explained in terms of reciprocal interaction of parts due to the struggle for existence and mutual interdependence. In 1893, he dismissed such conceptions as anthropomorphic. The cell-state model defined organisms in functional terms; Whitman sought principles of structural unity. "It is not division of labor and mutual dependence that control the union of the blastomeres. It is neither functional *economy* nor social instincts that bind the two halves of an egg together, but the constitutional bond of individual organization. It is not simple adhesion of independent cells, but integral structural cohesion."[18]

Cell division did not lead to organization, he argued; organization led to cell division. There was a structural foundation, a "grade of organization as the result of heredity," that was the starting point for every organism. It preceded the formation of cells and regulated it. That the organism dominates cell formation was indicated by comparative studies. One could find an organism using for the same purpose one, several, or many cells, massing its material, directing its movements, and shaping its organs, "as if cells did not exist, or as if they existed only in complete subordination to its will."[19] In protists, complex organizations were worked out within the limits of a single cell. The cell theory did not apply at all to "unicellular" organisms in which it was obvious that cell division was the result rather than the cause of organization or structural duplication. Therefore, Whitman argued, cells were not the primary unit of organic structure.

Whitman's objections were followed by protests against the cell state, made by a whole new generation of experimental embryologists that included Edwin Conklin (1863–1952) at Princeton, T. H. Morgan (1866–1945) at Columbia, Ross Harrison (1870–1959) at Yale, and Frank R. Lillie (1870–1947) at Chicago.[20] Lillie phrased the issue succinctly in 1906: "The organism is primary, not secondary, it is an individual, not by virtue of the cooperation of countless lesser individualities but an individual that produces these lesser individualities on which its full expression depends."[21] At Columbia University, E. B. Wilson, who had been so enthusiastic about the concept of the cell state, still insisted in 1925 that "the multicellular organism *in general* is comparable to an assemblage of Protista which has undergone a high degree of integration and differentiation so as to constitute essentially a cell-state."[22]

Epigenesis and Preformation

The principle of the organism as a whole was also upheld in opposition to the Weismannian model of development. The evidence and reasoning was similar. Weismann had proposed that cell differentiation occurred through the disintegration of the germ plasm of the nucleus. The first and most widely publicized experiment that contradicted that proposition was reported by Hans Driesch (1867–1941) in Leipzig in 1891.[23] He managed to separate the first two segmentation cells (blastomeres) of a sea urchin egg and grow them separately. If, during cell division, changes in the hereditary content of the cell had occurred, one would expect the separated cells to develop abnormally. But they did not. Instead two complete, though half-sized, larvae developed.

Weismann was able to point to a contradictory report by Roux three years earlier.[24] Roux's experiments were based on a different technique and on a different organism. He killed one of the two cells of a dividing frog's egg by pricking it with a hot needle; the other cell went on to produce what Roux considered to be a half an embryo. Weismann played the conflicting results off one another to argue that his own careful reasoning was more reliable than the results of experiments: "Other than experimental methods may lead us to fundamental views, and an experiment may not always be the safest guide, although it may at first appear conclusive."[25]

In the end, Weismann's rhetoric was no match for empirical data. Driesch's results with sea urchins were reproduced, making it clear that if cleavage cells are separated from one another in the two-cell stage, each of them may give rise to an entire animal. This was also true for the eggs of starfish, salamanders, and several other animals, including frogs. Roux's hot-needle experiments were deemed to have been faulty: the damaged cleavage cell remained and must have caused defects in the surviving embryo.

Roux was one of the great champions of a mechanistic approach to embryology. But what machine can be cut in half in which both halves continue to function as if nothing had happened? How can a cell that is to form a part of an embryo suddenly form additional parts? How can a part have a sense of the whole? The whole problem of development seemed to be brought into focus in Driesch's well-known axiom: "The relative position of a blastomere in the whole determines in general what develops from it; if its position be changed, it gives rise to something different. In other words, its prospective value is a function of its position."[26]

How a cell behaved and what it formed was determined by some powerful, mysterious field of forces pervading the organism. The fate of a cell was a function of its position in the whole. The principle of "the organism as a whole" became a central tenet of embryology; its capital problem was to find a physicochemical basis for it. Embryologists did not have the answers, but they did develop a common general conception: epigenesis. In epigenesis theory, the organism was conceived of in terms of the behavior of a particular protoplasm in a particular environment. The cells of an organism, for example, were not predetermined as different parts but could be primarily all alike in constitution. The morphological and biochemical differences that arose in the course of development were influenced by the action of the environment on whole groups of cells and on each member of them.

Epigenesis theory had considerable explanatory power. Many experiments showed that the development of a part is dependent upon the presence of another part. For example, if embryonic limb tissue is transplanted from its normal position to the middle of the back or belly, it will develop, and nerves and blood vessels may grow into it, that would have had very different positions if the limb had not been there. Hans Spemann (1869–1941) at the University of Freiburg was noted for his ingenious experiments demonstrating the regulative qualities of salamander embryos.[27] In 1924, his student Hilde Mangold reported that when she transplanted a part of an early embryo of one species of salamander onto a different part of an embryo of a different species, it modified the surrounding host tissue, and a new second embryo grew.[28] Spemann was awarded the Nobel Prize in 1935 (he did not share it with Mangold because she had died) for his detailed studies of the organizer effect.[29]

There were many illustrations of the regulative qualities of embryonic cells. But not all cells carried the ability to make a whole organism. All cells were not "totipotent." Driesch had assumed that any part is capable of any fate, and its actual fate is determined by its relation to other parts. By the end of the nineteenth century, embryologists knew this to be only a part of the truth. The fate of a cell was both a function of its internal organization and a function of its position. Cellular differentiation did occur at some stage of development. For example, if a right limb is transplanted to the left side of the body after it has begun to differentiate, it remains a right limb and is not modified by its new relationship to the rest of the body. Experiments on tissue cultures showed that some of the differences among somatic cells persisted when the cells were taken out of the body. Biochemical and morphological changes in a cell that occurred during development were inherited from one cell to the next. The fate of a cell or tissue was determined both by its internal organization and by its neighbors.

Embryologists were soon reminded of another way in which biology is not reducible to physics and chemistry. They obtained different results from different organisms. In some organisms, some differentiations occurred further back in development—in the fertilized egg itself before it divided. For example, the first cleavage cells of the eggs of some phyla such as mollusks (snails, squid, and octopus), annelid worms, and ascidians (sea squirts), give rise only to parts of an animal. One might get a right or a left half of an animal from right or left cleavage cells; or an anterior half or posterior half from anterior or posterior cleavage cells; or any one of the cells of the four-cell stage may produce the corresponding quarter of an entire animal. Such cases were referred to as "mosaic embryos." The difference between mosaic embryos and the regulative embryos of sea urchins and salamanders reflected the relative importance of cell-cell interactions in determining the fate of the cell. In mosaic embryos such cell interactions appeared to be quite limited, but all embryos were regulated to some extent by interactions between cells.

Embryologists thus used two concepts, epigenesis and preformation, to understand the emergence of the embryo from the egg.[30] Although the effects of the environment mediated through metabolic reactions in the cell could account for modifications in cells in the course of development, it was clear that some sort of organizational plan, a spatial principle, was required to bring parts into place in the proper way and at the proper time during development. As discussed in the following chapter, many embryologists investigated the organization of the egg as the primary basis of development.

10

The Egg

We are vertebrates because our mothers were vertebrates and produced eggs of the vertebrate pattern; but the colour of our skin and hair and eyes, our sex, stature and mental peculiarities were determined by the sperm as well as by the egg from which we came. There is evidence that the chromosomes of the egg and sperm are the seat of the differential factors or determiners for Mendelian characters while the general polarity, symmetry and pattern of the embryo are determined by the cytoplasm of the egg.

—Edwin G. Conklin, 1915

Without a structure in the egg to begin with, no formation of a complicated organism is imaginable.

—Jacques Loeb, 1916

EMBRYOLOGISTS' THINKING ABOUT HEREDITY and evolution was far removed from that of neo-Darwinian biologists, and from that of geneticists who studied chromosomal genes in the early twentieth century. The two main reasons for this stemmed from research begun in the nineteenth century. First, experimental embryologists were not concerned with adaptive changes to adult organisms. They were not concerned with the differences between individual varieties and species, but rather with the larger similarities and differences in the structural plans underlying different organismic types. They were interested in the form of the embryo and in the orderly changes during development—in how the parts of the organisms come together in space and time.

Second, while geneticists and neo-Darwinian evolutionists maintained that changes in the nuclear chromosomal genes of eggs and sperm were the basis of evolution, many embryologists insisted that the cytoplasm of the egg played the primary role in heredity and development. Although often overlooked by historians of evolutionary biology, embryologists argued that the large, "fundamental" organismic characteristics that placed the organism in its proper phylum, class, order, and family were determined by the egg cytoplasm, while the smaller differences—between, say, genera, species, varieties, and individuals—were due to changes in the chromosomes of egg and sperm.[1]

Embryologists' and geneticists' paths began to diverge when embyologists refuted Weismann's nuclear theory of development. When Hans Driesch separated the first two cells of a sea urchin egg, each cell formed a complete sea urchin larva (see chapter 9). This and other experiments showed that the fate of cells was determined in part by their relationship to other cells, indicating that development was not due to qualitative changes in the chromosomes, as Weismann had imagined. It was also clear from cytological investigations that the chromosomes were precisely divided at each mitotic division and evenly distributed to daughter cells. There seemed to be no qualitative nuclear differentiations except for the reduction division (meiosis) during the formation of sex cells. Each body cell, it seemed, inherited the sum total of hereditary qualities of the nucleus. Yet, biochemical and morphological differentiations in body cells did occur, and they were inherited by one cell generation from the previous. These problems led many embryologists to search for the basis of development in the egg cytoplasm.

The Body Plan in the Egg

Embryologists of the late nineteenth and early twentieth centuries showed that the cytoplasm of egg cells was highly organized, and unlike the nucleus, it did change during development. There was a characteristic organization of cytoplasmic materials in the eggs of many marine organisms. The cytoplasm in different egg regions often exhibited differences in pigmentation, viscosity, and other properties, which could be followed visually in the early stages of development. When the egg divided, each daughter cell obtained different amounts of the different types of cytoplasmic materials. In fact, by following cell lineages, embryologists were actually able to visually trace the early development of organs and tissues back to specific regions of the cytoplasm of the egg before it began to divide.[2]

The characteristics they traced to the egg were the fundamental organizational features of the developing organism. The fundamental nature of these characteristics is better appreciated when we consider what embryologists regard as pivotal episodes or stages in ontogeny. During the first stage of development, immediately following fertilization, the egg undergoes a series of divisions to produce a large number of cells that occupy the original volume of the egg, forming what is called a blastula. There is no growth of tissue or increase in cell mass during this period. The process by which cells are ordered spatially and temporally begins with the laying down of the body plan: the defining of the main axes of the embryo such that the anterior and posterior ends of animals are established. This is done through gastrulation, during which the early embryo undergoes an origami-like change, an indentation appears, and the mass of cells buckles to form a cup in frogs and sea urchins, for example.

As a result of gastrulation, the embryo is sorted into three distinct cell layers, or germ layers (endoderm, ectoderm, and mesoderm), which give rise to the basic body plan and differentiate into the many tissues and organs of the adult body. The endoderm (inner skin) gives rise to the lining of the digestive and respiratory tracts, and the lungs and liver in vertebrates. The ectoderm (outer skin) gives rise to the epidermis of the skin, the nervous system, the lens of the eye, and the inner ear. The mesoderm (middle skin) between the endoderm and ectoderm gives rise to muscles, skeleton, and other internal organs such as the kidney and heart.

Each of these cell layers, and therefore the organs and tissues to which they gave rise, were treaceable back to visible substances in specific locations in the cytoplasm of the fertilized egg. The arrangement of the cytoplasmic materials in eggs was so definite and so constant that characteristic organization patterns could be recognized for different phyla. In some sea urchins and certain species of sea squirts, for example, eggs had yellow and red pigment granules distributed only in a thin layer of cytoplasm near the cell surface, the cortex. In amphibian eggs, the pigment was brown and black and was located both in the cortical layer of the cytoplasm and in the interior of the egg. The distribution of these substances to different cleavage cells could be easily followed. Within a few minutes after the fertilization of the sea squirt egg, the anterior and posterior, dorsal and ventral, and right and left poles were clearly distinguishable in the egg, and the regions that would give rise to ectoderm, endoderm, and mesoderm were plainly visible in the characteristic positions.

The precise arrangement of such substances in the egg indicated the existence of some sort of submicroscopic structure in the egg cytoplasm. The egg cell was spatially organized in terms of two opposite poles: the substance in the vicinity of the "animal pole" usually gave rise to the ectoderm; the substance surrounding the "vegetal pole" usually became the endoderm. The axis connecting these poles, the chief axis of the egg, became the chief axis of the adult animal. Thus, the polarity of the developed animal was directly connected with the polarity of the egg from which it came. The bilateral symmetry of animals was also foreshadowed in the egg cytoplasm, and in many animals, such as cephalopods and insects, eggs were found to be bilaterally symmetrical while still in the ovaries. In most animals, bilateral symmetry is not perfect. In the bodies of all vertebrates, for example, the heart and most of the liver are on one side of the midline. But sometimes this asymmetry was found to be reversed: the heart was on the right side instead of the left.[3]

Thus embryologists had discovered that the early stages of development—the basic body plan and pattern of a jellyfish, starfish, worm, mollusk, insect, or vertebrate—resulted from a characteristic polarity, symmetry, and localized pattern or stratification in the egg cytoplasm.[4] They were also able to show that the sperm that fertilized the egg played no role in setting up those morphogenetic characteristics: when eggs developed without fertilization, either naturally or with development induced by a pinprick (artificial parthenogenesis), the characteristic polarity, symmetry, and pattern of the adult were still obtained just as if the egg had been fertilized. Therefore, the earliest and most fundamental differentiations that distinguished various phyla were not dependent upon the entering sperm. Offspring inherit these fundamental developmental features exclusively from their mother.

Maternal Inheritance

Embryologists also devised hybridization experiments to test the relative roles of the nucleus and cytoplasm, or sperm and egg, in heredity. They physically removed the nucleus from the egg of one species and inseminated that enucleated egg with the sperm of a different species. Because sperm cells are made up of little more than a nucleus, the resulting fertilized egg would possess a cytoplasm from the egg species and chromosomes from the sperm species. If the hybrid embryo showed only characteristics of the paternal species brought in by the sperm, the nucleus would be the "bearer of inheritance." But if characteristics of the maternal species appeared, then one could conclude that a cytoplasmic influence was operative. Eggs of echinoderms, mollusks, amphibians, and fishes that are inseminated in water outside the body were the best material. The eggs of sea urchins were especially useful because they readily united with spermatozoa of foreign species or genera.

These kinds of experiments were initiated in 1889 by Theodor Boveri, who in fact conducted them with the intention of proving that the nucleus was the sole bearer of inheritance. His results contradicted that presumption. They seemed to indicate that the general features of early development were due to cytoplasmic influences, whereas the influence of the sperm (nucleus) appeared later, in the sea urchin larva. In 1903, Boveri proposed a compromise between the roles of the nucleus and the cytoplasm in heredity. He distinguished between preformed and "epigenetic" developmental characteristics. The former—the general characteristics of the embryo—were blocked out or prearranged in the organization of the egg cytoplasm independently of the sperm. The epigenetic characteristics were due to reciprocal interactions between the nucleus and the cytoplasm in the course of development. Boveri thus proposed that the nucleus of egg and sperm would be responsible for "all the essential characteristics which distinguished individuals and species," and, by implication at least, he suggested that cytoplasmic qualities would distinguish higher groups.[5] One issue was perfectly clear: at the time of fertilization: the hereditary potencies of the two germ cells were not equal.[6] The sperm had no influence on early development.

Today, our ideas of heredity are based on observing *differences* between individuals of a species. The characteristics that embryologists traced to the egg cytoplasm involved the organizational features that make those individuals the *same*: all the early stages of development, including the polarity, symmetry, type of cleavage, and pattern, or relative positions and proportions of future organs. As the

Princeton embryologist E. G. Conklin remarked in 1920, "These characters are of such a general sort that they may not be recognized as phenomena of inheritance at all, and yet they from the background and framework for all the other characters."[7] When geneticists developed gene theory and the Mendelian chromosome theory during the first decades of the twentieth century, embryologists offered a compromise between views of the roles played by the nucleus and the cytoplasm. They suggested that while the cytoplasm determined the fundamental organismic features, the genes in the chromosomes of egg and sperm accounted for differences between individuals, species, or perhaps even genera. At the University of Cincinnati, Michael Guyer wrote in 1911: "we must restrict our assertion of equal inheritance to the sexual and specific differences which top off, as it were the more fundamental organismal features."[8] Similarly, the Oxford embryologist J. W. Jenkinson wrote:

> The characters, the determinants of which reside in the cytoplasm, are the large characters which put the animal in its proper phylum, class and order, which make it an Echinoderm and not a Mollusc, a Sea-urchin and not a Starfish; and these large characters are transmitted through the cytoplasm and therefore through the female alone. The smaller characters—generic, specific, varietal, individual—are equally transmitted by both germ-cells and the determinants of these are in the chromosomes of their nuclei.[9]

The Belgian embryologist Albert Brachet distinguished between what he called *l' héredité générale*, which had its seat mainly in the cytoplasm, and *l'hérédité spéciale*, that of the individual due to chromosomes.[10] Edwin Conklin at Princeton launched a rhetorically effective statement in 1915 when he argued that "we are vertebrates because our mothers were vertebrates and produced eggs of the vertebrate pattern but the colour of our skin and hair and eyes, our sex, stature and mental peculiarities were determined by the sperm as well as by the egg from which we came."[11] The next year Jacques Loeb also made the compromise plain:

> The facts of experimental embryology strongly indicate the possibility that the cytoplasm of the egg is the future embryo (in the rough) and that the Mendelian factors only impress the individual (and variety) characters upon this rough block. . . . In any case we can state today that the cytoplasm contains the rough preformation of the future embryo.[12]

This marked the beginning of a long and deep rift between embryologists and geneticists in the United States, who were most reluctant to concede that the cytoplasm played any role in heredity and evolution. Geneticists immediately raised the question of whether this cytoplasmic organization in the egg could be actually traced to the activity of chromosomal genes during the formation of the egg (oogenesis). In other words, were the characteristics of the egg perpetuated from one generation to the next independently of the nucleus or did they arise anew under the *direction* of the nucleus when the egg was formed? Would not the genes of the father and mother of the egg-bearing female have had some effect on her egg? Such characteristics may not be influenced by the nucleus of the sperm that fertilizes the egg, but if the biparental nucleus had been involved during egg formation, one would expect to see the effect of that biparental nucleus in subsequent generations. The embryonic character inherited maternally in the first generation might therefore show biparental inheritance in subsequent generations.[13]

Geneticists insisted that in order to disprove their claim that the nucleus is the sole basis of heredity and show that such cytoplasmic features were not due to the action of nuclear genes, one would have to follow egg characteristics through at least two generations. However, such experiments were not easy to do because normally the hybrid embryos from crosses between species did not develop fully into adults. But by the early 1920s, some evidence indicated that some egg characteristics, such as inverse coiling in snails, were influenced by nuclear genes.[14] These preliminary results encouraged geneticists to believe that the whole cell was a product of gene action. They insisted that the "maternal inheritance" observed by embryologists was not true cytoplasmic inheritance because all egg characteristics were ultimately traceable to the effects of genes. The Harvard geneticist L. C. Dunn wrote in 1917: "The whole case of the supporters of any theory which views the cytoplasm as determinative rests on either their refusal to go back and inquire the source of this cytoplasm, or on their refusal to give due emphasis to the source, even though they recognize it."[15]

These were harsh words. Certainly some embryologists conceded that it was highly misleading to say that "the embryo in the rough" was determined solely by the cytoplasm. Some egg characteristics may be developed under the influence of the nucleus.[16] Nonetheless, they continued to exclude genes from playing any significant role in early development and morphogenesis. They also insisted that there is a real difference in modus operandi between nucleus and cytoplasm: the characteristics in the egg cytoplasm make their appearance unaffected by the sperm that subsequently enters the egg. Parental egg and sperm do not contribute equally to the heredity of the offspring. As Boveri wrote in 1918:

> If one designates as heredity the totality of internal conditions which achieve the unfolding of characteristics of the new individual, this gives to the cytoplasm a much more specialized significance than one often has been inclined to assume; and more than ever one realizes the absurdity of the idea that it would be possible to bring a sperm to develop by means of an artificial culture medium.[17]

As late as 1945, the Cambridge zoologist V. B. Wigglesworth maintained that the nuclear genes were concerned with the details but not the form of the organism that had its basis in the egg cytoplasm:

> The essential organism is something apart from the cells which support it. It exists before the cells dispose themselves and define its form. The cells and their nuclei, as the vehicles of genes, play a great part in controlling the details of the form the organism will tak; but the framework which marks the main outlines of that form, which says that the organism shall be a vertebrate, an amphibian, a frog, or an insect, a dipteron, a *Drosophila*, which defines the head and the tail, the main regions of the body and the limbs—this framework exists even before the cells.[18]

Cellular Differentiation

There was still another problem confronting Mendelian genetics: cellular differentiation. The Mendelian chromosome theory was no more helpful than Weismann's nuclear theory. By growing different kinds of body cells in a test tube or petri dish, embryologists showed that at least some of the differences among cells persisted when they were taken out of the body—that is, they were hereditary (this is called cell heredity). Yet, the nucleus showed no signs of differentiation from one cell generation to the next. The most celebrated experiments indicating a lack of nuclear differentiation were carried out by Hans Spemann, who in 1928 managed to transfer a nucleus of a salamander embryo at the sixteen-cell stage to a cell without a nucleus. If that nucleus had undergone any irreversible differentiation, one would expect abnormal development. Instead, in a number of cases, a completely normal twin developed.[19]

Spemann had created clones; and when in 1938 he published his results in his book *Embryonic Development and Induction*, he called for the "fantastic experiment" of cloning from adult cells.[20] In the meantime, his experiments remained influential throughout most of the century. They indicated that the hypothesis of unequal division of the hereditary substance of the nucleus or genes was incorrect. At least up to the sixteen-cell embryo, nuclear differentiation did not occur: every nucleus had a complete set of hereditary factors. Mendelian geneticists also insisted that each cell inherited the whole germ plasm of the nucleus.[21]

The inheritance of cell differences with equivalent nuclei reinforced the argument that cytoplasm was the seat of somatic cell heredity. Conklin spoke for many embryologists when he argued in 1920 that if nuclear genes were the only agents of development, "these genes would of necessity have to undergo differential division and distribution to the cleavage cells; since this is not true, it must be that some of the differential factors of development lie outside the nucleus, and if they are inherited as most of these early orientations are, they must lie in the cytoplasm."[22]

Of course, as Spemann and others recognized, it was still possible that nuclear genes changed in some other way besides being sorted out in the course of development.[23] Nonetheless, in the absence of such evidence, the idea that the sorting out of cytoplasmic materials was the basis of cell heredity during development was perpetuated throughout most of the twentieth century (see chapters 12, 15, and 16).

The paradox of cellular differentiation in the face of nuclear equivalence was only one of the difficulties embryologists saw in the classical gene theory. It was also logically necessary to maintain some preformed spatial property in the cell—a problem that persists to the present day (see chapter 17). How could parts come together in an orderly fashion to make a cell or complex organism? This was a stumbling block for a materialistic outlook on life. Biologists of the late nineteenth century would never think of treating the processes of digestion and metabolism in any other way than in purely chemical or physicochemical terms. But when it came to the problem of how a harmonious organism developed "purposefully" from an egg, such leading biologists as Claude Bernard (1813–1878) and Hans Driesch turned to vitalism and invoked an unknowable "directive force" or Aristotelian "entelechy."[24]

Many nongeneticists found it difficult to believe that the cell and its parts were the by-products of genes. It seemed unlikely that mitochondria and chloroplasts were products of the nuclear genes. Cytological observations suggested that they arose by division of preexisting structures such as the nuclear chromosomes, and this seemed to be the case for centrioles as well. Was it also true for cell structure itself, as Rudolf Virchow had expressed it in his famous axiom *omnicellula e cellula*?

Physiologists viewed the cell as a chemical factory where enzymes carried out chemical reactions, but the orderliness of their activity was attributed to the structure of the factory. The cell was not a bag of enzymes; life was not governed by molecules or determinants. A. P. Mathews at the University of Chicago summarized this view concisely in 1915: "The orderliness of chemical reactions is due to the cell structure; and for the phenomenon of life to persist in their entirety that structure must be conserved."[25] "One cannot help assuming," wrote the physiologist L. Jost in 1907, "that the mode of arrangement of the ultimate parts of the organism is of greater importance than the chemical nature of these parts."[26]

Jacques Loeb, one of the great champions of mechanistic materialism, aptly compared the making of cell structure from genes to spontaneous generation. It was difficult enough to imagine the circumstances in the remote past that had led to the formation of the chemicals of living matter such as proteins. But to imagine how cell structure first evolved was beyond comprehension:

> It is at least not inconceivable that in an earlier period of earth's history radioactivity, electrical discharges, and possibly also the action of volcanoes might have furnished the combination of circumstances under which living matter might have formed. The staggering difficulties in imagining such a possibility are not merely on the chemical side—e.g., the production of proteins from CO_2 and N—but also on the physical side if the necessity of a definite cell structure is considered.[27]

Extrapolating the problem to eggs and animals, Loeb argued in 1916 that without a cell structure in the egg to begin with, no complex organism could form. Therefore, he reasoned, the cytoplasm must contain a primordial structure (which it in fact did), as indicated by polarity and symmetry and the visible stratification of the cytoplasm, which determined the first steps in morphogenesis.[28]

What was important was not just the pigment granules, but the underlying spatial organization they revealed.[29] What was the basis of this polar organization of the egg? How was it maintained? What would happen if the visible substances were artificially displaced? Would they return to normal? Here, too, experimental embryologists of the early twentieth century were resourceful. They placed fertilized eggs in a centrifuge, whirled them rapidly for a few minutes at moderate speeds, and subjected them to pressure several thousands of times that of gravity. Under such conditions, the heavier particles were thrown to one side of the egg, and the entire substance of the egg became stratified into layers or zones. In sea urchins, annelids, and gastropods, the nucleus, centrosome, and pigment granules could be displaced without modifying normal polarity, symmetry, and development.[30]

This experiment indicated that the fundamental organization of the egg resided in a comparatively immovable cytoplasmic structure, in a cortical ectoplasmic layer of the cell just inside the plasma membrane: the cortex: Thus, it was argued, the more rigid cortex played the role of a spatial principle.[31] It foreshadowed in some way the pattern of the future of the embryo and provided the basis of the orderly pattern of morphogenesis.[32]

To account for the regulative properties of the developing animal, embryologists of the early twentieth century adopted the notion of a "morphogenetic field." The term "field" had been used since the mid-eighteenth century to define the space to which observations of a telescope were limited. Beginning in the mid-nineteenth century, physicists used the word to refer to an area or space under a common influence. Similarly, embryologists adopted it to refer generally to a region of the developing organism within which developmental options are subjected to a common set of coordinating influences.[33] Hans Spemann likened it to a magnetic field and in 1938 argued, on the analogy of physical fields, that it was bound to a source.[34] He located it in the egg cytoplasm as manifested in cell polarity (animal pole and vegetal pole) and the stratification of egg materials. The inheritance of this spatial property or morphogenetic field associated with the cell cortex was not systematically investigated until the 1960s (see chapter 17).[35]

Throughout the twentieth century, embryologists protested against attributing the properties of life to one kind of molecule or one part of the cell. The terms "holistic," "holism," and "holon" (from the Greek *holos*, "whole") were coined by South African general Jan Christian Smuts (1870–1950) in his book *Holism and Evolution* (1926) for the tendency for nature to produce wholes (e.g., organisms) from the ordered groupings of unit structures. Holism was (and is) usually opposed to "reductionism," a term borrowed from chemists who used it to describe the reduction of a compound to a simpler substance by removing oxygen—for example, the removal of oxygen from metal ores, which "reduced" the metal ore to pure metal (e.g., $2\ Fe_2O_3 + 3\ C \rightarrow 3\ CO_2 + 4\ FE$). Reductionism designated the practice of describing a phenomenon in terms of an apparent, more "basic" or primitive phenomenon, to which the first is held to be equivalent. Embryologists added to holism the principle of "emergence." Physicists and chemists had shown how new properties could arise through the combination of smaller units into larger ones. New atoms with new properties are formed by new combinations of protons and electrons, new molecules by new combinations of atoms. Similarly, embryologists argued, the distinctive properties of life and the formation of new materials and qualities in the course of development arise or emerge from the interactions of parts that by themselves do not show these properties.[36]

Cytoplasmic Evolution

Experimental embryologists as a group did not develop a coherent theory for evolutionary change. Some adopted pluralist models that combined both mechanistic and metaphysical conceptions. When Boveri, for example, asked what caused large changes in organization, he turned to psychic explanations. Similarly, Spemann's holistic outlook led him to combine both Lamarckism and psychic forces.[37] He drew ideas from vitalists and from the idealist morphology of Goethe and the *Naturphilosophen*, still widely read in Germany.[38]

In the United States, C. O. Whitman opted for a blend of orthogenesis and micromutations when he argued that developmental patterns would constrain evolutionary directions. Natural selection was too chaotic a mechanism to account for complex organs. To believe that the eye developed by mutations and selections, he wrote, "either gradually or *per saltum*, would be hardly more satisfactory than appealing to a miraculous succession of miracles. . . . Without the assistance of some factor having more continuous directive efficiency," he argues, "selection would fail to bring out of the chaos of chance variation, or kaleidoscopic mutation, such progressive evolution as the organic world reveals."[39]

Others saw a basis for macromutations in the organization of the egg cytoplasm. When Conklin declared in 1938 that "the characteristics of the phylum are present in the cytoplasm of the egg cell," he did so to emphasize how macroevolutionary change might occur.[40] One of the main difficulties in explaining the origin of different phyla was the dissimilar localizations of corresponding organs or parts. How could vertebrates be derived from annelids or from any other invertebrate type? If the overall pattern of development was determined by the egg, Conklin reasoned, then only a slight modification in the localization of the formative substances of the egg would produce profound modifications of the adult in which the relative positions of the parts were changed.[41] Whether such changes actually occurred was not known.

But by the 1920s, even some leading geneticists had defected from gene theory. By that time, the genetics school headed by Thomas Hunt Morgan at Columbia University, using the fruit fly, *Drosophila melanogaster*, had forged a great synthesis of genetics and cytology by locating genes on chromosomes and had mapped their spatial relationships, like beads on a string. However, geneticists failed to produce new species. Many of the mutations in *Drosophila* were small changes that made a part a little longer or a little smaller, for example. When the mutations were larger, they seemed only to produce organisms less capable of surviving outside the laboratory. Geneticists studied primarily defects. The purebred strains in Morgan's laboratory included *Drosophila* that were eyeless, had abnormal abdomens, or displayed some other abnormality. To biologists who had originally turned to Mendelian genetics and championed it in the hope that it would be the basis for evolution by macromutations, it seemed clear that gene changes were not the stuff of evolution after all. The Danish geneticist Wilhelm Johannsen spoke for several lapsed Mendelians in 1923 when he commented:

> Is the whole of Mendelism perhaps nothing but an establishment of very many chromosomal irregularities, disturbances or diseases of enormous practical and theoretical importance but without deeper value for an understanding of the "normal" constitution of natural biotypes? The Problem of Species, Evolution, does not seem to be approached seriously through Mendelism nor though the related modern experiments in mutations. . . . Chromosomes are doubtless vehicles for "Mendelian inheritance" but Cytoplasm has its importance too.[42]

Mendelian genetics, as Johansen saw it, was "mostly operating with 'characters' which are rather superficial, in comparison with the fundamental Specific or Generic nature of the organism."[43] *Drosophila* geneticists seemed to be dealing with trivial differences between individuals. As Johannsen put it: "the Pomace-flies in Morgan's splendid experiments continue to be pomace-flies even if they lose all 'good' genes necessary for a normal fly-life, or if they be possessed with all the 'bad' genes, detrimental to the welfare of this little friend of the geneticists."[44] He speculated about the existence of "a great central 'something' as yet not divisible into separate factors" located in the cytoplasm.[45] In England, William Bateson discussed the limits of the chromosome theory in 1926 when emphasizing its failure to account for cellular differentiation: "pending that analysis, the chromosome theory, though providing much that is certainly true and of immense value, has fallen short of the essential discovery.[46]

At the same time, neo-Lamarckian naturalists repeatedly emphasized that they could see no connection between the gene mutations reported by geneticists and the evolutionary processes leading to species and higher taxa.[47] As Maurice Caullery at the Sorbonne wrote in 1935, "The properties of the characters to which Mendelism applies are limited, in an almost absolute way, to variations which do not extend beyond the framework of the species."[48] Many neo-Lamarckians pinned their hopes for the inheritance of acquired characteristics on the cytoplasm. In Germany, during the 1920s, several botanists, including Carl Correns, one of the early champions of Mendelism, turned to investigate the role of the cytoplasm in heredity (see chapter 15).

Morgan and other American geneticists remained steadfast in their advancement of the chromosome theory. They modified their views on evolution to match their experimental observations. By 1920, Morgan maintained that the small gene mutations he observed in the laboratory together with selection theory would account for evolution (see chapter 13). He denied any distinction between hereditary traits corresponding to "fundamental" and "trivial" differences and argued that there was little genetic evidence for cytoplasmic heredity:

> Mendelian workers can find no distinction in heredity between characteristics that might be ordinal or specific, or fundamental, and those called "individual." This failure can scarcely be attributed to a desire to magnify the importance of Mendelian heredity, but rather to experience with hereditary characters. That there may be substances in the cytoplasm that propagate themselves there and that are outside the influence of the nucleus, must, of course, be at once conceded as possible despite the fact that, aside from certain plastids, all Mendelian evidence fails

> to show that here are such characters. In a word, the distinction set up between generic versus specific characters or even "specificity" seems at present to lack any support in fact.[49]

Supported by a lack of genetic evidence of non-Mendelian heredity, Morgan declared in 1926, "Except for the rare cases of plastids (chloroplasts) inheritance all known characters can be sufficiently accounted for by the presence of genes in the chromosomes. In a word the cytoplasm may be ignored genetically."[50]

Mendelian genetics could not address problems of development. Nonetheless, it quickly overshadowed embryology, comparative morphology, and natural history to become one of the most enterprising disciplines in the history of science (see chapter 12). There is more literature about the history of genetics than about any other aspect of twentieth-century biology. I turn next to that historiography, beginning with the discovery of Mendel's laws.

III

GENETICS AND THE CLASSICAL SYNTHESIS

11

Mendel Palimpsest

> Each generation perhaps found in Mendel's paper only what it expected to find; in the first period a repetition of the hybridization results commonly reported, in the second a discovery in inheritance supposedly difficult to reconcile with continuous evolution. Each generation, therefore, ignored what did not confirm its own expectations.
>
> —R. A. Fisher, 1936

GENETICS RESOLVED ONE of the major difficulties confronting the theory of natural selection. According to Darwinian theory, inherited variations between individuals were the source of all evolutionary innovations. Variations that improve chances of survival and reproduction are favored in each generation; those that reduce chances of reproduction become less common. Thus, a population changes over time—it evolves (see chapter 13). One of the main criticisms of this view was that it did not fit with prevailing conceptions of inheritance (see chapter 6). It was commonly assumed that a father's traits "blended" with a mother's traits at fertilization. And if such variations blended with each other in each generation, it was argued, this fuel for evolution would soon be washed out and would disappear altogether.

Geneticists' experiments indicated that there were hereditary factors that separate cleanly in the germ cells, even when the inherited visible trait in an offspring was intermediate in a character. Although visible traits (phenotypes) often blended, say a pink offspring from red and white flowers, the germinal material, the "genes" (the genotype), did not blend. The genes representing red and white traits would be inherited intact, as distinct particulate entities, and transmitted from one generation to the next, and both could later reappear in offspring from a pink flower. Natural selection could in principle act on any heritable trait.

This discovery is generally attributed to the Austrian monk Johan Gregor Mendel (1822–1884), one of the greatest legends in the history of science. Three moments in this legend are extraordinary. The first is that in the 1860s he discovered the laws governing the inheritance of individual characters; second, that the scientific world failed to recognize the monumental importance of these finding in his lifetime; and third, the remarkable rediscovery in 1900 of what came to be called Mendel's laws.

Mendel's Laws

Mendel changed his name to Gregor in 1843 when he entered the Augustinian monastery at Brünn (now Brno), an Austrian village now part of the Czech Republic. He was ordained a priest in 1847, and in 1851 he went to the University of Vienna to be educated as a teacher of mathematics and natural sciences. After repeatedly failing the exams to become a teacher, he spent most of his life in the monastery of St. Thomas in Brünn, where, shortly after he graduated in 1856, he began experiments with the garden pea plant, *Pisum sativum,* which he continued for eight years.

Textbooks of genetics relate that Mendel had the idea that in each generation a plant inherits two factors for a trait, one from each parent. To test his notion, he tracked several single traits through two generations. For instance, in one series of experiments, he crossed plants with true-breeding purple flowers with plants with true-breeding white flowers. All plants grown from the seeds that resulted from this cross had purple flowers. When he let these purple-flowered plants self-fertilize and grew the seeds, some plants had white flowers again. He called these traits that did not show up in the first generation of offspring recessive traits, because they appeared to recede or disappear in the hybrid. If his hypothesis was correct—if each plant had inherited two factors for flower color—then the purple trait would be dominant, because it had masked the white trait in first generation of offspring (F_1).

Mendel crossed hundreds of such hybrid F_1 plants, tracked thousands of offspring, and counted the relative number of plants showing dominance and recessiveness for a trait. He reported that in the second generation (F_2), on average three of every four plants had the dominant trait and one had the recessive trait. This 3:1 ratio suggested that the possible outcomes of crosses were a matter of chance combinations of two different factors. Mendel had an understanding of probability: the chance of each outcome occurring is proportional to the number of ways it can be reached. Of course, his reported ratios were not exactly 3:1. The result of F_1 crosses, for example, resulted in 705 purple flowers and 224 white flowers (3.15:1).

Mendel reported his findings at a meeting of the Brünn Natural History Society in 1865, and they were published in the society's journal the following year.[1] However, he gave up his experiments when he was was made abbot in 1868. Although respected by his fellow monks, he did not gain recognition as a great scientist in his lifetime. Three botanists, Hugo de Vries (1848–1935) Carl Correns

(1864–1935), and Erich Tschermak (1871–1962), were credited with independently rediscovering his laws in 1900.[2] Subsequently, geneticists summarized Mendelian theory as comprising two laws. Mendel's first law, or the law of segregation, states that only one form of a gene (allele) specifying an alternative trait can be carried in a particular germ cell (egg or sperm or pollen), and that germ cells combine randomly in forming offspring. His second law, called the law of independent assortment, states that each trait is inherited independently of any other. The purple flower factor, for example, may be inherited with another factor, say, for seed shape. This principle was later modified when geneticists discovered linkage, the inheritance of two or more genes situated close to each other on the same chromosome.

Neglect and Rediscovery

Thus, after being eclipsed for thirty-five years, Mendel's experiments came to be universally hailed as providing the foundation for a chain of scientific research that has led to the Darwinian evolutionary synthesis of the 1930s and 1940s and to the spectacular accomplishments of contemporary molecular biology. Loren Eisley summarized the Mendel legend beautifully:

> Mendel is a curious wraith in history. His associates, his followers, are all in the next century. That is when his influence began. Yet if we are to understand him and the way he rescued Darwinism itself from oblivion we must go the long way back to Brünn in Moravia and stand among the green peas in a quiet garden. Gregor Mendel had a strange fate: he was destined to live one life painfully in the flesh at Brünn and another, the intellectual life of which he dreamed, in the following century. His words, his calculations were to take a sudden belated flight out of the dark tomblike volumes and be written on hundreds of university blackboards, and go spinning through innumerable heads.[3]

If Mendel's experiments were neglected for thirty-five years, today they show no signs of dwindling in curiosity or significance. More historical attention has been given to analyzing them and commenting on them than any other experiments in biology. But here we meet with an apparent paradox. Although everyone agrees that Mendel's experiments are central to modern biology, there has been little agreement about what their exact significance is. His motives, his experimental protocols, and his own beliefs about heredity and evolution have been the subject of controversy for a century. In fact, just about every possible scenario has been offered to account for them:[4]

1. Mendel was a non-Darwinian. Although Mendel was an evolutionist, he did not entirely agree with Darwin's views and set out to disprove them.[5]
2. Mendel was a good Darwinian. His experimental protocols and reported results can be explained on the assumption that he had no objections to Darwinian selection theory.[6]

3. Mendel was not directly concerned with evolution at all. He placed it on the back burner while he investigated the laws of inheritance for agricultural purposes.[7]
4. Mendel believed in the fixity of species.[8]
5. Mendel laid out the laws of inheritance that justifiably carry his name (this is the standard view).[9]
6. Mendel was no Mendelian. He was not trying to discover the laws of inheritance, and some Mendelian principles are lacking in his papers.[10]
7. Some of Mendel's data were falsified.[11]
8. None of Mendel's data were falsified.[12]
9. Mendel's reported experiments set out in his paper of 1866 are wholly fictitious.[13]

Most interpretations of Mendel's objectives and thought processes aim to explain why his experiments were neglected for so long. Simply put, the problem is this: if his experiments provide a foundation for genetics and evolutionary theory, why was this not recognized in his day? Why was there not a meeting of the minds, so to speak, between Darwin and Mendel? The question was raised by geneticists, who offered a variety of reasons for the long neglect of his work. Some writers have pointed to social issues and suggested that Mendel was professionally an outsider, an amateur, and a monk, (anticlerical attitudes may have interfered with the proper evaluation of his work), and that the journal in which he published his results was obscure.[14] Most commentators emphasize other reasons maintaining that conflicting theories and competing research interests may have been partly responsible, and that his work on inheritance was overshadowed by larger questions of evolution when Darwin published his *Origin of Species*.[15] Still others have argued that Mendel's methodology was unorthodox—that his statistical manner of analysis was ahead of its time, and as a result, his contemporaries were simply unprepared to appreciate his results.[16] Yet, each of these suggestions has been contradicted, and there is still no consensus over why his work went unrecognized.[17]

Over the past few decades some scholars have painted a radically different picture. They suggest that Mendel was not so far "ahead of his time."[18] They suggest that he did not develop the concept of paired hereditary factors equivalent to the alleles of classical genetics; he did not clearly announce a "law of segregation" that he thought might be applicable to all plant hybrids. And the way he understood his work may have been quite different from that of twentieth-century geneticists. Contrary to accepted opinion, Mendel was not trying to discover laws of inheritance of variations between individuals that might be the source of evolutionary change. Instead, he belonged to Linnaean tradition of hybridists who were examining the possibility that species hybridization might be a source of evolutionary change. Linnaeus had supposed, in his *Fundamenta Fructificationis* of 1762, "that the Creator at the actual time of creation made only one single species for each natural order of plants[,] . . . [that] all genera were primeval and consisted of a single species," and that from these species all others arose by hybridization.[19]

The central question for Mendel and his fellow hybridists was whether or not hybrid offspring bred true and therefore actually could initiate new species.[20] The laws of inheritance were of concern to Mendel only inasmuch as they related to the role of hybrids in the genesis of species. He approached this problem with the conception of constant and independently transmitted characters. This problematic differs from that of geneticists at the turn of the century, who were not interested in hybridization as a means of speciation, but used cross-breeding analysis as a means to determine the nature of hereditary variability that in turn provided the fuel for evolution.

If we accept this historical reinterpretation, Mendel was not the mythical lonely pioneer running ahead of his contemporaries. He was not someone who made an intellectual leap so great that it could not be understood by them, but rather someone whose work was firmly situated in the context of mid-nineteenth-century hybridization research. As one commentator put it, "If anything, Mendel's reputation was modest not because he was so radically out of line with his times but because his identity with his contemporaries was so complete!"[21] Later, at the turn of the century, his work was reinterpreted by geneticists to produce the legend of the long neglect.

Making a Discoverer

How, then, has this Austrian monk, with his experiments on garden peas, come to move so many people? The answer to this question perhaps lies more in what is written *about* Mendel than in what is written *by* Mendel. How was the story about his discovery, neglect, and rediscovery invented? Where did it originate? In 1977, the geneticist Alexander Weinstein argued that the belief that Mendel's work was virtually unknown before 1900 dates back to statements made at the turn of the century by two of the rediscoverers of Mendel's laws, de Vries and Correns. (It is now generally understood that the third "rediscoverer," Tschermak, did not actually understand the significance of Mendel's work when he first referred to it.) De Vries and Correns both insisted that they had read Mendel only after they had conducted their own experiments and reached their own interpretations. Thus each was "anxious to protect [his] priority, and have his work regarded as independent of the work of Mendel and of other rediscoverers."[22] In fact, there was a widespread belief among commentators that de Vries at first intended to suppress any reference to Mendel, and that his plans were interrupted when he found that Correns was going to refer to the monk.[23] This inference is based on de Vries's failure to mention Mendel when he first announced his discovery in a short abstract before Correns's paper. De Vries mentioned Mendel's work only later in two longer papers, in which he remarked that it was *trop beau pour son temps*.[24]

In 1979, the sociologist August Brannigan took this suggestion further to argue that Correns, realizing that he had lost priority to de Vries, referred to Mendel's work as a strategy to minimize his loss and effectively undermine the priority of de Vries's claim to the discovery. Thus, Brannigan concluded that "Mendel's revival in 1900 took place in the context of a priority dispute between Correns and de Vries, and that this dispute led scientists to overlook the original intent of the earlier research." The labeling of the discovery as "Mendel's laws" was a strategy to neutralize the dispute. This, Brannigan asserts, "is perhaps the single most important fact in the reification of Mendel as the founder of genetics."[25]

Why Multiple Meanings?

But did this dispute really lead scientists to overlook the original intent of Mendel's work? One might think this plausible, for scientists are often concerned only with those who precede them insofar as they see in past work elements of what they take to be the truth. Looking at the past, they often impose their own framework of understanding on the work, irrespective of the author's intention. From this perspective, all would agree that Mendel's work was superior to that of his contemporaries. But Brannigan's suggestion that there was no real interest in Mendel's intentions is belied by the fact that Mendel's intentions were addressed by Bateson in 1902 and by many geneticists through to the present day. Moreover, it implies that Mendel's intentions would be obvious had anyone bothered to discern them. Yet, despite the many attempts to reconstruct Mendel's thought processes, there have been and continue to be different opinions on what Mendel thought he had discovered.[26]

Mendel's private papers were burned, and he wrote little about himself.[27] He published only two papers. The main source for reconstructing his thought processes is his paper of 1866, "Experiments in Plant Hybrids," a concise transcript of two reports given at the Brünn Natural History Society in 1865. The whole story of the development of the new theory is usually claimed to be given in these forty-four printed pages. But scientific papers are not diaries. Moreover, they are usually designed to give a veneer of objectivity and "matter of factness" to published claims. Their rhetorical style often obscures the intentions and biases of the author and the process by which results are produced. Mendel's remarks concerning his experiments and the nonevolutionary views of the hybridist C. F. Gaertner illustrate the point. Conflicting interpretations of the following passage (in both English and original German) have been offered:

> Gaertner, by the results of these transformation experiments, was led to oppose the opinion of those naturalists who dispute the stability of plant species and believe in a continuous evolution of vegetation. He perceives in the complete transformation of one species into another an indubitable proof that species are fixed within limits beyond which they cannot change. Although this opinion cannot be unconditionally accepted, we find on the other hand in Gaertner's experiments a noteworthy confirmation of that supposition regarding variability of cultivated plants which has already been expressed.[28]

Does this statement mean that Mendel was an evolutionist or a nonevolutionist? The Darwinian evolutionist R. A. Fisher stated in 1936, "It will be seen that Mendel expressly dissociates himself from Gaertner's opposition to evolution, pointing out on the one hand that Gaertner's own results are easily explained by the Mendelian theory of factors."[29] Similarly, Gavin de Beer commented, "This passage comes as near to the acceptance of the mutability of species as anyone could wish."[30] Yet, Callender offers the exact opposite interpretation: "If this statement is to be taken literally, as Mendel most assuredly intended it to be taken, then it says quite simply that he gave *conditional acceptance to the view, expressed by Gaertner, 'that species are fixed within limits beyond which they cannot change.'* Nothing could be clearer."[31] Surely many things have been clearer. We do not have to decide here which interpretation is the correct one. The point is that Mendel's literary style, his attempts to sound objective in his evaluation of Gaertner's views, and his use of double negatives obscure his own intentions. It is not surprising that there is no consensus about the meaning he gave to his own experimental work.

I am not suggesting that the priority dispute between Correns and de Vries was not important to the initial recognition accorded Mendel. But Mendel's place in the history of genetics is not simply the result of a priority dispute that supposedly led scientists away from examining his intentions. Much more than this underlies the Mendel legend. Mendel and the meaning of his experiments have come to be clothed in various social and intellectual guises. Any understanding of the significance of Mendel's experiments would have to recognize the cultural importance of "founding father" mythologies in science. Historians have suggested that stories about the heroic insights of great geniuses who were ahead of their time and other myths of origins play important roles in defining and strengthening scientific disciplines.[32] The "long neglect" theme portraying Mendel as a creative genius clothed in monastic virtues, pursuing the truth undauntedly on the lonely frontiers of knowledge, unappreciated by his contemporaries, has been important in keeping Mendel's experiments alive. It is a tragedy that appeals to our sense of moral indignation, comparable to the suppression of Galileo by the church.[33]

The cultural importance of founding fathers in scientific disciplines may help us to understand why accounts of Mendel's discovery and neglect are repeated over and over again. But geneticists' stories about Mendel do not simply reappear over and over again. They have also changed in such a way that Mendel's thoughts and motivations are altered; he has been undressed and redressed in new colors of allegiance. To understand why, we need to appreciate that scientists' accounts of history play important roles in the knowledge-making process; they surround experimental evidence and constitute part of the art of persuasion in science. Geneticists' stories about Mendel's discovery and neglect vary, and we need to know their specific rhetorical functions. (There is a similar story of discovery, neglect, and rediscovery in the origins of biochemical genetics; see chapter 14.)

What is often at stake in geneticists' reconstructions of Mendel's thoughts and motivations is a definition of the concepts and movements that should be associated with the genetics tradition. Throughout the twentieth century, the significance of Mendelian genetics has changed. For example, the first generation of geneticists viewed Mendelism as being in direct conflict with Darwinian theory. By the 1930s, Mendelism was held to be compatible with Darwinism. Of course, the meaning of many experiments can be, and is, continually renewed as science proceeds. However, it is not just the meaning scientists place on Mendel's experiments that have changed with the development of Mendelian genetics; the inferences as to the meaning Mendel himself gave to his experiments and the reasons his work was neglected have also changed.

The history of genetics is marked by various conflicts between competing groups over evolutionary theory and approaches to biological research. And we can find these issues reflected in geneticists' early accounts of Mendel's discovery and the reasons for its neglect. To examine these stories, therefore, is at the same time to explore early controversies with geneticists over evolutionary theory.

Geneticists versus Statisticians

The first interpretation of Mendel's views about evolution occurred in the context of a now-famous conflict between Mendelian geneticists and statisticians or biometricians.[34] The dispute revolved around the question of whether new species emerged gradually by natural selection, or suddenly through large mutations. Biometricians in England, led by Karl Pearson (1857–1936) and W. F. L. Weldon, were Darwinians; they supported continuous evolution. Mendelian geneticists, led by William Bateson (1861–1926), were non-Darwinians; they supported discontinuous evolution.

Bateson was the champion of Mendelism in Britain; he wrote the first Mendelian textbooks. In 1905, he coined the word "genetics" in a letter to Adam Sedgwick regarding his candidacy for a new professorship at Cambridge, which he hoped would be in his specialty. He commented that there was no word in common usage that carried the combined meaning of heredity and the correlated phenomenon of variation: "Such a word is badly wanted, and if it were desirable to coin one, 'GENETICS' might do."[35] He did not get that professorship; but he was appointed professor of Biology at Cambridge in 1908.[36] Two years later, he became the first director of the new John Innes Horticultural Institute at Merton Park in southwest London.

Bateson also championed the saltationist view that evolution occurred by leaps, and Mendel's own experiments on such marked traits as height and dwarfism in peas seemed to confirm that this was indeed the case. As Bateson wrote in the first textbook of genetics in 1909, "The concept of evolution as proceeding through the gradual transformation of masses of individuals by the accumulation of impalpable changes is one that the study of genetics shows immediately to be false."[37]

The clash between Bateson and the Darwinian biometricians fumed in private correspondence and in published journals throughout the first decade of the century. Bateson found it impossible to believe that the biometricians had "made an honest attempt to face the facts" about the nature of hereditary variations. He doubted that they were "acting in good faith as genuine seekers of the truth."[38] Weldon, for his part, doubted that one could reproduce Mendel's results with further pea experiments. As he wrote to Pearson in 1901, "If only one could know whether the whole thing is not a damned lie!"[39]

There was still more to this. Mendelians and biometricians based their studies of heredity on different methods. The biometricians' approach was developed by Darwin's cousin Francis Galton (1822–1911), who championed quantitative studies in biology. Galton believed that just about everything could be measured: he attempted to develop a quantitative scale for beauty and even studied the effectiveness of prayer by examining the mortality rates of royalty whose subjects prayed for their health, and by comparing shipwrecks for vessels that carried missionaries to those that did not. In his book *Natural Inheritance* (1889), Galton formulated what came to be known as his "law of ancestral inheritance."[40] The idea behind the law was that an individual contains contributions from all its ancestors, the amount of the contribution being larger as the ancestor is nearer. As it concerns such a trait as height, for example, an individual inherits ¼ of his characteristics from each parent, 1/16 from each grandparent, 1/64 from each great-grandparent, and so on. When the trait did not blend—as in the case of, say, whether an individual would have blue or brown eyes—ancestral inheritance would have an effect on the ratio of traits in a population. Thus, on average each parent would determine the eye color of ¼ of their offspring, each grandparent 1/16, and so on. Galton's approach opened the door to statistical studies of the frequency of traits in populations—what is now called population genetics (see chapter 13). But during the first decades of the twentieth century, Bateson, as well as Pearson and Weldon, insisted that the parent-offspring correlations of biometricians were quantitatively in flat contradiction to Mendelism.[41]

Bateson cast Mendel himself as a non-Darwinian ally, as he told stories about why Mendel's work was neglected. In 1902, he suggested that Darwin's principle of natural selection "had almost completely distracted the minds of naturalists from the *practical* study of evolution. The labors of hybridists were believed to have led to confusion and inconsistency, and no one heeded them anymore."[42] Later, in his book *Mendel's Principles of Heredity* (1909) he claimed that, like himself, Mendel had carried out his work in virtual conflict with Darwinians, and that this was partly responsible for its neglect: "While the experimental study of the species problem was in full activity the Darwinian writings appeared. . . . *The Origin* was published in 1859. During the following decade, while the new views were on trial, the experimental breeders continued their work, but before 1870 the field was practically abandoned."[43]

Bateson not only suggested that Mendel's work was in virtual conflict with Darwinian thinking; he also imposed a non-Darwinian motive on Mendel's experiments, asserting that "with the views of Darwin which were at that time coming into prominence Mendel did not find himself in full agreement, and he embarked on his experiments with peas, which as we know he continued for eight years."[44] If Mendel's work had come into the hands of Darwin, he declared, "the history of the development of evolutionary philosophy would have been very different."[45]

Mendel Made Darwinian

By the 1930s, Darwinian gradual evolution and the statistical study of variations in populations had merged with Mendelian genetics, and Bateson and the first generation of geneticists were considered to have seriously erred in allying Mendelism with discontinuous evolution (see chapter 13). Ronald Fisher (1890–1962), a celebrated population geneticist and one of the architects of the new synthesis, aimed to pound the last nail in the coffin of the controversy between Darwinian statisticians and non-Darwinian geneticists. In his now-famous paper of 1936, "Has Mendel's Work Been Rediscovered?," he offered a detailed examination of Mendel's experiments, a new interpretation of Mendel's views on evolution, and the reason for his neglect. He recast Mendel as a good Darwinian and charged that Bateson had deliberately misrepresented Mendel and deceived scientists into believing that Mendel was a non-Darwinian evolutionist like himself. In effect, he argued, Bateson had distorted history to suit his own interest:

> It cannot be denied that Bateson's interest in the rediscovery was that of a zealous partisan. We must ascribe him two elements in the legend which seem to have no other foundation: (1) The belief that Darwin's influence was responsible for the neglect of Mendel's work, and of all experimentation with similar aims; and (2) the belief that Mendel was hostile to Darwin's theories, and fancied that his work controverted them.[46]

Fisher asserted that Mendel worked squarely within a Darwinian framework, but that his intent had been misrepresented because his paper was not examined with sufficient care, that biologists had imposed their own meanings on Mendel's work, and that their interpretations were influenced by the theory of their times.[47] In his view, the biases of others were obstacles to the recognition of Mendel's true intentions, or, as he put it, "Each generation perhaps found in Mendel's paper only what it expected to find."[48] It is striking, however, that he excluded his own interpretation from any biases—especially because Mendel scholars today agree that Mendel was a non-Darwinian in the tradition of hybridists.

Is the Scientific Paper Fiction?

Mendel's intentions and evolutionary beliefs aside, Fisher addressed still another aspect of the Mendel legend in 1936: the question of whether his published results were too good to be true, that is, whether his reported ratios of recessive and dominant traits were too close to theoretical expectation. In short, did Mendel fake his data? When examining Mendel's data using statistical tests, Fisher calculated that only once in 30,000 times would there be so close a fit with prediction. But, in raising this issue, he did not intend to discredit Mendel; his aim was to reveal Mendel's power of abstract reasoning: to show that his experimental protocols made sense only if he had his theory in mind before doing his experiments, and only if he was a Darwinian.[49] Fisher considered the possibility that Mendel unconsciously biased his data by misclassifying some of the results—say, scoring a "round" as a "wrinkled" so as to favor expected ratios. However, he suspected that it was actually a case of conscious fiddling—but not by Mendel: "Mendel was deceived by an assistant who knew too well what was expected." Since the centenary of Mendel's paper in 1965, many geneticists have jumped to Mendel's defense to reconsider unconscious bias.[50] His honor continues to be well defended.

Still another question emerged from Fisher's and Bateson's discussions of Mendel's experiments: should scientific papers be taken literally? Although it is generally overlooked by commentators, Bateson had suggested that perhaps all of the experiments Mendel reported were fictitious. After summarizing his seven experiments, Mendel wrote, "In the experiments described above plants were used which differed only in one essential character." Bateson found it difficult to believe that Mendel actually possessed plants that differed in only one of the seven contrasting pairs of traits he examined.[51] One would expect that some or all the crosses would have involved more than one contrasting pair of traits. Fisher agreed. He also recognized that Mendel did not provide all the relevant data, and that this behavior would be easily intelligible if the "experiments" reported in the paper were fictitious. Such oversimplification is often used when teachers illustrate principles to students in a lecture. Nonetheless, he believed that Mendel's paper should be taken literally. The issue hinged on what Mendel would have counted as an experiment. Fisher suggested that Mendel might have assembled the data for each of the seven traits from the different crosses in which it was involved, but then reported the results for each trait as if it were a single experiment.[52]

In recent years, many scholars have examined the structure and rhetoric of scientific papers.[53] The Nobel Prize–winning biologist Peter Medawar was one of the first to do so when, in 1963, he posed the question, "Is the Scientific Paper a Fraud?" He did not mean to suggest that the scientific paper misrepresents "facts" or that the interpretations found in a scientific paper are wrong or deliberately mistaken. What he meant was that "the scientific paper may be a fraud because it misrepresents the thought process that accompanied or gave rise to the work that is described in the paper."[54] The typical scientific paper embodied "a totally mistaken conception, even a travesty, of the nature of scientific thought":

> First, there's a section called the "introduction" in which you merely describe the general field in which your scientific talents are going to be exercised, followed by a section called "previous work" in which you concede, more or less graciously,

> that others have dimly groped towards the fundamental truths that you are now about to expound. Then a section on "methods"—that's O.K. Then comes the section called "results." The section called "results" consists of a stream of factual information in which it's considered extremely bad form to discuss the significance of the results you're getting. You have to pretend that your mind is, so to speak, a virgin receptacle, an empty vessel, for information which floods into it from the external world for no reason which you yourself have revealed. You reserve all appraisal of the scientific evidence until the "discussion" section, and in the discussion you adopt the ludicrous pretense of asking yourself if the information you've collected actually means anything; of asking yourself if any general truths are going to emerge from the contemplation of all the evidence you brandished in the section called "results."

Admittedly, this description is somewhat of an exaggeration, for certainly many scientific papers do not follow this structure. But we can agree with Medawar that there is "more than a mere element of truth in it." The conception underlying this style of scientific writing is that scientific discovery is an inductive process.[55] The scientific paper gives the illusion that discovery begins with simple, unbiased, unprejudiced, naïve, and innocent observation. Out of this unbridled evidence and tabulation of facts, orderly generalizations emerge. Yet, scientists know full well that discoveries do not occur in this way. They know what meaning to place on their results before they conduct their experiments.

Medawar traced the inductive structure of scientific papers to the nineteenth-century philosopher John Stuart Mill. However, it would be naive to believe that scientists are the dupes of philosophers. The narrative of the scientific paper, which presents discoveries as an inductive process, plays an important persuasive role in science, giving a veneer of objectivity to scientific interpretations.[56] But there are other problems. Often, the "methods" section of scientific papers is *not* O. K. Seemingly trivial yet vital information concerning procedures is often left out. Much of modern science involves special technical skills, what Michael Polanyi called tacit knowledge, an adeptness that is difficult to articulate—like riding a bicycle.[57] The result, as Mendel's case illustrates, is that interpretations of a scientist's conduct and procedures sometimes involve considerable speculation and conjecture.

Commenting on Mendel's paper, one writer remarked: "All geneticists admitted that it was written so perfectly that we could not—not even at present—put it down more properly."[58] Yet, it is this very "perfection" that has made Mendel's conduct so difficult to ascertain. Mendel's experiments thus become a flexible resource, and as a founding father whose intentions are vitally important, he is adaptable indeed. In a sense, he also "had the luck to please everyone who had an axe to grind." The tale of Mendel's discovery, lost and found, is a parable so satisfying that few biologists can bear to let it go. Even if a little wrong and a little oversimplified, some will say, they should stick with the traditional textbook story anyway because a more complex one would only confound students.

Examining the rhetoric of scientific papers and the nature of experiments is crucial to understanding the development of science. As the issues raised in biologists' discussions of the neglect of Mendel's experiments imply, ideas and data do not win out solely on their own strength, no matter how powerful they may later appear to be. The social context, specialization, divergent interests, and institutional strategies—so central in accounts of Mendel's neglect and rediscovery—are equally crucial for understanding the development and success of genetics as a discipline during the early twentieth century.

12

Emerging Genetics

> If we are to progress fast there must be no separation made between pure and applied science. The practical man with his wide knowledge of specific natural facts, and the scientific student ever seeking to find the hard general truths which the diversity of nature hides—truths out of which any lasting structure of progress must be built—have everything to gain from free interchange of experience and ideas.
>
> —William Bateson, 1911

THE HISTORY OF GENETICS has two aspects: its foundations in theory and method, and its effect on our world. Consideration of both is vital to understanding its remarkable growth. During the first half of the twentieth century, genes were abstract entities. No one was certain of their physical nature, how they were reproduced from one generation to the next, how they affected traits said to be under their control, or what role they played in development. Despite these limitations, genetics, especially in the United States, was extraordinarily successful and rose to an authoritative position in the field of heredity by the 1930s. To understand the relationship between genetics and other disciplines, I first outline the state of the field of heredity in the first decades of the last century. In so doing we see that several distinct conceptions of heredity were produced from within several specialties. Subsequently, we shall examine the rhetoric between specialties and trace the course that led genetics in the United States to a central position in biology.

The Field of Heredity

During the first decade of the century, heredity was a pivotal issue for biologists. The problems that depended upon it were many. What is the nature of the variations that fuel evolution? Do variations come in big jumps or small steps? Do new traits first appear in an individual as adaptations to the environment, as neo-Lamarckians expected, or do they occur randomly as "mistakes," some (very few) of which are beneficial? How does an organism grow and develop from an egg? How does cellular differentiation occur? Why are certain cells of the embryo capable of producing the whole (a property called totipotency). What is the physical basis of heredity? What kinds of differences between individuals are inherited, that is, which are due to nature and which are due to nurture?

These problems were the concern of several specialties, including cytology, embryology, physiology, agricultural breeding, eugenics, comparative anatomy, paleontology, and biometry. Indeed, how an individual understood the term "heredity" resulted largely from the problems, techniques, and theories of his or her speciality. To begin with, Mendelian genetics, based on cross-breeding analysis, involved statistical examination of the reappearance of visible differences between individuals. Initially championed by William Bateson and Wilhelm Johannsen, and later by Thomas Hunt Morgan, it supported the notion of discontinuous evolution. Biometry, on the other hand, led by W. F. R. Weldon and Karl Pearson, who supported continuous evolution, was based on the Galtonian theory of ancestral heredity and on statistical examinations of visible characters within populations. Naturalists (paleontologists, comparative morphologists, etc.) compiled data illustrating genealogical relationships between species and between living and extinct forms, brought to light by field trips and expeditions sent out from museums and universities. For many neo-Lamarckian naturalists, the habits and environments of organisms played a direct role in shaping heredity and evolution.

Cytologists investigated the physical basis of heredity—the link between generations—by microscopic examination of the cell. By the first decade of the century, they were emphasizing the importance of chromosomes in cell division, the physical continuity of the cell, the nucleus, and other cytoplasmic organelles, including mitochondria, chloroplasts, and perhaps centrioles. "Heredity" for cytologists remained somewhat uncertain, as E. B. Wilson remarked in 1914: "Our conceptions of cell organization, like those of development and heredity, are still in the making. The time has not yet come when we can safely attempt to give them very definite outlines."[1]

Experimental embryologists investigated the causes of development, which they held to be the same as heredity. The fact that the egg of a rat gave rise to a rat, and that of a frog to a frog, represented the first aspect of the phenomena of heredity. For them, heredity was a process of production and reproduction and included concepts of integration, organization, regulation, and differentiation. While geneticists focused on the chromosomes, embryologists focused on the egg cytoplasm as the basis of heredity and development (see chapter 10). E. G. Conklin described heredity from the embryological point of view in 1908:

> Indeed, heredity is not a peculiar or unique principle for it is only similarity of growth and differentiation in successive generations. The fertilized egg cell undergoes a

> certain form of cleavage and gives rise to cells of a particular size and structure, and step by step these are converted into a certain type of blastula, gastrula, larva and adult. In fact, the whole process of development is one of growth and differentiation, and similarity of these in parents and offspring constitutes hereditary likeness. The cause of heredity are thus reduced to the causes of successive differentiations of development, and the mechanism of heredity is merely the mechanism of differentiation.[2]

Another group that was concerned with heredity was breeders situated outside the universities. During the early years of Mendelism, breeders came to view heredity as an important "economic force." Breeders could use cross-breeding analysis to detect whether a certain variety was a purebred or hybrid and thereby improve the quality and/or quantity of their stocks. Major breeding work aimed at improving pedigreed plants and animals was carried out in public and private institutions in Europe and the United States. Breeders emphasized that the hereditary values of specially bred strains of plants and animals were as real as the seemingly more concrete values of land or goods. Some considered the value of the "unseen carriers of heredity" to be "far above that of gold." An article in the first issue of the *American Breeder's Magazine* in 1910 stated that "heredity is a force more subtle and more marvelous than electricity. Once generated it needs no additional force to sustain it. Once new breeding values are created they continue as permanent economic forces."[3]

Finally, there were those concerned with improving the stock of humanity through selective breeding. Francis Galton coined the word "eugenics" (from the Greek for "well born") in 1883.[4] He founded the English Eugenics Society in 1907, and he provided funds for a chair of eugenics at University College, London University.[5] Eugenics societies mushroomed in many countries.[6] Ostensibly, eugenics was aimed at improving the "fitness" of human populations by decreasing the reproduction of those deemed unfit and increasing that of those deemed fit. Historians have argued that it often functioned as a middle-class ideology that, like much of social Darwinism, was aimed at legitimating existing social order as "natural." Eugenicists often argued that social status reflected differences in fundamental abilities due to innate, biological inherited differences, and therefore that a just society is necessarily unequal, and necessarily hierarchical.

Eugenics was popular in the United States as elsewhere, especially after the First World War. Eugenicists laid great stress on sterilizing defective persons: the insane, the mentally retarded, and epileptics. As a result of their efforts, by the mid-1930s sterilization laws had been passed in many states and in many European countries. The United States had instituted restrictions on immigration from nations with "inferior" stock, such as Italy, Greece, and the countries of eastern Europe. In Germany, the race hygiene movement began around the turn of the century.

The greatest horror occurred when eugenics was used as a rationale for producing a "master race." The mass murder and torture of millions of innocent people—Jews, gypsies, mentally ill persons, and homosexuals, among others—by the Nazis doused enthusiasm for the "eugenics movement."[7] But eugenics was far from an historical aberration of Nazism or extreme right-wing politics. Historians have shown that the eugenicists in many countries were not inherently racist or classist.[8] Critics today argue that eugenicists failed to recognize the sizable role of the environment or culture and education in establishing human characteristics. Moreover, geneticists later showed that all races are populations with mixtures of many different groups, and that the genetic diversity between individuals is far greater than it is between races. The differences between races are literally skin deep.[9]

At first glance one might think that the growth of these various groups interested in various aspects of heredity, from cytologists to eugenicists, was simple a matter of specialties dividing up the work, posing different questions, and using different approaches. This then would be the intellectual division of labor and mutualistic interchange working for the benefit of the whole that philosophers and biologists had predicted (see chapter 8). At the surface this might seem to be the case, but underneath was conflict and competition. To the extent to which members of each specialty defined and explored heredity with their own methods and theories, each claimed the value of his or her approach to be greater than that of others."[10]

The biometricians Pearson and Weldon maintained that "the problem of animal evolution is essentially a statistical problem" and that a statistical knowledge of the changes going on in a number of species was the "the only legitimate basis" for understanding evolution.[11] In 1898, Pearson stated his views about Galton's laws in no uncertain terms: "If Darwin's evolution be natural selection combined with *heredity*, then the single statement which embraces the whole field of heredity must prove almost as epoch-making to the biologists as the law of gravitation to the astronomer."[12]

In opposition to the views of biometricians, Conklin asserted in 1908 that embryology offered the most effective way to study heredity:

> Heredity is today the central problem of biology. This problem may be approached from many sides—that of the breeder, the experimenter, the statistician, the physiologist, the embryologist, the cytologist—but the mechanism of heredity can be studied best by investigation of the germ cells and their development.[13]

Yet, leading Mendelian geneticists considered all approaches to heredity other their own to be either wrong or trivial. Bateson proclaimed the importance of genetics above all other approaches. He discarded his old specialty of embryology as moribund. "Formerly," he wrote, "it was hoped that by simple inspection of embryological processes, the modes of heredity might be ascertained, the actual mechanism by which the offspring is formed from the body of the parent."[14] But these expectations were not realized. "With the existing methods of embryology," he asserted, "nothing could be analyzed further than the physiological events themselves." Alternatively, he rationalized, "we at least can watch the system by which the differences between the various kinds of fowls or the various kinds of sweet peas are distributed among their offspring. By thus breaking the main problem up into its parts we give ourselves fresh chances."[15]

Bateson dismissed biometricians' views about gradual Darwinian evolution (see chapter 11), and he also ridiculed paleontologists and comparative anatomists, advising biologists to turn away from the study of evolution until the genetic facts were in: "Naturalists may still be found expounding teleological systems which would have delighted Dr. Pangloss himself, but at the present time few are misled. The student of genetics knows that the time for the development of theory is not yet. He would rather stick to the seed pan and the incubator."[16]

Genotype and Phenotype

In addition to the term "genetics" that he had coined for the new discipline, Bateson offered genetics much of its lexicon: "zygote" (from the Greek *zugōtos*, "yoked") to describe the fertilized egg, homozygote (from the Greek *homo*, "the same"), and heterozygote (*heteros*, "different") to describe pure breeds and hybrids; and allelomorph (*allēlēn*, "another," plus *morphe*, "form") to describe two different versions of the same trait, such as height and dwarfism. These were important steps in the development of the new science of genetics. But no conceptual distinction did more to disqualify nonexperimentalists from the study of heredity than that between the genotype and the phenotype, introduced by Wilhelm Johannsen in 1909 and 1911.[17] He began his discussion with the metaphor of "inheritance" itself.

Biologists had borrowed the terms "heredity" and "inheritance" from jurisprudence, in which the meaning of these words is the transmission of money or things, rights or duties from one person to another: the heirs or inheritors. Darwin spoke of characters that may be attributed to inheritance from a common progenitor.[18] But the metaphor was imperfect and misleading, because biologists had to distinguish between the characteristic, say, the purple color of a flower, which was not inherited, from something else that stood for the characteristic that was inherited. Not surprisingly, embryologists were quick to see the problem. As E. G. Conklin commented in 1908, "The comparison of heredity to the transmission of property from parent to children has produced confusion in the scientific as well as the popular mind."[19] Similarly, T. H. Morgan wrote in 1910: "When we speak of the transmission of characters from parent to offspring, we are speaking metaphorically; for we now realize that it is not characters that are transmitted to the child from the body of the parent, but that the parent carries over the material common to both parent and offspring."[20]

Johannsen's genotype conception of heredity articulated the distinction between the social and biological meanings. The genotype, which he defined as "the sum total of all the 'genes,'" lay hidden in the germ cells.[21] The phenotype, on the other hand, could be seen readily in variations all around us. "All 'types' of organisms," Johannsen wrote, "distinguished by direct inspection or only by finer methods of measuring or description, may be characterized as 'phenotypes.'"[22]

Genetics was in a transition period, and Johannsen drew on the history of chemistry to argue that the traditional phenotype conception of heredity was the reverse of the real facts, just as the famous "phlogiston" theory was an expression diametrically opposite to chemical reality.[23] Naturalists and biometricians, whose views of heredity were based on direct inspection of an organism, were seriously mistaken, because they assumed that an adult individual was a good representation of its genotype. Mendelian geneticists showed that an individual might have hidden recessive traits, and therefore adult traits were not at all fair representations of its genotype. By observing phenotypes, one also could not distinguish between what was inherited and what was due to the environment. Moreover, whereas phenotypes blended, genotypes did not.

As Johannsen saw it, the noninheritance of acquired characteristics was "an expression of the fact that the external conditions may easily mold phenotypes in a more or less adaptive manner, but can hardly or rarely induce changes in the genotype."[24] He emphasized that the genotype conception was an ahistorical conception of heredity, and he dismissed the biometricians' "laws of ancestral influence" (which state that an individual contains contributions from all its ancestors, the amount of the contribution being larger as the ancestor is nearer) as superstition. "Ancestral influence! As to heredity, it is a mystical expression for a fiction. The ancestral influences are the 'ghosts' in genetics, but generally the belief in ghosts is still powerful." Genetics, he asserted, should be pursued "*with* mathematics not as mathematics. . . . Certainly, medical and biological statisticians have in modern times been able to make elaborate statements of great interest for insurance purposes, for the 'eugenics-movement' and so on. But no profound insight into the biological problem of heredity can be gained on this basis."[25]

He also decried breeders on grounds of their lack of professional training:

> The practical breeders are somewhat difficult people to discuss with. Their methods of selection combined with a special training and "nurture" in the widest sense of the word are most unable to throw any light upon questions of genetics, and yet, they only too frequently make hypotheses as to the nature of heredity and variability. Darwin has somewhat exaggerated the scientific value of breeders' testimonies, as if a breeder *eo ipso* must be an expert in heredity.[26]

Johannsen further dissociated Mendelian factors from all such discredited self-reproducing vital entities or determinants, such as those proposed by Weismann (see chapter 8). "Of all the Weismannian army of notions and categories," he wrote, Mendelism, "may use nothing."[27] He introduced the term "gene" as a neutral term to replace other theory-laden, deterministic terms. Johannsen's "gene" did not have the power to reproduce itself, to grow, and to assimilate; it did not possess an independent life of its own. As he later put it, "The Mendelian units as such, taken *per se* are powerless."[28]

Disciplinary Design

Implicitly, the distinction between genotype and phenotype entailed recognition of developmental processes and the role of the environment in the formation of characteristics. "Between the characters that furnish the data for the theory, and the postulated genes to which the characters are referred," Morgan wrote in 1926, "lies the whole field of embryonic development."[29] The distinction effectively offered geneticists the conceptual route by which to bypass the organization of the cell, regulation by the internal and external environment of the organism, and the temporal and orderly sequences during development. However, there was a problem. In practice geneticists often ignored the influence of those developmental conditions. As a result, they often referred to characteristics as being the direct result of genes, as they spoke of "a gene for this" and "a gene for that." The notion that the gene determines the characteristic (that genotype determines phenotype) was pure tautology in the typical breeding experiments because the presence of genes was inferred by experimental manipulation of phenotypes. The problem of identifying genes with characteristics was raised by the embryologist Frank Lillie in a letter to Julian Huxley in 1928, "If you will excuse a paradox," Lillie wrote, "gene theory is essentially a theory of phenotypes, i.e., something always static for as soon as it changes it is already another phenotype."[30]

The first generation of geneticists turned inward to make and solve their own problems. Bateson summarized the institutional strategy for genetics in 1911, soon after he was appointed director of the John Innes Horticultural Institution, in Merton. He argued that no separation should be made between pure and applied science, and that geneticists should develop close relationships with practical breeders.[31] His statements about fostering relationships with breeders were heeded, but Bateson was reluctant to take the next step in the Mendelian program—of situating the genes in chromosomes.

During the First World War, leadership in genetics changed hands from Britain to the United States. T. H. Morgan (1866–1945) and his talented *Drosophila* group of Calvin B. Bridges, Alfred H. Sturtevant, and Hermann J. (Joe) Muller at Columbia University combined Mendelian analysis with cytological studies to establish the Mendelian chromosome theory.[32] Morgan synthesized the results of genetics in several books, including *The Mechanism of Mendelian Heredity* (1915), *The Physical Basis of Heredity* (1919), and *Embryology and Genetics* (1934). He became the most famous geneticist of his generation and in 1933 was awarded the Nobel Prize in Medicine, the first time the prize had ever gone to a nonmedical scientist.[33]

When Morgan began his work on *Drosophila* genetics in 1910, a landmark series of cytological studies on sex determination had already convinced many skeptics of the importance of the chromosomes in heredity. During the first decade of the century, it was established that the sex of almost all plants and animals was determined at the moment of fertilization by the combination of X and Y chromosomes. The crucial step in this discovery was taken in 1905 by Nettie Stevens (1861–1912) and E. B. Wilson.[34]

Geneticists lined genes up along chromosomes like beads on a string. Genes were inherited in groups, but they could also be rearranged when sex cells are formed. This phenomenon made it possible to determine the relative location of individual genes in the chromosomes. Morgan's school ingeniously learned to construct maps of genes based on recombination frequency: the farther two genes were from each other on chromosomes, the more inclined they would be to recombine with other genes. Cytological studies later showed that when the interchange of genes takes place, large blocks of the chromosomes were actually exchanged: crossing over occurred between homologous chromosomes at the first stages of meiosis during the formation of egg and sperm cells. Morgan wrote about the immediate aims of genetics in 1919: "Our study of the germ-plasm is largely confined . . . for the present to the study of the transmission of the genes, to the kinds of effects they produce on the organism, and to the special relations of the genes in the chromosomes where they are located."[35]

American geneticists formed their own discipline, with its own techniques, theories, journals, and societies, and they infiltrated departments of biology, zoology, and botany. Genetics grew rapidly for a number of reasons. One cannot overstate the importance of model organisms. Mice, rats, and especially the fruit fly (*Drosophila*) were the effective tools for genetics, analogous to the marine organisms of embryology. The simplicity of Mendelian analysis and its use on rapidly reproducing *Drosophila* for producing data was important for recruiting young scientists into a specialty where their chances of immediate success were good. Genetics had a highly competitive capacity to produce results and students. Mendelism was devised by botanists, and various plants, especially corn, were also crucial to its development. In addition to the "fly room" at Columbia was the famous corn genetics group at the Department of Plant Breeding at Cornell, led by R. A. Emerson and the cytologist Leslie W. Sharp.[36] Emerson developed one of the largest schools of genetics in the United States and had made corn second only to *Drosophila* as an object of genetic investigation. Many of Emerson's students would become distinguished geneticists, including George Beadle, Barbara McClintock, Marcus Rhoades, and Milislav Demerec (see chapter 14). There were also schools led by E. M. East and W. E. Castle at Harvard's Bussey Institution. The groups at Columbia, Cornell, and Harvard had no difficulty graduating one Ph.D. a year between 1910 and 1930.[37]

The early association of genetics with practical breeding was vital.[38] Major breeding work was carried out by the United States Department of Agriculture, in state experimental stations, and under other public and private auspices. Clubs devoted to the study of heredity and eugenics were organized throughout the United States, and philanthropists were encouraged to dedicate generous sums of money to foster the development of genetics in universities. In fact, the relationship between Mendelism, practical breeding, and eugenics and was so intimate that the word "genetics" covered all of them in the United States during the first three decades of the century.

In 1903, the American Breeder's Association was founded "to bring the practical breeder into close contact with scientists" and "to achieve scientific and economic results of the highest order." The American Breeder's Association was later renamed the American Genetics Association (1914), and *American Breeder's Magazine* carried the subtitle *A Journal of Genetics and Eugenics*. In 1915 a new journal was initiated, *Genetics*, distinct from the periodicals for eugenics and practical breeding. And in 1932, a new genetics society, the Genetics Society of America, was founded as an offshoot of the American Society of Naturalists, with aims distinct from those of the American Genetics Association.

Biology out of Balance

While genetics, with its well-defined methods, aims, and institutional developments, thrived, older disciplines floundered, and some feared the encroachment of onto their own domains. Leading physiologists, embryologists, and paleontologists hoped that genetics was a fad. Joseph Needham spoke for many nongeneticists when he commented in 1919,

> We are all out of balance. Some of our laboratories resemble up-to-date shops for quantity production of fabricated genetic hypotheses. Some of our publications make a prodigious effort to translate everything biological into terms of physiology and mechanism—an effort as labored as it is unnecessary and unprofitable. Why not let the facts speak for themselves? They go from one extreme to another. In my high school days we did nothing but dissecting; later came morphology and embryology, then experimental zoology, then genetics, and the devotees of each new subject have looked back upon the old with something like that disdain with which a debutante regards last year's gown. Natural history and classification are perhaps long enough out of date, so that interest in them may again be revived.[39]

Comparative anatomists and paleontologists felt the pinch of genetics and the derogatory remarks of its leaders.[40] The paleontologist W. K. Gregory, at the American Museum of Natural History in New York, responded in 1917 that although paleontologists may reserve judgments about the mechanisms of evolution, they had their own and well-deserved place within the life sciences:

> As long as museums and universities send out expeditions to bring to light new forms of living and extinct animals and new data illustrating the interrelations of organisms and their environments, as long as anatomists desire a broad comparative basis for human anatomy, as long as even a few students feel a strong curiosity to learn about the course of evolution and the relationships of animals, the old problems of taxonomy, phylogeny and evolution will gradually reassert themselves even in competition with brilliant and highly fruitful laboratory studies in cytology, genetics and physiological chemistry.[41]

The American naturalist William Morton Wheeler saw his field as being on the endangered list. In his diatribe of 1923, "The Dry-Rot of Our Academic Biology," he confessed that his "mental condition is, no doubt, partly due to the disappointing spectacle of our accomplishments as more or less decayed campus biologists in increasing the number, enthusiasm, and enterprise of our young naturalists."[42] He turned to the history of science to argue for the central position of natural history, and the need for intellectual breadth in the life sciences.

> History shows that throughout the centuries . . . natural history constitutes the perennial rootstock or stolon of biologic science and that it retains this character because it satisfies some of our most fundamental and vital interests in organisms as living individuals more or less like ourselves. From time to time the stolon has produced special disciplines which have grown into great, flourishing complexes. . . . More recently another dear little bud, genetics, has come off, so promising, so self-conscious, but alas, so constricted at the base.[43]

Wheeler may have expressed a common viewpoint of nongeneticists, but American geneticists saw the situation differently. They believed that genetics would become the stolon that would bind all of biology into a unified field, which, would one day rival the physical sciences. Indeed, genetics had come to such a prominent position by 1930 that the idea of converting the American Society of Naturalists to a "genetics society" was seriously considered. The former was made up of members of various specialties and had as its scope of interests "organic evolution."[44]

American embryologists also felt their field threatened. As F. R. Lillie lamented in 1927, "Genetics has become quite a unitary science and the physiology of development is at most a field of work."[45] Although geneticists could say nothing about development, they still claimed full authority over heredity. Their strategy was simple: focusing on the sexual transmission of genes, they simply redefined the term "heredity" operationally to suit their practice. As the Harvard geneticist Leslie Dunn wrote in 1917,

> The working of the effective method is known for heredity, if heredity be properly only concerned with the way in which hereditary factors are distributed in the germ cells. For development, the mechanism is but grossly known, but we have learned enough . . . to foster a suspicion that one day the governance of the chromosomes over development will be explained in physico-chemical terms.[46]

As genetics grew institutionally, so did its narrow view of heredity. E. G. Conklin commented in 1919, "Development is indeed a vastly greater and more complicated problem than heredity, if by the latter is meant merely the transmission of germinal units from one generation to the next."[47] F. R. Lillie was more reluctant to concede the geneticists' restricted version of heredity and maintained what he called "the physiological conception of heredity as repetitive life histories."[48] Similarly, the Belgian embryologist Albert Brachet wrote in 1935:

> For the embryologist the word heredity takes on a very broad meaning; heredity is the totality of the developmental potentialities in the fertilized egg; it is the ensemble of the causes which make the egg produce, when in adequate environmental conditions, following a succession of well-defined processes, a new organism having all the characters of the species to which it belongs. Thus understood, heredity is the real object of embryology—to know it at the same time in its origin, in its manifestations and in the mechanisms which it puts in place in order to realize its final goal.[49]

Embryologists found it more than difficult to reconcile embryology and genetics, because there was no sorting out of nuclear genes in the course of development. Morgan insisted that "each cell inherits the whole germ plasm."[50] As long as this remained a necessary part of the gene theory, then it seemed to them that differentiation of cells was an environmental relationship mediated through the cytoplasm, not the nucleus. Despite these problems, in 1926 Morgan asserted that the application of genetics was a most promising method of attack on the problem of development.[51] Embryologists were not long in responding. The next year Lillie wrote, "Those who desire to make genetics the basis of physiology and development will have to explain how an unchanging complex can direct the course of an ordered developmental stream." As he saw it,

> the progress of genetics and of development can only result in a sharper definition of the two fields, and any expectation of their reunion (in a Weismannian sense) is in my opinion doomed to disappointment. . . . Instead of distorting our workable conceptions to include that which they can in no wise compass, may it not be profitable, for a while, to admit that more lies without than within our confines of mechanism and statistics?[52]

Five years later, the Yale embryologist Ross Harrison continued to express his deep concern about reducing development to the action of genes. Gene theory, he argued, may even be an impediment to understanding development:

> The prestige of success enjoyed by the gene theory might easily become a hindrance to the understanding of development by directing our attention solely to the genome, whereas cell movements, differentiation and in fact all developmental processes are actually effected by the cytoplasm. Already we have theories that refer the process of development to genic action and regard the whole performance as no more than the realization of the potencies of the genes. Such theories are altogether too one-sided.[53]

Morgan responded to critics with his own historical interpretation that embryology had simply failed as a progressive research program:

> If another branch of zoology that was actively cultivated at the end of the last century had realized its ambitions, it might have been possible to-day to bridge the gap between gene and character, but despite its high-sounding name of *Entwicklungsmechanik* [developmental mechanics] nothing really quantitative or mechanistic was forthcoming. Instead philosophical platitudes were invoked rather than experimentally determined factors. Then, too, experimental embryology ran for a while after false gods that landed it finally in a maze of metaphysical subtleties.[54]

Are Genes Real?

Genes were certainly useful as a way of speaking about the results of genetic crosses, but it was far from certain whether they really had physical reality as a material particle or were simply instrumental entities, hypothetical factors. Some early geneticists suspected that genes controlled biochemical reactions by way of enzymes, and some even suspected that genes themselves might be enzymes (see chapter 14). However, Morgan was reluctant to speculate about the nature of the gene. In his book *The Theory of the Gene* (1926), he attributed five principles to the gene, all derived from purely numerical data: segregation, independent assortment, crossing over, linear order, and linkage groups. He raised the question about the physical reality of genes again in his Nobel Prize lecture in 1934:

> What are genes? Now that we locate them in the chromosomes, are we justified in regarding them as material units; as chemical bodies of a higher order than molecules? . . . Frankly, these are questions with which the working geneticist has not much concerned himself, except now and then to speculate as to the nature of the postulated elements. There is no consensus amongst geneticists as to what the genes are—whether the gene is a hypothetical unit, or whether the gene is a material particle.[55]

H. J. Muller (1890–1966), who would develop a following of his own, was more speculative.[56] He formulated a theory of the gene as a discrete physical unit comparable to a virus. As he put it in 1922, "Perhaps we may be able to grind genes in a mortar and cook them in a beaker after all."[57] In his 1926 paper "The Gene as the Basis of Life," he argued that genes possessed the unique fundamental properties of identical reduplication and mutation:

> What is meant in this paper by the term "gene" material is any substance which, in given surroundings—protoplasmic or otherwise—is capable of causing the reproduction of its own specific composition, but which can nevertheless change repeatedly—"mutate"—and yet retain the property of reproducing itself in its various new forms. There is clear evidence that such material is to be found in the chromatin.[58]

This ability of genes to vary (mutate) and to reproduce themselves in their new form meant to Muller that they were the building blocks of evolution:

> Genes (simple in structure) would, according to this line of reasoning, have formed the foundation of the first living matter. By virtue of their property (found only in "living" things) of mutating without losing their growth power they have evolved even into more complicated forms, with such by-products—protoplasm, soma, etc.—as furthered their continuance. Thus, they would form the basis of life.[59]

Muller's theory of the gene as a naked virus was later considered prophetic by geneticists of the 1950s.[60] Muller also provided one of the most important techniques for probing the gene and revitalizing genetics. In 1927, he reported that X-rays could increase the frequency of gene mutations in *Drosophila* by a factor of 1500.[61] The following year, L. J. Stadler published the results of his similar studies with barley.[62] The artificial production of mutations gave genetics a new lease on life. It provided genetics with one of its most important analytic devices and one of its most important sources of material for investigation. Geneticists no longer had to wait for mutations to arise spontaneously. The study of mutations was soon extended from X-rays to gamma rays, beta rays, cathode rays, and ultraviolet light. Muller himself argued that the mutations produced were of two kinds: chromosomal rearrangements and gene mutations, which he saw as "reconstructions of the gene." The kinds of mutations produced were neither specific nor directed by the mutagenic agents. The types of phenotypic changes were the same as those known to occur spontaneously. As Muller put it, "the artificial building blocks of evolution" were as good as "the natural stones."[63] He was awarded the Nobel Prize in 1946.

Whether or not genes were hypothetical entities, Mendelian gene mutations and recombination emerged at the center of evolutionary thinking of the 1930s and 1940s, coupled with a population-statistical approach. Since Darwin, biology had become increasingly fractionated and fractious. Certainly, there was a great deal of cooperation within groups—individuals collaborated and shared methods, concepts, and results—but there was much less cooperation between specialties. During the 1930s and 1940s there were calls for a reunification of biology and for specialties to put aside their differences and misunderstandings. A new Darwinian synthesis was possible.

13

Darwinian Renaissance

Evolution may lay claim to be considered the most central and the most important of the problems of biology. For an attack upon it we need facts and methods from every branch of the science—ecology, genetics, paleontology, geographical distribution, embryology, systematics, comparative anatomy—not to mention reinforcements from other disciplines such as geology, geography, and mathematics.

Biology at the present time is embarking upon a phase of synthesis after a period in which new disciplines were taken up in turn and worked out in comparative isolation. Nowhere is this movement towards unification more likely to be valuable than in this many-sided topic of evolution; and already we are seeing the first fruits in the reanimation of Darwinism.

—Julian Huxley, 1942

THE EVOLUTIONARY SYNTHESIS, the coming together of naturalists and geneticists over Darwinism and Mendelism, is a hallmark of twentieth-century biology. It emerged in the 1930s and 1940s to resolve many of the large issues that had confronted Darwin's theory, and it constitutes what many biologists today regard as biology's basic evolutionary paradigm. The phrase "the evolutionary synthesis" was introduced by Julian Huxley (1887–1975) in his book *Evolution: the Modern Synthesis* (1942) to indicate two generally accepted conclusions: that evolution can be explained by natural selection acting on variations resulting from gene mutations and recombination, and that phenomena observed by paleontologists, systematists, and field naturalists can be explained in a manner consistent with known genetic mechanisms.

As the expression "evolutionary synthesis" connotes, this unification was generated not by the complete overthrow of one specialty's paradigm by another, but by a merger of different viewpoints and approaches. Huxley's aim was to show

how diverse specialties working together might resolve long-standing issues. His take on the antagonisms to overcome depicts the conflicts described in chapters 6 and 12:

> It was in this period, immediately prior to the war, that the legend of the death of Darwinism acquired currency. The facts of Mendelism appeared to contradict the facts of paleontology, the theories of the mutationists would not square with the Weismannian views of adaptation, the discoveries of experimental embryology seemed to contradict the classical recapitulation theories of development. Zoologists who clung to Darwinian views were looked down on by the devotees of the newer disciplines, whether cytology or genetics, *Entwicklungsmechanik* or comparative physiology, as old-fashioned theorizers; and the theological and philosophical antipathy to Darwin's great mechanistic generalization could once more raise its head without fearing too violent a knock.[1]

The new "unification" would be based on a functional division of labor, each specialty providing something another lacked. The circumstances of the Second World War provided favorable conditions for such a synthesis. In England, biologists who contributed books on evolution, genetics, and cytology were forced to put aside their research when laboratory and fieldwork was impeded. Several of the leading books appeared during the war years.[2] The scope of the synthesis and how it occurred has been intensively investigated by scholars.[3] Their aim has been to explain how conceptual misunderstandings were mended and to sort out the specific contributions of individuals and specialties.

Merging Mendelism

The architects of the synthesis emphasized that understanding Mendelian heredity was crucial.[4] As Huxley remarked:

> What would Darwin or any nineteenth-century biologist say to facts such as the following . . . ? A black and albino mouse are mated. All their offspring are grey, like wild mice: but in the second generation greys, blacks and albinos appear in the ratio 9:3:4. . . .
>
> To the biologist of the Darwinian period the product of the grey mice would not have been inheritance but "reversion" to the wild type, and the reappearance of blacks and whites in the next generation would have been "atavism," or "skipping a generation." . . .
>
> In reality, the results are in both cases immediately explicable on the assumption of two pairs of genes, each transmitted from parent to offspring by the same fundamental genetic mechanism. The "reversions," "atavism," and "sports" are all due to new combinations of old genes.[5]

Mendelism was key. But we must keep in mind that the early leaders of genetics, including Hugo de Vries, Wilhelm Johannsen, William Bateson, and Thomas

Hunt Morgan were inspired by the thought that new species and varieties appeared suddenly, by leaps, not by creeps, perhaps by sudden mutation of a single hereditary unit. Three kinds of evidence supported this view, each of which required refutation before the synthesis:

1. De Vries had discovered that new types of evening primrose *Oenothera lamarckiana* could arise at a single step: in a single generation, plants suddenly differed markedly from their parents in a number of characteristics, including differently configured leaves and flowers. He believed that the new types arising by large "mutations" were new species; their sudden occurrence meant that selection had little or nothing to do with the *origin* of species. He articulated his views in his influential treatise *The Mutation Theory* (1901–3).
2. Early cross-breeding work seemed to support a view of evolution by leaps. After all, when Mendel had crossed a tall with a dwarf strain of peas he found in the second generation a ratio of three tall to one dwarf plant. This was a case of a significant disjunction, not a continuum.
3. Johannsen's experiments seemed to indicate that the individual variations always present in populations were not the basis of evolutionary change.[6] In 1903, he tested Darwin's theory by trying to alter the genetic nature of the garden bean through artificial selection. But he was unable to do so. He selected the largest and the smallest beans from the seed lot of a single variety and raised the offspring from them in successive generations, creating what he called pure lines. In the first generation, the larger beans produced slightly larger offspring than the smaller ones, but in later generations selection for large and small size had no effect whatsoever on the offspring. Thus, Johannsen and others concluded, the visible continuous variations that Darwin observed in natural populations, and that he considered to be the basis of natural selection, were actually not hereditary at all, but due to the effect of the environment on individual organisms.[7]

Each of these three conclusions was refuted by the 1920s. First, genetic studies indicated that the types de Vries had found in the evening primrose were not new species, and they were not due to single-gene mutations. They involved many genes and were due to a peculiar of kind of genetic segregation; *Oenothera lamarckiana* has a rather unusual type of chromosome behavior.[8] Second, among the many traits studied by geneticists by the 1920s, examples of large changes caused by single genes were relatively rare. No simple Mendelian ratios could be found in the offspring derived from humans differing in height, or when a tall race of wild yarrow is crossed with a dwarf alpine plant of the same species. The difference between tallness and shortness is governed by many genes, each of which has only a slight effect on size. Not only did geneticists find that one characteristic could be influenced by many genes (polygeny), they also found that one gene could affect many traits (pleiotropy).

Third, Johannsen's pure-line experiments were faulty. Selection experiments conducted for size and other characteristics with mice, flies, corn, and other organisms showed that selection could produce cumulative changes over as many as fifty to a hundred generations.[9] Johannsen's results were later understood to be due to the fact that he mistook an artificial population for a natural one. The garden beans he used had been selected by breeders over many generations for uniformity and constancy. Thus, differences due to small effects of genes had been artificially eliminated before he began his experiments. Plenty of heritable variations existed in the wild for natural selection.

The Importance of Sex

According to the new Darwinian synthesis, evolution was not driven by mutations, large or small, but by the size of populations, or "gene pools," and the kind of variations within them. Because every individual born from sexual reproduction contains a novel genetic mix, a population contains an enormous diversity due to gene recombination, which would fuel evolution. This view contradicted that of leading geneticists well into the 1930s. Even though every geneticist knew that during meiosis, chromosomes often exchange parts to form new combinations of genes and so change the chromosomes in their germ cells, some geneticists insisted that this was not the stuff of evolution. T. H. Morgan, for example, abandoned the notion of large-scale heritable changes and accepted the notion that subtler, more gradual gene mutations were the basis of evolution. But he could not accept the idea that recombination of genes would create anything new, and that the common variations found in the wild would provide the fuel for evolution. He insisted that evolution was essentially driven by new gene mutations.[10]

In neo-Darwinian theory, natural populations would contain a large pool of hereditary variation, and therefore new mutations would rarely, if ever, be the direct source of variation upon which evolutionary change is based. Instead, they replenish the supply of variability in the population, which is constantly being reduced by selective elimination of unfavorable variants. But in any one generation, the variation contributed to a population by mutation is minuscule compared to the variation brought about by recombination of preexisting genetic differences. Thus, one would not expect to find any relationship between mutation rate and direction of evolution.[11] This was the view that emerged during the 1940s.

One can point to geneticists such as H. Nilsson-Ehle in Sweden and Erwin Baur in Germany, who had emphasized the importance of recombination for evolution. In the United States, geneticists were very familiar with hybrid vigor in corn resulting from genetic recombination, and the Harvard corn geneticist E. M. East and his students were well known for their Darwinian views. A. H. Sturtevant had also adopted a Darwinian conceptual framework.[12] But none of these geneticists developed a comprehensive theory or an evolutionary research program.

Population Genetics

The union of Darwinism and Mendelism meant that important aspects of evolution change could be studied as a branch of applied mathematics in which one could assess the changes in gene frequencies within a population.[13] The mathematical treatment of evolution meant that one could check deductions and make quantitative predictions. Beginning in the 1920s, population geneticists, led by R. A. Fisher and J. B. S. Haldane in England and Sewall Wright in the United States, established the basis for population genetics by weaving together biometric methods, Mendelian inheritance, and the effects of selection into quantitative models of evolutionary processes.[14] They were the founding fathers of theoretical population genetics, but they had fierce disagreements.

Ronald Fisher (1890–1962) was educated in mathematics at Cambridge during the 1910s. An accomplished mathematician at age twenty-two, he became interested in genetics and evolution when he discovered the biometrician Karl Pearson's paper "Mathematical Contributions to the Theory of Evolution." But unlike Pearson, he saw no inherent conflict between Mendelism and natural selection, and he set out to demonstrate how little basis there was for the opinion that the discovery of Mendel's laws of inheritance was unfavorable, or even fatal, to the theory of natural selection. In his famous book of 1930, *The Genetical Theory of Natural Selection*, he aimed to construct a mathematical and deterministic analysis that would establish the laws of evolutionary change on a par with those of physics. The second part of the book was concerned with his eugenics aims. He surveyed the collapse of empires and the decline of British peerage, and he advocated the proper distribution of family allowance to people of superior beauty, intellect, health, and talent, whose breeding, he argued, should be promoted.

The first part of the book focused on a mathematical treatment of the slow and deliberate effects of selection on individual genes. Far from being opposed to Darwinian selection, Fisher emphasized that Mendelism actually made the theory intelligible by refuting the old blending inheritance theory. If the latter were true, he argued, heritable variation would be approximately halved in every generation, and therefore Darwin's theory would require a colossal amount of new variation each generation.[15] Mendelism conserved the variance in the population. He also rejected the non-Darwinian arguments of those who believed in macromutations and who had argued that small imperceptible variations would have no selective advantage. As he commented:

> If a change of 1 mm. has selection value, a change of 0.1 mm. will usually have a selection value approximately one-tenth as great, and the change cannot be ignored because we deem it inappreciable. The rate at which a mutation increases in numbers at the expense of its allelomorphs will indeed depend upon the selective advantage it confers, but the rate at which a species responds to selection in favour of any increase or decrease of parts depends on the total heritable variance available, and not on whether this is supplied by large or small mutations.[16]

Fisher put "fitness" on a quantitative basis: he defined it solely in terms of the number of offspring an individual leaves behind. The centerpiece of his book was the fundamental theorem of natural selection, which he arrived at by combining Mendelism with certain principles of population ecology. According to the theorem, natural selection in itself would always drive a population toward greater fitness. However, since the environment was always changing, the population is always deteriorating in fitness. Therefore, the population never reaches optimum fitness, and is condemned to race forever against an ever further receding destination. This situation has been dubbed "the Red Queen's dilemma" in reference to the character in Lewis Carroll's *Through the Looking-Glass*.

In his calculations of fitness and mathematical treatment of evolutionary change, Fisher simplified the concept of the gene, treating each as a particulate, noninteracting entity upon whose individual effect selection could act. By such simplifications, if a certain gene bestowed an advantage that resulted in enhanced reproduction, he could calculate how rapidly it would increase in frequency. Indeed, he showed that small changes could have large effects in relatively short time periods. For example, if a gene change increased the likelihood of leaving offspring by only 1 percent, the genetic constitution of a species would be greatly modified in only a hundred generations.

John Burdon Saunderson Haldane (1889–1988) was the polymath son of the Oxford physiologist John Saunderon Haldane. He took up the breeding of guinea pigs with his sister Naomi as a teenager when he discovered the phenomenon of genetic linkage—that two or more genes situated close together on the same chromosome are inherited together. After serving in the First World War, he studied at Oxford, was a reader in biochemistry at Cambridge (1922–32), and then was professor of genetics and of biometry at the University of London (1933–57). He subsequently immigrated to India, adopted Indian nationality, and worked in Calcutta and Orissa. Although he at first viewed evolution as driven by mutations, he adopted the Darwinian perspective about the importance of preexisting variation, the significance of recombination, and the evolution of populations driven by natural selection.[17] In a series of papers published between 1924 and 1931, he constructed theoretical population models, plotting the factors that influence fitness and the frequencies of specific phenotypes.

Haldane thought evolution worked much faster than Fisher did, and he searched for an example in the real world. In 1924, he published the now-infamous story of the peppered moth, *Amphidasys betularia*, told as one of the strongest examples of the power of natural selection. The peppered moth's mutant black form was known to have undergone a spectacular increase in the second half of the nineteenth century. Haldane supposed that if the first black moth had been sighted in Manchester in 1848, it would have had a frequency lower than 1 percent of the population, and yet by 1901 it had virtually replaced the typical form in the region. Based on this conjecture, he showed that the black form would have a selective advantage of 50 percent. Such intense selection pressure would show that natural selection was a powerful force propelling evolution. Though the case of the peppered moth is today seriously doubted as a real case of evolution by natural selection, it was polished as the jewel in the crown of neo-Darwinian evolutionary biology of the twentieth century.[18]

The conclusion of population geneticists that even a slight genetic difference, when introduced into a population, could rapidly accumulate and result in evolutionary change was important in winning over such neo-Lamarckians as Ernst Mayr and Bernhard Rensch. Mayr became one of the great champions of Darwinian gradualism, but he was a staunch critic of what he dubbed "beanbag genetics" of population geneticists who were concerned solely with changes in the frequencies of individual genes and ignored the interaction between genes.[19]

Sewall Wright (1892–1964) had relatively little formal mathematical training; he considered himself more of a physiological geneticist than a population geneticist. Taught genetics at Harvard by W. E. Castle, he worked as an animal breeder for the U.S. Department of Agriculture before being appointed professor at the University of Chicago (1926–1954).[20] His quantitative studies of changes in the genetic composition of rat populations corroborated Fisher's about the importance of Mendelism and natural selection and the effectiveness of selection acting on small random variations.[21] However, he also investigated another factor that was not due to selection: the random effects in small populations, which he called random genetic drift.

Random Drift and Nonadaptive Change

The size of the variations did not matter, but the size of the population did. Whether a small or large population best favored evolutionary change divided Fisher and Wright for many years. For Fisher, natural selection would occur rapidly and most effectively in large populations, because more variant genes were stored there. For Wright, evolution would occur rapidly and most effectively in small populations because a genetic novelty would spread quickly in a relatively small population without selection due to genetic drift.[22] The chance isolation of a group in which, say, 75 percent of the individuals happened to be purple and not yellow would mean that the purples would pass on a higher percentage of their genes to the next generation, and the shift in gene frequency would become more pronounced in subsequent generations.

Based on his view that organisms actually live in small scattered populations, Wright thus advanced his "shifting balance" theory of evolution in natural populations—the theory that genes can be "fixed" in populations by a random process, that is, the chance partitioning of environments. Not all evolutionary changes were adaptive. Though anathema to the deterministically minded Fisher, this nonselectionist turn in evolutionary theory was, initially at least, taken up by other architects of the synthesis.

In Stalinist Russia, population genetics was begun by Sergie Chetverikov (1880–1959), who, in the 1920s, developed a major school of evolutionary biology at the Kol'tsov Institute in Moscow.[23] In 1922, H. J. Muller visited the Kol'tsov Institute and brought with him cultures containing thirty-two mutants of *Drosophila* from Morgan's laboratory, a gift that Chetverikov's school subsequently used to do the first population genetic studies of *Drosophila*. Chetverikov was exiled in 1929 because, it was rumored, he was denounced to the political police by one of his students.[24]

In the United States, population genetics of *Drosophila* were pioneered by the Ukrainian-born Theodosius Dobzhansky (1900–1975), who immigrated to the United States in 1927. There he worked in Morgan's lab at Columbia University. Subsequently, he taught at Columbia University (1940–62) and then at Rockefeller University (1962–71). In his pathbreaking book *Genetics and the Origin of Species* (1937), Dobzhansky articulated the concept that evolution was an affair of populations, not of individuals—or, as he put it, "Evolution is a change in the genetic constitution of populations."[25] "Populations evolve" because the traits of organisms within those populations are modified in kind and frequency. "The rules governing the genetic structure of a population," he wrote, "are distinct from those governing the genetics of individuals."[26]

Soon after arriving in the United States, Dobzhansky became an enthusiast for Wright's shifting balance theory and put it to the test by studying wild populations of fruit flies in remote mountains and deserts. Random genetic drift played a significant role in maintaining variation. The available variations spread quickly in small populations, he argued; successful variants could then prevail in larger populations by natural selection.[27] In his studies high in the Sierra Nevada, he studied sixteen strains of the fruit fly *Drosophila pseudoobscura*. The strains differed in particular inversions of large segments of chromosomes, and he recorded spectacular booms and busts in the population of different strains over several years. He ruled out chance, mutation pressure, and internally directed changes, and concluded that natural selection was almost certainly involved, although the selective factors involved remain unknown. The relative importance of random drift and natural selection was actively debated in the 1950s, at which time Dobzhansky and other leading evolutionists moved away from drift.[28] The importance of such random processes as genetic drift relative to adaptive change through natural selection remains controversial among evolutionists to the present day.[29]

The architects of the synthesis introduced a whole new lexicon for population genetics: *balanced polymorphism, climatic rules, founder principle, gene pool, gene flow, introgression, isolate, sibling species, stabilizing selection, allopatric speciation*, and *taxon*.[30] But there was much more to the problem of evolution than statistical studies of gene frequencies.[31] In his books *Recent Advances in Cytology* (1931) and *The Evolution of Genetic Systems* (1939), the British cytogeneticist Cyril Darlington showed that many chromosomal mechanisms and genetic systems regulate the amount of inbreeding or outbreeding—that is, the amount of recombination.[32] Darlington's books were complemented by Michael White's *Animal Cytology and Evolution* (1945).[33] Naturalists focused on the major temporal and spatial problems of evolution, such as adaptation, geographical variations, the multiplication of species, and the origin of higher taxa. Huxley's *Evolution: The Modern Synthesis* (1942) and the Harvard ornithologist Ernst Mayr's *Systematics and the Origin of Species* (1942) dealt with animals; G. Ledyard Stebbins' *Variation and the Evolution of Plants* (1950) dealt with plants.[34]

The Species Problem

In *Genetics and the Origin of Species*, Dobzhansky defined species as "that stage of the evolutionary process at which one actually or potentially interbreeding array of forms becomes segregated into two or more separate arrays which are physiologically incapable of interbreeding."[35] Mayr (b. 1904) subsequently revived what he referred to as the "biological species concept." A species, he argued, is the lowest group of animals or plants that, at least potentially, forms an interbreeding array of populations unable to breed freely with other sorts of animals and plants. "The major intrinsic attribute characterizing a species is its set of isolating mechanisms that keeps it distinct from other species."[36] Mayr's definition allowed for the idea that species themselves, not just individuals, may be targets of selection. The paleontologist George Gaylord Simpson (1902–1984) went further, defining species as "a phylogenetic lineage (ancestral-descendant sequence of interbreeding populations) evolving independently of others, with its own separate and unitary evolutionary role."[37]

Neo-Darwinian evolutionists considered two competing conceptions of speciation. Darwin, it will be recalled, held that new species could emerge from the splitting of a single population in a single geographical area; he understood this as a division of labor (see chapters 4 and 6). Mayr argued that geographic isolation was necessary.[38] Speciation, in his view, was not so much about the origin of new variations or evolutionary changes within populations, but rather a matter of isolating mechanisms, such as geography, which led to sexual incompatibility, which would then prevent inflow of new genes into gene pools: "A new species develops if a population which has become geographically isolated from its parental species acquires during this period of isolation characters which promote or guarantee reproductive isolation when the external barriers break down."[39]

Accordingly, a geographical barrier of some sort—a river or isolation on an island—would split a species into two groups on which selection could act. Since it is unlikely that the population on one side of this barrier would have exactly the same frequencies of genes when first separated, selection would have two very different sets of variations from the beginning. Mutation and selection would also gradually push the two groups in different directions, and eventually each group would be recognizable as a different species. Mayr called this hypothetical form of speciation "allopatric" in contrast to "sympatric" speciation (splitting of a single population in the same geographical area). He rejected the latter, arguing that it would be "subject to so many limitations that it is surely rare, if not exceptional."[40] Plants were an exception. Instantaneous speciation within a single population of plants could occur by chromosomal changes such as polyploidy (that is, the occurrence of three or more complete sets of chromosomes). Polyploidy is virtually unknown in animals.[41]

In the United States, evolutionary biology took on some of the trappings of a new discipline with its own Society for the Study of Evolution, founded in 1946 by Mayr, to discuss common evolutionary problems of geneticists, ecologists, systematists, and paleontologists, and its own journal, *Evolution: An International Journal of Organic Evolution*, launched in 1947. Edited by Mayr, its aim was to stimulate evolutionary research and bring scattered literature from several specialties together in readily accessible form.[42] Evolution by adaptation—by the gradual accumulation of traits with even the "slightest selective value"—became the central core of the evolutionary synthesis.[43] Selection explained the arrival of the fittest as much as it did the survival of the fittest. In its most important components, the synthesis was remarkably similar to Darwin's original theory of 1859.

Within the boundaries of the modern synthesis, one could quibble about details of speciation; the importance of random processes such as genetic drift relative to the force of adaptive change through natural selection was hotly debated; there could be disagreement about the relative importance of contributions of individuals and specialties. But one point all the founding fathers agreed upon was that evolution did not occur by large, discontinuous, macromutational changes.

Microevolution as Macroevolution

The macromutation ideas of Richard Goldschmidt (1885–1968) represent a notorious polemic in the history of the evolutionary synthesis.[44] Goldschmidt was one among many leading German scientists who, as Jews, fled Nazi Germany; he immigrated to the United States in 1933. He argued for the importance of macromutations in his text of 1940, *The Material Basis of Evolution*: "I cannot agree with the viewpoint of the textbooks that the problem of evolution has been solved as far as the genetic basis is concerned," particularly with respect to the origins of new structures or organs.[45] He acknowledged that natural selection could lead to races, but maintained that it could not lead to new species.[46] "The decisive step in evolution, the first step toward macroevolution, the step from one species to another, requires another evolutionary method than that of sheer accumulation of micromutations."[47] Thus, he spoke of "hopeful monsters," perhaps resulting from chromosomal rearrangements.

For the architects of the evolutionary synthesis, there was no distinction to be made between microevolution and macroevolution.[48] The difference between the two was arbitrary and merely quantitative. As Dobzhansky put it in 1937, "We are compelled at the present level of knowledge reluctantly to put a sign of equality between the mechanisms of micro- and macro-evolution, and, proceeding on this assumption, to push our investigations as far ahead as this working hypothesis will permit."[49]

Paleontology represented all that biologists really knew directly about the course of life's history. Few paleontologists had accepted natural selection as the only direction-giving mechanism in evolution, and most had considered macroevolutionary processes to be different from microevolution. Many were neo-Lamarckians, and some argued for orthogenesis—a linear direction in macroevolution due to some kind of intrinsic force. At Harvard, Percy Raymond commented in 1941:

> The school of experimental biologists has so firmly convinced them that they can never prove anything that, whatever their beliefs, paleontologists have generally chosen to remain silent. Probably most are Lamarckians of some shade; to the uncharitable critic it might even seem that many out-Lamarck Lamarck. . . . Recently orthogenesis has become popular, particularly with the younger generation. This idea is natural for the paleontologist, whose lines of descent are necessarily straight.[50]

More than anyone else, George Gaylord Simpson brought the disparate disciplines of paleontology and genetics together under a common explanatory framework.[51] Simpson worked at the American Museum of Natural History (1927–59), and he also taught at Columbia University (1945–59) and Harvard (1959–70). He was educated primarily as a geologist, but in the 1930s, he realized that a grounding in population genetics was essential for an evolutionist, and he integrated it with his detailed knowledge of paleontology in his influential book *Tempo and Mode of Evolution* (1944).[52] In the introduction, he summarized the hostile rift between the two cultures he aimed to reconcile:

> The attempted synthesis of paleontology and genetics, an essential part of the present study, may be particularly surprising and possibly hazardous. Not long ago, paleontologists felt that a geneticist was a person who shut himself in a room, pulled down the shades, watched small flies sorting themselves in milk bottles, and thought that he was studying nature. . . . On the other hand, the geneticists said that paleontology had no further contributions to make to biology, that its only point had been the completed demonstration of the truth of evolution, and that it was a subject too purely descriptive to merit the name "science." The paleontologist, they believed, is like a man who undertakes to study the principles of the internal combustion engine by standing on a street corner and watching the motor cars whiz by. . . . It is not surprising that workers in the two fields viewed each other with distrust and sometimes with the scorn of ignorance.[53]

As Stephen Jay Gould observed, Simpson's book broke the paleontological mold: far from being descriptive, it was based on heavy use of quantitative work such as graphs, frequency distributions, and pictorial models from population genetics that he applied to large-scale patterns of diversity in the history of life.[54] He confronted those supporting orthogenesis and insisted that "the history of life, as indicated by the available fossil record, is consistent with the evolutionary process of genetic mutation and variation, guided toward adaptation of populations by natural selection." Much of the evidence for internally directed, nonadaptive linear trends in evolution was refuted as further discoveries of the fossil record revealed more complex branching. The phylogeny of horses, for example, was argued to be an adaptively oriented bush: many of the fossils represented extinct side branches rather than steps on the way toward the modern horse. Even very large and apparently burdensome structures such as the gigantic antlers of the Irish elk, it was argued, had some adaptive purpose.[55]

There still remained the notorious lack of evidence in the fossil record for transitions between higher taxa of fundamentally different body plans. As Simpson saw it in 1944, an imperfect fossil record could account for part of the problem, but not all of it. Some of the fossil evidence had to be taken at face value, and that meant that the typical gradual mode of phyletic change by selection would not be sufficient. He rejected macromutations and suggested instead that perhaps macroevolution sometimes occurred in a small part of the population, which would be expected to leave behind little fossil evidence. Thus, he considered Wright's notion of genetic drift, whereby inbreeding in a small population and selection would cause a population to change relatively rapidly. In 1944, Simpson referred to this as "quantum evolution," and described it as "the dominant and most essential process in the origin of taxonomic units of relatively high rank, such as families, orders, and classes." Not all evolutionary change would be adaptive:

> The aspects of tempo and mode that have now been discussed give little support to the extreme dictum that all evolution is primarily adaptive. Selection is a truly creative force and not solely negative in action. It is one of the crucial determinants of evolution, although under special circumstances it may be ineffective, and the rise of characters indifferent or even opposed to selection is explicable and does not contradict this usually decisive influence.[56]

Although Simpson entertained a pluralistic view of evolutionary mechanisms in 1944, as Gould emphasized, the evolutionary synthesis hardened over subsequent years to exclude virtually everything but evolutionary adaptations caused by natural selection (a position called panselectionism). In 1953, Simpson rejected discontinuity in the pace of evolution, and he discounted the idea that genetic drift could trigger major evolutionary events leading to the origins of higher categories such as classes or phyla.[57]

Lessons of Synthesis

After a long "eclipse," as Julian Huxley called it, Darwinian theory was finally shining its evolutionary light on biology. No other field of science had such an explanatory theory uniting so many different specialties. Those who reflected on what had led to this transformation in evolutionary thinking noted that it differed from the model of scientific change described in Thomas Kuhn's influential book *The Structure of Scientific Revolutions* (1962).[58] Kuhn argued that science does not evolve in a linear manner by the gradual accumulation of scientific results—not from disproving hypotheses and refuting theories in the manner philosophers had imagined. It was more like putting in pieces to a puzzle according to an accepted conceptual and methodological framework. During "normal science," scientists work within cognitive structures he called paradigms. The paradigm was a way of both seeing and doing.

During revolutionary periods, Kuhn argued, one paradigm is replaced by a dramatically different one, due largely to an intolerable accumulation of anomalies. For example, an earth-centered universe was replaced with a sun-centered one, and then with Einsteinian astrophysics. Creation theory was replaced by evolutionary theory; the phenotype conception of heredity was replaced by the genotype conception. All of these changes were revolutionary. Science then would be punctuated by stops and new beginnings.

The Structure of Scientific Revolutions had far-reaching ramifications affecting virtually all aspects of academic thought. Scientists themselves often appeal to Kuhn's book to argue that they themselves are revolutionaries confronting the traditional paradigm and meeting resistance from an inherently conservative scientific establishment. No one denies today that scientists do indeed work within structural frameworks that shape how they observe and understand. But historians and philosophers of science have noted a number of problems with Kuhn's theory of scientific change.[59] One monolithic paradigm is seldom replaced soundly by another monolithic paradigm; often there are several competing paradigms (specialties). Rarely are revolutions so complete, and rarely is the knowledge produced so incommensurate with the previous paradigm, as Kuhn had presupposed. Revolutions occur by various means, and the stories about them all carry lessons.

Biologists' discussions about the construction of the evolutionary synthesis are exemplary. There was no scientific revolution in the sense of one paradigm's being dramatically replaced by another through an accumulation of anomalies. And there were many competing specialties or paradigms, none of which were entirely overthrown and replaced by an incommensurable one. The model for scientific change was a synthesis, a breaking down of disciplinary boundaries, resolving misunderstandings, mistrust, feelings of superiority, and intolerance between members of different specialties. The evolutionary synthesis is a lesson about the hazards of specialization and about the need for intellectual breadth and willingness to learn outside one's own specialty. Mayr summarized this well when writing about the conflict he perceived between naturalists and experimentalists (geneticists):

> When I read what was written by both sides during the 1920s, I am appalled at the misunderstandings, the hostility, and the intolerance of the opponents. Both sides display a feeling of superiority over their opponents "who simply do not understand what the facts and issues are." How could they have ever come together? Just as in the case of warring nations, intermediaries were needed, evolutionists who were able to remove misunderstandings and to build bridges between hierarchical levels. These bridge builders were the real architects of the synthesis.[60]

The synthesis required bridge builders to unite the competing camps. But it also required gatekeepers and guardians of the faith. Biology was far from over, and the neo-Darwinian revolution of the 1940s and 1950s was far from complete. The bridge builders of the hardened synthesis became its chief defenders during the latter twentieth century, as a new generation of biologists from several new specialties—molecular evolutionists, developmental evolutionists, and symbiologists—challenged its central tenets.

Many long-standing questions remained unresolved, none more so than the relation between development and evolution. There were a number of issues. How did the embryo evolve? Why have so few structural plans of animals evolved? One had to consider developmental constraints and the basis of fundamental body plans. Why, over the past 500 million years of evolution, following the Cambrian explosion, have no new phyla or body plans appeared in the course of evolution?[61] Embryologists generally did not participate in the evolutionary synthesis, not because they were excluded in any institutional manner, but because the two paradigms were so dramatically different. Ontogeny is controlled and purposeful; evolution, according to neo-Darwinism, was stochastic and random. Many continued to believe that early development and macroevolutionary changes had their seat in the materials and organization of the egg cytoplasm, whereas Mendelian genes showed their effects only later in development (see chapter 10).[62] Others emphasized that evolution operated through modifications of developmental processes. The synthesis did not include the problems of morphogenesis or the possibility of mechanisms of inheritance other than chromosomal genes (see chapter 17).

The modern synthesis did not deal with the great transitions in life: the genesis of bacteria and of protists, and the multicellularity of plants and animals. It did not touch on microbial evolution and the role of microbial symbiosis in development and heredity (see chapters 18 and 19). These aspects of biology required a new molecular comparative approach to examining changes occurring at the level of DNA, RNA, and proteins. A new field, molecular evolution, emerged the 1960s and 1970s that conflicted with the aims, methods, and doctrines of classical neo-Darwinism. Molecular biology had a major impact on evolutionary biology. It emerged during and after the Second World War, when a new generation of geneticists, working in collaboration with chemists and physicists, set out to discern what genes are and what they do in the cell.

14

Genes, Germs, and Enzymes

> The war is over here, but as you know, we never did any suffering to speak of. The suffering is with the soul as it is everywhere. . . . The upswing in science, though, is sensational. If it lasts, science in the United States will receive much support. It is having its effects on those that are making plans for the future. Big ideas are hatching and they certainly look good on paper. The genetics groups are feeling it too. You undoubtedly know that Beadle and his whole group are moving to Cal. Tech. this coming July. . . . The set-up at Cal. Tech will be splendid. Much emphasis will be put on chemical aspects of genetics with the cooperation of the department of chemistry there—Pauling in particular. . . . The emphasis in genetics seems to be swinging into genic action.
>
> —Barbara McClintock to Boris Ephrussi, December 29, 1945

THE FIRST BREAKTHROUGH in understanding how genes actually function in the cell begins with a tale of discovery, neglect, and discovery no less remarkable than that of the birth of genetics itself (see chapter 11). The achievement, it is said, was made by an outsider, the British physician Archibald Garrod (1857–1936), who, in 1908, made the astonishing discovery that a gene controls biochemical reactions by directing the formation of a single enzyme. (Enzyme-proteins are catalysts that increase the rate at which a chemical reaction occurs.) Geneticists have maintained that Garrod's insight was unappreciated and overlooked for several decades until it was rediscovered in the 1940s by George Beadle (1903–1989) and Edward Tatum (1909–1975).[1] They proposed that a gene acts by determining the specificity of a particular enzyme and thereby controls, in a primary way, enzymatic synthesis and other metabolic reactions in the organism. Their proposal, known as the "one-gene: one-enzyme" hypothesis, played a directive role in the development of biochemical genetics throughout the 1940s and 1950s. Beadle and Tatum were awarded the Nobel Prize in 1958, which they shared with Joshua Lederberg, a founder of bacterial genetics.

The Garrod Tale

Beadle himself popularized the story of Garrod's neglect in the 1950s. He referred to Garrod as the "father of chemical genetics," and in his Nobel lecture he insisted that he had only "rediscovered what Garrod had seen so clearly so many years before."[2] Since that time, the story of Garrod's neglect and subsequent discovery has been repeated in hundreds of classrooms and asserted in biology texts, in historical writings, and in journalists' sensationalist accounts of gene theory.[3] Geneticists have been perplexed by it. Bentley Glass wrote in 1965: "The work of Garrod was certainly known among men of medicine. Nevertheless, its pregnant insight into the nature of gene action was disregarded among geneticists. . . . Why was this development, so surely foreshadowed in Garrod's work, postponed for nearly forty years?"[4] Beadle offered an explanation in 1966: "It seems to me that, like Mendel, Garrod was so far ahead of his time that biochemists and geneticists were not ready to entertain seriously his gene-enzyme-reaction concept. Like Mendel's, Garrod's work remained to be rediscovered independently at a more favorable time in the development of biological sciences."[5]

Like the neglect of Mendel, the tale about Garrod has been told by geneticists as a lesson to illustrate a host of conditions impeding the progress of reason and the recognition of profound discoveries: the "prematurity" of Garrod's publications, the inadequately prepared biological world, his inclusive data, the technical difficulties of doing biochemical genetic work on the organisms of classical genetics, barriers between scientists resulting from specialization, and the demand that human genetics produce immediate practical applications, a demand that interfered with the fundamental research.[6]

Yet, recently, scholars have offered a dramatically different story. They have shown that Garrod did not make a conceptual advance in biochemical genetics that was neglected for decades; he did not offer a theory of gene action, and his work was simply not "rediscovered" by Beadle and Tatum.[7] As in the case of Mendel, geneticists had taken Garrod's work out of context and imposed their own meanings on it, and the significance of his work changed over time with a shift in context. A brief overview of these changes will help us to understand the great metamorphosis in genetics that occurred during the 1940s and 1950s.

Physiology and Genetics

Garrod was a lecturer in chemical pathology at St. Bartholomew's Hospital and senior physician at the Hospital for Sick Children in London. In the early twentieth century, he was interested in a group of congenital metabolic diseases in humans, which he referred to in the title of his book as *Inborn Errors of Metabolism* (1909).[8] Alkaptonuria, one of the first inborn errors he studied, has as its most obvious symptom blackening of the urine. Although a rare condition, it had long been recorded medically, and important aspects of its biochemistry were understood. The substance responsible for blackening of the urine is alkapton or homogentisic acid, which oxidizes to a black pigment when exposed to air. Little was known about the underlying mechanisms of its heritability. Garrod consulted William Bateson, who in 1902 diagnosed alkaptonuria as a Mendelian recessive character.[9]

In 1908, Garrod discussed the case in more detail and suggested that the condition was due to the blocking of an enzymatically controlled reaction; alkapton is not further oxidized, and as a consequence accumulates and is excreted in the urine. He drew similar conclusions concerning other congenital abnormalities such as albinism, cystinuria, and porphyrinuria. It is amply clear that Garrod recognized that the lack of an enzyme was blocking a metabolic step. However, he made no statements about gene action. By 1923, when he published the second edition of *Inborn Errors of Metabolism*, numerous suggestions of gene action in terms of enzymes had appeared in the genetics literature. But Garrod referred to none of them. He was interested in metabolism and pathology, not in gene theory.

Garrod championed the concept of chemical individuality—the idea that every species and every individual had different proteins.[10] Inherited chemical individuality, in the form of variable enzymes, could explain certain diseases in humans. His writings were part of a conceptual change in the newly emerging discipline of biochemistry. During the first decades of the century, biochemists began to recognize that each metabolic reaction is specifically catalyzed by a single enzyme.[11] In 1908, Garrod explained this shift in thinking:

> It was formerly widely held that many derangements of metabolism which result from disease were due to a general slackening of the processes of oxidation of the tissues. The whole series of catabolic changes was looked upon as a simple combustion, and according as the metabolic fires burnt bright or burnt low, the destruction of the products of the breaking down of food and tissue was supposed to be complete or imperfect . . . Nowadays, very different ideas are in ascendance. The conception of metabolism in block is giving place to that of metabolism in compartments. The view is daily gaining ground that each step in the building up and breaking down, not merely of proteins, carbohydrates, and fats in general, but even of individual fractions of proteins and of individual sugars, is the work of special enzymes set apart for each particular purpose.[12]

Garrod's announcement was among many that heralded the importance of enzymes in physiological processes. But physiologists and biochemists of Garrod's day were reluctant to relate their work on metabolism to genes. Indeed, they generally doubted the very reality of genes, viewing them as hypothetical elements useful for explaining the results of breeding but of little explanatory value for physiology. As H. J. Muller lamented in 1926 in his famous paper "The Gene as the Basis of Life":

> Just how these genes thus determine the reaction-potentialities of the organism and so its resultant form and function is another series of problems; at present a closed book in physiology and one which physiologists as yet seem to have neither the means nor the desire to open.[13]

Early Gene-Enzyme Associations

Bateson mentioned Garrod's work in his two books *Mendel's Principles of Heredity* (1909) and *Problems of Genetics* (1913). However, what he saw as important in Garrod's work was not a theory of gene action by way of enzymes, but a means for understanding gene mutations and the underlying nature of dominant and recessive traits.[14] Bateson noticed that almost all Mendelian mutations were recessive and often resulted in the phenotypic loss of some structure or function. He suggested that recessive traits resulted from the absence of a factor that was present in dominant traits. The alternative hypothesis at the time was that separate factors were responsible for dominant and recessive traits. Bateson's "presence and absence theory" was simpler, and he pointed to alkaptonuria, the first example of Mendelism in humans: it was a recessive trait that occurred due to the loss of an enzyme necessary to break down alkapton.

Geneticists ridiculed Bateson's presence and absence theory for leading to absurd evolutionary conclusions: if gene mutations were a principal source of evolutionary change, and most mutations were recessive lacking some factor that the dominant had, then evolution must proceed by the loss of genetic elements. And if evolution proceeded by loss of genes, then simpler organisms such as worms or microbes would have a more complex genetic makeup than humans.[15]

Bateson did not overlook the significance of Garrod's work for biochemical genetics. In fact, the idea that genes controlled metabolic reactions by way of enzymes was ubiquitous in genetics from the beginning. Geneticists in France, Germany, and England used it to explain coat color in mice and guinea pigs and pigmentation in moths and plants. One of Bateson's own associates, Muriel Whedale Onslow, interpreted plant pigments in terms of genes and enzymes, and in the 1930s the gene-enzyme relationship was also extensively investigated in plant pigments by J. B. S. Haldane and his colleague Muriel Scott-Moncrieff.[16] The gene-enzyme relation was discussed in the United States in the 1920s—by H. J. Muller, C. O. Bridges, and Morgan, who insisted that enzymes were products of genes—rather than being the genes themselves.[17] In 1917, Sewall Wright also interpreted the results of his studies of the coat color of guinea pigs in terms of genes and enzymes.[18]

That same year, the chemist Leonard Troland developed the idea of a one-to-one relationship between genes and enzymes. He argued that genes were enzymes, and that his "enzyme theory of life" solved many of the fundamental problems of theoretical biology: the origin of life, the source of variations, the mechanisms of heredity, and development. Indeed, Troland was optimistic that someday one might be "able to show how in accordance with recognized principles of physics a complex of specific, autocatalytic, colloidal particles in the germ-cell can engineer the construction of a vertebrate organism."[19]

Although such pronouncements have a familiar ring today, to many biologists of the 1920s, no less than today, they seemed hyperbolic and simplistic. After all, geneticists had shown that one gene may effect the production of many characters (pleiotropy), and a single character may require the action of many genes (polygeny). The United States' leading cell biologist, E. B. Wilson, pointed to this evidence to argue that not only was Troland deceived about the construction of an organism out of autocatalytic (or self-reproducing) particles, but that his one-to-one theory was absurdly naive: "I believe it is not a great overstatement," he wrote in 1923, "when I say that every unit may affect the whole organism and that all units may affect each character. We begin to see more clearly that the whole cell-system may be involved in the production of every character."[20]

The One-Gene: One-Enzyme Hypothesis

Following the Second World War, the face of genetics was transformed as biochemists and geneticists teamed up to explore what genes do in the cell, and what they are. What was new was not the concept that genes function by way of enzymes, but the use of microorganisms. Single-celled microbes allowed geneticists to avoid the complexity of tissue differentiation and cellular integration when searching for a bridge between gene and character. The geneticist isolated mutants that were found to be unable to grow, or that grew poorly on a well-defined growth medium, and the biochemist sought the reason for this inability. Thus, a suitable organism for biochemical genetics was one whose sex life *and* growth could be brought under meticulous control. The model organisms of classical genetics, *Drosophila*, corn, and mice, so useful for establishing the chromosome theory of inheritance, were quickly outcompeted by rapidly reproducing microorganisms: fungi, yeast, algae, protists, and bacteria.

In their landmark paper of 1941 on the genetics of the bread mold *Neurospora crassa,* Beadle and Tatum argued that because the cell was a highly integrated system, "there must be orders of directness of gene control ranging from simple one-to-one relations to relations of great complexity."[21] However, by the mid-1940s, Beadle's views had hardened. The numerous instances in which single-gene mutations resulted in a block of a metabolic step led him to hypothesize that many or all genes have single primary functions. The idea that one gene produce one metabolic block soon became well known as the "one-gene: one-enzyme hypothesis."[22] Although this hypothesis offered the hope that one could know what genes actually do in the cell, many geneticists were reluctant to accept such a simple one-to-one relationship. As Beadle commented years later, "I recall well the year 1953 when, at the Cold Spring Harbor Symposium on Synthesis and Structure of Macromolecules there appeared to be no more than three of us who remained firm in our faith."[23]

This was the context in which Beadle "rediscovered" Garrod's work. He had first learned of it in reviews of physiological genetics by Sewall Wright and J. B. S. Haldane in the early 1940s.[24] He saw in Garrod's writings the idea that each inborn error of metabolism in humans, like those in *Neurospora*, could be interpreted as a block at some particular point in the normal course of metabolism that resulted from a congenital deficiency of a specific enzyme. In other words, genes control single metabolic functions. But, unlike many of those who followed him, Beadle fully recognized that Garrod himself had made no such announcement about gene action. "Thus, while he did not express it exactly so, the relation gene-enzyme-specific chemical reaction was certainly in his mind. It is proper that he should be recognized as the father of chemical genetics."[25]

Garrod had said just enough for Beadle. The story of Garrod's creative insight and subsequent neglect was important for bolstering his one-gene: one-enzyme hypothesis. The notion of the independence and the inevitability of the truth is embedded in the notion of rediscovery: the true concept of gene action, lost and found, would be vindicated, just as Mendel's laws had been. Thus began, in 1950, the legendary myth about the long neglect of Archibald Garrod.

At the centenary of Mendel's paper, geneticists embellished the Garrod story further, referring erroneously to him as a "geneticist" and asserting that he had been ignored by his colleagues. Bentley Glass wrote in 1965, "The first big steps forward in understanding the nature of gene action were those made by Garrod, so ignored by the fraternity of geneticists who perhaps were too engrossed in their own experiments to read anything not published by another recognized geneticist."[26] Beadle compared the neglect of Garrod to that of Mendel, commenting that "I strongly suspect that an important component of the unfavorable climate for receptiveness in these two instances is the persistent feeling that any simple concept in biology must be wrong."[27]

Beadle's suspicion that Garrod's one-to-one theory was doubted because of the feeling that it was too simple holds some truth. However, we should remember two points. First, ample experimental evidence had accumulated over previous decades indicating that one gene may be concerned with many characters, and one character may depend on many genes. Secondly, no satisfactory proof of the one-gene: one-enzyme theory existed. This limitation was addressed by prominent scientists of the 1940s and 1950s. The physicist-turned-geneticist Max Delbrück offered the most significant criticism of the one-gene: one-enzyme hypothesis in 1946, when he pointed out that the data were only compatible with the interpretation; they did not prove it. He emphasized that the procedures of isolating mutations in the *Neurospora* work precluded the results by restricting alternative possibilities. If in fact one gene normally controlled many enzymes, no mutations in such genes could be detected. He challenged geneticists to devise methods by which the hypothesis could be disproved. "If such methods are not available, then," he argued, "the mass of 'compatible' evidence carries no weight whatsoever in supporting the thesis."[28]

Delbrück's challenge was easy to make but difficult to meet. *Neurospora* geneticists argued that since the data could be accounted for on a one-to-one basis, there was little value in making up a more complex interpretation.[29] Beadle himself confessed that he knew "no way of proving it in a single instance." Nevertheless, he argued that the hypothesis "served a useful purpose" and that there were "no compelling reasons for abandoning it," even though it might be later found to "err in the direction of oversimplification."[30] Indeed, it was a useful heuristic. The idea that each gene specified one enzyme was genetic orthodoxy for decades, and the gene itself was generally defined in terms of specifying a single protein (see chapter 18).

Domesticating Microbes

The revolution in biochemical genetics was not the result of a new theoretical insight, but of a technical breakthrough. The new microbial techniques' profound importance for the genetic analysis of metabolic pathways was indisputable. No previous studies of gene action using plants or animals had come close to matching the *Neurospora* technology in scope and detail. Microbiology and genetics had developed apart prior to the Second World War. Microbiology was strong on the applied side, being of considerable importance to medicine and industry. Biochemists had studied their metabolism and enzyme activities and defined their nutritional requirements, their ability to use certain compounds as sources of carbon, and their sensitivity to antibiotics. In all biochemical respects, microbes were fundamentally similar to plants and animals, but their genetics had not been explored. In most cases, they seemed to reproduce solely vegetatively; their small size and lack of obvious sexual differentiation made it difficult to distinguish between sexes, and therefore to carry out genetic studies.

Beadle shifted to the bread mold *Neurospora* after he had studied two classical genetics model organisms: corn and *Drosophila*. In 1934, following his doctoral research on the cytogenetics of corn with R. A. Emerson at Cornell University, he worked as a postdoctoral student in T. H. Morgan's laboratory, which had relocated to the California Institute of Technology in Pasadena a few years earlier. Several other leading cytogeneticists were former students of Emerson, including Marcus Rhoades (1903–1991) and Barbara McClintock (1902–1992), who was awarded the Nobel Prize in 1983 for her discovery of transposable elements or "jumping genes."[31] In Morgan's lab, Beadle met Boris Ephrussi (1901–1979), a postdoctoral student from Paris. Ephrussi's research had been on experimental embryology and tissue culture, and he had arrived in Morgan's laboratory with the mission of closing the gulf between embryology and genetics. His aim was to build up "the chain of reactions connecting the gene with the character."[32]

When Ephrussi returned to Paris in 1935, Beadle accompanied him. Together, they developed a transplantation technique for analyzing the chain of reactions leading to eye pigment formation in *Drosophila*.[33] They removed embryonic eye tissue from larvae of two strains with eye color mutants (vermilion and cinnabar) and injected them into the body of larvae of normal flies. Their technique was crude, but they were able to conclude that two substances (eye hormones) intervened in the formation of normal eye pigment.

The next task was to try to establish the chemical identity of the two substances. When the German army invaded, Ephrussi fled to the United States as a refugee scholar at Johns Hopkins University, where he remained until 1944, when he became active in Les Forces Français Libres. Beadle had returned to the United States to take up a position at Stanford University, where he worked with a new collaborator, Edward Tatum, who had just completed his Ph.D. work in biochemistry on the physiology and nutrition of bacteria. Beadle and Tatum were able to identify one of the eye color hormones in *Drosophila* as the amino acid tryptophan. But after three years, they had failed to isolate and characterize the second hormone. As Beadle later remarked, this failure was a "blessing in disguise" as far as he and Tatum were concerned, for, in frustration, they abandoned *Drosophila* and turned their attention to the bread mold *Neuropora crassa*.[34]

The idea of switching to *Neurospora* occurred to Beadle in the middle of a series of lectures that Tatum gave on comparative biochemistry.[35] In the course of his lectures, Tatum recounted the nutritional requirements of yeast and fungi, some of which exhibited well-defined blocks in vitamin biosynthesis. It occurred to Beadle that one should be able to select mutants in which known chemical reactions were blocked—that is, select mutants unable to synthesize known metabolites, such as vitamins and amino acids—that could be supplied to the medium. A mutation unable to make a given vitamin could be grown in the presence of that vitamin and identified as such when grown in media lacking it. *Neurospora* seemed to be ideal. Its growth requirements had already been worked out, and so had its genetics. A study of its life cycle and genetics had been begun by B. O. Dodge (1927) of the New York Botanical Garden. In 1931, Morgan had assigned Carl Lindegren to work on the genetics of the mold.[36]

By the spring of 1941, Beadle and Tatum were X-raying *Neurospora* or exposing them to ultraviolet light and seeking mutants. Despite the war, a number of graduate students and postdoctoral fellows hurried to Stanford to learn the new *Neurospora* techniques. The team of Beadle and Tatum soon grew into the *Neurospora* school with a talented group of researchers that included M. B. Mitchell, Norman Horowitz, and David Bonner. By the end of the war, their work resulted in a library of about a hundred genes affecting vital syntheses in *Neurospora*. Almost immediately, others began to apply similar procedures to other microbes.

The Chosen Few

The *Neurospora* work established standardized procedures for investigating the biochemical effects of genes, and it proved to be of great importance in the drug industry and biochemistry, providing a powerful tool for the analysis of the pathways in which vitamins, amino acids, and other compounds are synthesized. But even with regard to this technological revolution, its origins are murky and were a subject of heated controversy in the 1950s. Although textbooks assert that Beadle and Tatum were the first to use microorganisms for biochemical genetic studies, the German biologist Franz Moewus and his collaborators preceded them. Seen by some as one of the most important figures in the new genetics, Moewus was ultimately dismissed as the perpetrator of one of the most ambitious frauds in the history of science. He was subsequently expunged from that history.[37]

Moewus began to work on the unicellular green algae *Chlamydomonas* in the 1930s, and studying its sexuality in particular. His research attracted Richard Kuhn, director of the Chemical Institute of the Kaiser Wilhelm Institute for Medical Research in Heidelberg. In 1938 Kuhn was awarded the Nobel Prize for his work on carotenoid pigments and vitamins, and later he was appointed the director of science under the Nazi regime. Working in collaboration with Kuhn, Moewus soon developed what some considered to be the basic concepts and methods for biochemical genetics of microorganisms.

In 1940, a year before Beadle and Tatum's first paper on *Neurospora*, and before scientific communication between the United States and Germany was cut off, Moewus and Kuhn published a series of four remarkable papers dealing with the general theory of biochemical genetics as applied to *Chlamydomonas*.[38] These papers represented the first systematic investigations of biochemical genetics in any microorganism, and the first studies purporting to demonstrate the precise manner in which biochemical activities of the cell were controlled by genes and enzymes. Moewus reportedly mapped more than seventy genes in *Chlamydomonas*, which would make it one of the genetically best-known organisms of that time. The influential protozoologist H. S. Jennings at Johns Hopkins commented in 1941 that "the work of Moewus has placed the genetics of protozoa on a new footing. It has brought the phenomenon of inheritance in these organisms into the same system that is manifested in the Mendelian inheritance of higher organisms."[39] Beadle commented on the biochemical side in 1945: "Without doubt the most remarkable series of studies in biochemical genetics is that of the German investigators Moewus, Kuhn, and co-workers on the flagellate *Chlamydomonas*."[40]

During the 1940s and 1950s, Moewus's work rose to the center of an intense controversy. While some leading geneticists argued for the priority of Moewus's work over that of Beadle and Tatum, others raised serious questions about its authenticity. There were reports that some of his data were statistically too good to be true and that his biochemical results were dubious. Some biologists found it impossible to believe that the number of genetic crosses Moewus reported could ever have been carried out as described. Others told of how they caught him rigging his "experiments." By the mid-1950s, after extensive attempts to confirm crucial aspects of Moewus's results by himself and others, Moewus was dismissed as a fraud. In the end, his dubious data overshadowed any technological originality that might be attributed to him. Tracy Sonneborn (1905–1981) summarized the issues in 1955:

> Moewus certainly anticipated a great deal of important work, priority for which is usually given to others. The situation is terribly complicated because the date that formed the basis for the important contributions in the way of methodology and ideas are subject to very serious questions. It is hard to disassociate the two aspects of Moewus, namely the ideas and methodologies which he clearly grasped and forcefully presented and which have been proven to be sound by the work of others; and

on the other hand, the experimental data relating to them which are, to say the least, quantitatively dubious.[41]

Chlamydomonas genetics had to begin anew in the 1950s and 1960s. Led by the research of Ruth Sager, Nicholas Gillham, and others, *Chlamydomonas* proved to be an especially suitable tool for genetic studies of chloroplasts and their genes.[42]

There were several other model organisms of microbial genetics. Each had specific attributes, and each found its own research niche. In the 1930s, H. S. Jennings directed his students at Johns Hopkins University to the genetics of protists with the expectation that novel mechanisms of heredity might be found. Tracy Sonneborn succeeded admirably. In 1937, he discerned sex-like mating types in *Paramecium aurelia* and showed that it possessed genes that were inherited in the classical Mendelian way.[43] *Paramecium* feeds on bacteria, and attempts to define an artificial chemical culturing media to do biochemical genetic studies were unsuccessful. But *Paramecium* proved to be extraordinarily valuable for investigating the role of the cytoplasm in heredity. During the 1940s and 1950s, Sonneborn built up a prominent school of ciliate genetics at Indiana University.[44] *Paramecium* genetics greatly expanded genetics beyond Mendelism (see chapters 15 and 18).

Baker's yeast (*Saccharomyces cerevisiae*) proved to be extraordinarily useful for biochemical genetics. Ever since Pasteur's day, yeast had been one of the most frequent objects of biochemical studies because of its importance for the beer industry. But its life cycle was poorly understood. Genetic studies of baker's yeast were carried out by Otto Winge at the Carlsberg Laboratories in Copenhagen before the Second World War, and by Carl Lindegren working in Morgan's laboratory at Caltech.[45] After the war, yeast genetics was led by Lindegren and Sol Speigelman at Washington University, in St. Louis, and by Boris Ephrussi and collaborators at the Rothschild Institute for Physico-Chemical Biology, in Paris. Most investigations of yeast focused on genetic "anomalies" exhibiting non-Mendelian inheritance. Yeast later became the organism of choice for studies of the mitochondrial genome (see chapter 19).

Bacterial genetics, pioneered by Joshua Lederberg (b. 1925), entailed a change in geneticists' assumptions as well as techniques.[46] By the early 1940s it was clear that the genetic systems of animals, plants, and protists exhibited great similarity. Their nuclear hereditary machinery was organized into discrete chromosomes of definite size, shape, and genetic make-up. The chromosomes divide normally by mitosis, and at a certain point in the life cycle they undergo meiosis, which is accompanied by crossing over. This applied to plants, animals, and protists. But bacteria seemed to be forever excluded from this generalization. They had no recognizable nucleus and no chromosomes. They also seemed to lack sexual reproduction and consequently would exhibit evolutionary mechanisms altogether different from other organisms. In *The Evolution of Genetic Systems* (1939), the British cytogeneticist C. D. Darlington referred to "asexual bacteria without gene recombination"[47] and "genes which are still undifferentiated in viruses and bacteria." Julian Huxley summarized what everyone "knew" about bacteria in 1942:

> Bacteria (and *a fortiori* viruses if they can be considered to be true organisms), in spite of occasional reports of a sexual cycle, appear to be not only wholly asexual but pre-mitotic. Their hereditary constitution is not differentiated into specialized parts with different functions. They have no genes in the sense of accurately quantized portions of hereditary substances; and therefore they have no need for accurate division of the genetic system which is accomplished by mitosis. The entire organism appears to function as soma and germplasm, and evolution must be a matter of alteration in the reaction system as a whole. That occasional "mutations" occur we know, but there is no ground for supposing that they are similar in nature to those of higher organisms, nor since they are usually reversible according to conditions, that they play the same part in evolution. We must, in fact, expect that the processes of variation, and evolution in bacteria are quite different from the corresponding processes in multicellular organisms. But their secret has not yet been unraveled.[48]

Lederberg had developed a strong interest in genetics as a teenager, and in 1946, two years after entering medical school at Columbia College, he dropped out to pursue microbial genetics. He had become convinced that bacteria at times exhibited the phenomena of "sex," or, more specifically, the phenomenon of gene transfer between different strains, and he mapped out a plan to demonstrate it. Francis Ryan invited Lederberg into his laboratory at Columbia College. He would have remained there, but he soon learned that Tatum was interested in similar problems with bacteria and was then in the process of moving to Yale. Lederberg arrived at Yale in March 1946, and by early May he had experimentally demonstrated genetic recombination in *Escherichia coli*.[49]

Bacteria possessed genes arranged along structures similar to the chromosomes of flies, guinea pigs, or humans. In addition to a main circular "chromosome," bacteria possessed smaller rings of genes, which Lederberg named plasmids (see chapter 19). However, bacterial "sex" did not involve pairing of chromosomes or producing offspring. It was not linked to organismic reproduction per se, as it is in plants and animals. In bacterial conjugation, genetic material of the "male" plasmids (and sometimes bits of the main chromosome) is transferred to the "female" recipient, and some genes may recombine with the female's chromosome.

The very properties that had hitherto excluded bacteria from classical genetics made them extremely valuable to the emerging field of molecular biology. Small and fast-growing, bacteria double in number every thirty minutes, and one could grow populations of billions of microbes in a few cubic centimeters. Geneticists of the 1950s confirmed that bacteria also had other means besides sex for transferring genetic material from one organism to another. Viruses or bacteriophages could also act as vehicles to transfer genes between bacteria (a process known as transduction). Certain kinds of bacteria can also absorb and incorporate into their own chromosome the genes released by dead bacteria (transformation). These processes were important for identifying the genetic material as DNA (see chapter 16); their importance in evolution has only recently begun to be recognized (see chapters 18 and 19).

Since the 1920s, a number of scientists had had a hunch, promoted by H. J. Muller, that viruses were in all respects similar to genes, and that their investigation might prove important to understanding what a gene was and how it was duplicated.[50] Thought to occupy a place of their own between living and the nonliving entities, viruses consist primarily of a protein coat surrounding an inner core primarily of nucleic acid. They depend on host cells for all phases of their life cycle. The simplest viruses were the phages of bacteria. Some bacteria can harbor phages indefinitely. In others, phages destroy the bacteria as they replicate. After several minutes, the bacterial cell ruptures, releasing 100 to 200 progeny viruses that can infect other bacteria. The viruses that infect our cells operate in much the same way. During the 1940s and 1950s, a growing number of scientists turned to bacteriophages in order to elucidate the nature of the gene (see chapter 16).

After completing his Ph.D. in 1947, Lederberg moved to the University of Wisconsin, where he quickly emerged as one of the most important leaders in the genetics of microorganisms. His investigations had three general aims: to test the one-gene: one-enzyme hypothesis; to use viruses to study the nature of the gene (was it a nucleic acid or protein?); and to elucidate the nature of mutations in bacteria, especially drug resistance, a problem with practical importance for medicine and immunology and, theoretically, for fundamental problems in genetics and evolution. This nascent field of molecular biology was fostered by both government and private organizations in the United States, especially the National Institutes of Health, the National Science Foundation in the 1950s,[51] and the Rockefeller Foundation, which had begun earlier.

The Rockefeller Foundation

The Rockefeller Foundation played an instrumental role in establishing microbial genetics and integrating physicochemical approaches into genetics. Founded in 1913 and reorganized in 1928, the Rockefeller Foundation mastered the art of conducting a large program of research grants for individuals and projects.[52] In 1932, Warren Weaver became director of its natural science division. A former physicist, his principal aim was to develop those aspects of biology that seemed ripe for applying the techniques of physics and chemistry, a field that, in 1938, he named molecular biology.

Weaver and his associates played an active part in selecting research programs, initiating collaborations, and establishing research centers. The Rockefeller Foundation had partly funded research in Paris at the Rothschild Institute for Physico-Chemical Biology since the 1920s; it supported Ephrussi's visit to Morgan's laboratory, and it funded research at the Institute of Genetics in France during the 1950s.[53] The Rockefeller Foundation also helped build up the biology department at Indiana University from obscurity, eventually making it one of the major genetic centers in the world. It attracted talented students from around the country, including James D. Watson, who shared the Nobel Prize in 1962 with Francis Crick and Maurice Wilkins for discovery of the structure of DNA. By 1950, the biology department hosted at Indiana such leading geneticists as Sonneborn, Muller, the plant physiologist Ralph Cleland, the corn cytogeneticist Marcus Rhoades, and Salvador Luria (1912–1991), a refugee from fascist Italy, who focused on bacteria and their viruses (phages). Luria, together with Max Delbrück (1906–1981) and Alfred D. Hershey (1908–1997), received a Nobel Prize in 1969 for medicine and physiology "for discoveries concerning the replication mechanism and the genetic structure of viruses." Officials at the Rockefeller Foundation also claimed responsibility for training Tatum and teaming him up with Beadle. As Weaver wrote in 1941:

> A brilliant young geneticist, namely Beadle, has turned up one of the most important and exciting leads which has developed for a long time in the ultra-modern field lying between genetics and biochemistry. This discovery has resulted in major part from circumstances which we helped create; and also in major part from the fact that we put with Beadle several years ago a very able young biochemist [Tatum] whom we had specifically trained for this sort of work.[54]

The Rockefeller Foundation allowed recipients to investigate fundamental problems without immediate applied economic value.[55] Beadle's new techniques for investigating how the gene works at the biochemical level promised immediate solutions to practical problems in medicine: it could uncover new vitamins and amino acids. Beadle estimated that there were "several thousand genes" in the human genome; one gene would serve as a template for the manufacture of a specific protein, one would provide the mold for insulin, another that for pepsin, another that for albumin, and another that for the protein that forms antibodies (see also chapter 17).[56]

The importance of genetics for medicine seemed monumental. However, Beadle had no interest in patents or funding from the drug industry.[57] Grants from the Rockefeller Foundation freed him of all obligations other than to work hard and publish his results freely. In 1945, he moved his group from Stanford to the California Institute of Technology, where he succeeded Morgan as director of the biological division.[58] Beadle's group became one part of the research triumvirate that characterized much of the emerging field of molecular biology. In addition to Beadle's group, there was the phage group, led by Max Delbrück, which focused on the nature of the genes; and Linus Pauling and his collaborators, concentrating on the structure of macromolecules.

Delbrück had arrived at the California Institute of Technology in 1937 on a Rockefeller Foundation fellowship to work with Morgan, and after the war, he emerged as the leader of a band of young people who investigated viruses as "naked genes."[59] His interest in genes had been sparked by a lecture in 1932 by the Danish physicist Niels Bohr (1885–1962) who insisted that life was as an elementary fact beyond physicochemical explanation.[60] Delbrück did his biological work on the side while working as an assistant physicist at the Kaiser Wilhelm Institute for Chemistry in Berlin. In 1935 he coauthored a famous paper, "The Nature of Genetic

Mutations and the Structure of the Gene," in which he suggested that "the gene is a polymeric entity that arises by the repetition of identical atomic structures."[61]

Delbrück's ideas about the gene and the search for the secret of life were popularized by the Austrian Nobel Prize–winning theoretical quantum physicist Erwin Schrödinger in his famous book *What Is Life?* (1944). The book was written for the layperson, much of it was not original, and much of what was original was known to be erroneous even at the time.[62] Nonetheless, it was enough to arouse interest among young physicists, leading some of them, such as Seymour Benzer, Maurice Wilkins, and Gunther Stent, to turn to the study of the molecular biology of the gene.[63] Certainly, expectations were high that new physical laws might be discovered. But, after the Second World War, some were shocked by the military use made of atomic energy, while others were dissatisfied by the experimental direction in the field. As Stent remarked, physics suffered from a "general professional malaise."[64]

Linus Pauling (1901–1994) was arguably the most important chemist of the twentieth century.[65] His work, style, and successes were crucial for the development of research on the fine structure of large molecules such as proteins and nucleic acids, information that was very important for understanding their activity (see chapter 16). Collectively, the three groups led by Pauling, Beadle, and Delbrück were dedicated to solving some of the great problems of biology and biomedicine: the structure and nature of proteins; nucleic acids and other constituents of life; the structure of the gene and its mechanisms of inheritance; and the structure and properties of antibodies, enzymes, viruses, and bacteria. They presented an extensive research proposal, titled "The Fundamental Problems of Biology and Medicine," to the Rockefeller Foundation in 1946:

> In general our program involves an attempt to uncover basic principles rather than to attack specific practical problems. We want to determine the structure of genes and the mechanism of their action rather than to develop commercially profitable mutants; to obtain a fundamental understanding of viruses and of antibodies rather than to prepare an antiserum effective therapeutically against a particular disease; to learn the basis of the physiological activity of drugs in terms of their molecular structure rather than to find a new bacterio-static substance: but it would be expected that practical discoveries useful in specific fields would be made incidentally in the course of fundamental investigations, and these discoveries will not be ignored. It would in particular be hoped that in the course of the fundamental studies new ideas would be developed which would provide the basis for clinical research on such great medical problems as those presented by neoplastic and cardiovascular diseases.[66]

By 1953, there were about sixty Ph.D. candidates in the Division of Biology at Caltech and about seventy postdoctoral research fellows working in various branches of chemical biology. By that time, there was evidence that the molecular structure of the gene had been solved, with an implied system of gene replication. I shall turn to that work in chapter 16, but first I examine another, more turbulent aspect of genetics during the cold war.

15

Genetic Heresy and the Cold War

> Of the many previous attempts to demonstrate experimentally the inheritance of acquired characteristics, all have failed. In most cases, the attempts yielded negative results. When positive results were claimed, the work later proved to be fraudulent, indecisive, or incompetently performed; repetition with unobjectionable methods always failed to establish the claims. No wonder most geneticists consider the matter closed.
>
> . . . That there is a strong political and philosophical element in the controversy cannot be denied, as I shall later show. . . .
>
> However, we do find, in the work on cytoplasmic inheritance, evidence for the inheritance of acquired characters, but only when the characters belong to that very small class which is determined by migratory plasmagenes (or viruses) or when the characters occur in unicellular organisms. It is conceivable, but not yet demonstrated, that similar (but not identical) phenomena could occur in plants.
>
> —T. M. Sonneborn, 1950

MENDELIAN GENETICS WAS OUTLAWED in the Soviet Union during the cold war. It was denounced as abstract and idealist, and incompatible with Soviet science and dialectical materialism. Under legislation from the Ministry of Education, all Mendelian doctrines were systematically rooted out of schools, universities, and libraries. Eminent Soviet geneticists, cytologists, and evolutionists were subjected to condemnation and suppression as dangerous bourgeois reactionaries, removed from their positions, deprived of their laboratories, and, in some cases, imprisoned.

Led by Trofim Denisovich Lysenko (1898–1976) and supported by the Communist Party, antigeneticists in the Soviet Union denied that the basis of heredity was in some special substance such as chromosomal genes, and they asserted that evolutionary and agricultural advances were achieved through the inheritance of acquired characteristics. With the rhetorical aid of the philosopher I. I. Prezent,

Lysenko and his followers claimed that their beliefs were based on practical experience and on the philosophical system of dialectical materialism. They proclaimed that theirs was a proletarian science, a science for the people, and they condemned Mendelian genetics as academic and useless.[1]

Rumblings over the usefulness of Mendelian genetics had begun in the 1920s, when Bolsheviks called for a proletarian biology that, like every other science, should be based on dialectical materialism. Not much came out of such debates until 1927, when Lysenko was popularized in the press as a scientist close to his peasant roots, a young agronomist from the Ukraine, who had made a major discovery called vernalization. Some plant seeds need to be exposed to cold for extended periods in order to germinate. Lysenko used this as the basis for an agricultural technique that allowed winter crops to be obtained from summer planting by first soaking and chilling the germinated seed. He was not the first to discover this technique, but during the 1930s, he organized a boisterous campaign, making extravagant claims for it based on modest evidence.[2]

The impression that Lysenko had indeed achieved marvelous results came at a time of growing impatience for increased agricultural productivity. After two years of famine, the Ukrainian Commissariat of Agriculture instituted the widespread use of vernalization, and Lysenko moved to a newly created department for vernalization at the Institute of Genetics and Plant Breeding in Odessa. He subsequently advanced a vague theory to explain his technique, arguing that knowledge of the various stages of plant development would open the way for direct manipulation through control of the environment. Nature could be sculpted at will, and this, for Lysenko, meant that Mendelian genetics was groundless.

Lysenko's real practical success was difficult to assess because his methods were lacking in scientific rigor: he reported only successful results, and they were based on small samples, inaccurate records, and almost total absence of control groups. In the 1930s, he also incorporated the work of Ivan Vladimirovich Michurin (1855–1935) into his theory and made him out to be a hero of Soviet science.[3] Michurin was last in the line of an impoverished aristocratic family in central Russia who, during the first decades of the century, cultivated fruit trees and experimented with grafting and hybridization. Like many others of his generation, he believed that the environment exercised a direct adaptive influence on heredity. Lysenko embraced Michurin's ideas as part of his theory of heredity. Lysenkoism, born out of Soviet agronomic practice and based on the inheritance of acquired characteristics, denied a distinction between germ plasm and somatoplasm.

During the 1930s, Lysenko and his followers launched a massive attack on academic science, pressuring geneticists to abandon research that was unlikely to lead to immediate practical results for agriculture. Lysenko denigrated the esteemed plant geneticist Nikolai Ivanovich Vavilov (1887–1943), who, in 1940, was arrested and falsely tried for treason; he died in prison of malnutrition in 1943. He was not the only one. Ivan Ivanovich Schmalhausen, professor of evolution at Moscow University and director of the Institute for Evolutionary Morphology, was removed from his academic positions and his career and books were destroyed, because he supported the Mendelian gene theory.[4] Lysenko replaced Vavilov as director of the Institute of Genetics of the Academy of Sciences, became president of the Lenin Academy of Agricultural Sciences, and set out to purge science of Mendelian geneticists.

In 1948, the Communist Party and the Lenin Academy of Agricultural Sciences, backed by Joseph Stalin, pledged support for strengthening Lysenkoism. The Mendelian chromosome theory was condemned as "an antiscientific reactionary, idealistic-metaphysical trend, divorced from life and sterile in practice, in contrast to the Michurin trend, which represents the creative development of Darwin's teaching, and is a new and higher stage in the development of materialist biology."[5] Lysenkoists maintained their power and official government status after Stalin's death in 1953 until Krushchev was ousted in 1965. Far from making any practical breakthroughs, Lysenkoism set Soviet agriculture back a generation.[6]

By 1950, it was evident to many leading Western biologists, both communists and noncommunists, that Lysenkoists had made no real new discoveries. Western geneticists requested that Lysenkoists provide detailed descriptions of their technical procedures so that their reports could be tested by others. They asked them to repeat their experiments with controls demanded by critics and to publish their data so others could analyze them. Lysenkoists responded with arguments based on Soviet ideology and dialectical philosophy. In the West, Lysenkoism came to represent the breakdown of scientific professional freedom and political control over the beliefs of scientists and the direction of research. Historians and biologists have extended considerable effort to analyzing how Lysenkoism arose and maintained its power—how dogmatism, authoritarianism, and the abuse of state power can help create and sustain erroneous theories.[7]

Non-Darwinian Development

The lessons of what is called "the Lysenko affair" continue to ring loudly. One consequence has been a reinforcement of the view that Morganist doctrines were correct: that Mendelian gene changes were the only basis of evolution and heredity, and that there was no real evidence for the inheritance of acquired characteristics. But these views are incorrect. Indeed, it would be a profound error to think that all disagreements with the principles of Mendelian genetics were due to ideology, to frauds, or to institutional power that interfered with the progress of reason. Many embryologists who advocated "the cell as a whole" as the basis of heredity and development opposed neo-Darwinism, as did the famed British biologist Conrad H. Waddington (1905–1975), who developed a model of development that involved the "genetic assimilation of acquired characters."[8]

In the 1940s, Waddington attempted a rapprochement between genetics and embryology, calling the gap between them "so frequently lamented as one of the main flaws in the structure of biological theory." In keeping with the classical theory of epigenesis, he coined the word "epigenetics" for the study of the processes that operate between genotype and phenotype.[9] Throughout his career, Waddington insisted that population genetics and neo-Darwinism did not account for the "real guts" of evolution—"how you come to have horses and tigers."[10]

To offset the random effects of natural selection and to emphasize a two-way relationship between genotype and phenotype, he introduced the concept of canalization. In his now-famous geographical metaphor of the "epigenetic landscape," Waddington compared the development of an organism and its relationship to the environment to a ball rolling through valleys and hills. In the landscape, there are well-worn trails that represent the ontogenetic pathways a particular species has followed in the past. The landscape is active, like the shifting environment. It may jostle the ball off its path and force it to make a detour, but the ball pushes back to restore its internal energy balance and still reaches the same place: an individual of the species. However, Waddington suspected that the individual's development had subtly worn down a detour. If enough embryos are pressed by circumstances to make this detour in their own genetic structure, the detour may become a formal rerouting for the species. This is how he imagined genes and environment push against each other to create evolutionary change.

During the 1940s and 1950s, interest in mechanisms of inheritance other than chromosomes was also growing in the West: non-Mendelian, cytoplasmic heredity. And there was also new genetic evidence for environmentally directed adaptive hereditary changes through the cell cytoplasm. Interest in the cytoplasm had its roots in nineteenth-century embryology and cytology (see chapter 10). Recall that many experimental embryologists maintained that Mendelian genes were concerned only with relatively trivial adult traits and not with the fundamental organismic characteristics that distinguished higher taxonomic groups, and that the more plastic egg cytoplasm was largely responsible for development.

Plasmon to Plasmagenes

The idea that the cytoplasm may be largely responsible for macroevolution and for early development was widely discussed among paleontologists and embryologists in many countries. If true, the cytoplasm's effects would usually be detected only by crossing widely different groups. This would be difficult to test, because matings between distantly related animals were sterile. But it was possible in some plants. Between the two world wars, leading botanists in Germany, including Carl Correns and his student Fritz von Wettstein, Otto Renner, Friedrich Oehlkers, and later Peter Michaelis turned to investigate the role of the cytoplasm in heredity and to challenge what some called the "nuclear monopoly."[11]

In 1926, von Wettstein coined the term "plasmon" for the "genetic element of the cytoplasm," in contrast to what Hans Winkler in 1920 had dubbed the "genome": the whole collection of genes in the chromosomes.[12] The evidence for the plasmon was merged with that which Renner called plastidom: the genome of the chloroplasts, as detected by non-Mendelian inheritance of chlorophyll characters in *Oenothera*. In 1937, von Wettstein rejected the rigid distinction between nuclear and cytoplasmic affects and asserted that both genome and plasmon work together as a cooperative hereditary system:

> Growth and gastrulation, chlorophyll formation and pigmentation, hairiness and habitus, all of these traits are the product of the cooperation between the genome and the plasmon.
>
> One should therefore dispense with the entirely wrong opinion that race and species characteristics are determined by nuclear genes and more profound characteristics of organization (=traits of higher taxonomic groups) by the cytoplasm. This is basically wrong and shouldn't be discussed over and over again.
>
> Cooperation (between the plasmon and the genome) is the essential point.[13]

Leading American geneticists remained intransigent, however. The plasmon was a relatively vague concept and, as A. H. Sturtevant and George Beadle saw it in 1939, all so-called cytoplasmic traits could easily be explained by the delayed effect of the nucleus during the genesis of the egg.[14] The Harvard corn geneticist E. M. East further suggested that even the non-Mendelian inheritance of chlorophyll characters could be due to a cytoplasmic disease affecting the chloroplasts, a possibility Correns himself had first considered. "Weighing all the evidence for and against the plasmon," East concluded in 1934, "one is forced to the Scotch verdict 'not proven.'"[15]

After the Second World War, the debate arose anew with the emergence of microbial genetics and studies of gene action. The idea that the cytoplasm might control the fundamental traits of the organism was revitalized. Several leading European biologists, including Boris Ephrussi, André Lwoff, Philippe L'Héritier in France, and Jean Brachet in Belgium, combined their work with that of T. M. Sonneborn in the United States in support of the importance of cytoplasmic genetic entities in heredity, development, and evolution.[16] They linked their results with cytological studies of cytoplasmic bodies, long neglected by geneticists. Since the late nineteenth century, simple microscopic observations had suggested to cytologists that cells contain in their cytoplasm a variety of particles that perpetuated themselves from one cell generation to the next (see chapter 8). These included mitochondria, present in cells of all organisms with the exception of bacteria; chloroplasts, found in plants and many protists; centrioles, which in animal cells could be seen at the poles of the mitotic spindles; kinetosomes or "basal bodies," found at the base of cilia and flagella of many microbes and of ciliated and flagellated cells of animals. In addition, there were "infectious" genetic entities due to symbiotic bacteria and viruses.[17]

During the 1940s and 1950s, all these entities were discussed under the rubric of "plasmagenes."[18] Some geneticists such as Sol Speigelman initially developed the plasmagene concept as part of a theory of gene action, as the means for understanding the gene–enzyme relationship. Like de Vriesian theory of old, he postulated that genes (nucleoproteins) from the nucleus would migrate to the cytoplasm and thus determine cell structure and function.[19] But others, including Sonneborn, Ephrussi, Lwoff, and Brachet, envisaged plasmagenes as independent of Mendelian genes in terms of their origin and function.[20] Plasmagenes were generally thought to vary in size from microscopically visible particles down to submicroscopic particles of the same order of size as viruses and genes. Some of the larger plasmagenes, such as chloroplasts, were thought to contain smaller plasmagenes within them. The existence of plasmagenes not only contradicted the doctrines of Morganist geneticists about the almost exclusive role of nuclear genes in heredity, but it also contradicted Weismannian views about the noninheritance of acquired characteristics.

The Inheritance of Acquired Characteristics

Geneticists' doctrine of the noninheritance of acquired characteristics was based in part upon the distinction between germ plasm and somatoplasm. It was supported by observations that in animals, shortly after the egg is fertilized by the sperm, cells are set aside from which sex organs are formed. It was a zoocentric concept and did not apply at all to single-celled microbes (and even in animals, the gametes were not completely isolated from environmental forces that effect changes in the body.) It also did not apply to plants. Many or all parts of the plant body may give rise to germ cells or to a new individual. A vegetative cutting, grown by simply putting a twig into the soil, could produce flowers and seeds.

Genetic studies of cytoplasmic inheritance further showed that, at least in some cases, acquired characteristics were indeed hereditary.[21] First, there were cases of the inheritance of acquired microbes. "Hereditary symbiosis" will be discussed in more detail in chapter 19, but as one example, consider the famous work of Sonneborn and his colleagues on a "killer trait" in the ciliated protist *Paramecium aurelia*. This was the exemplar of the interaction of plasmagenes and the environment during the 1940s and 1950s. Some strains of *Paramecium* produce a poison called paramecin that kills *Paramecium* of certain other strains, called sensitives. This killer trait was due to cytoplasmic particles Sonneborn called kappa. These were thought to be either infectious viruses or (as is now recognized) symbiotic bacteria. Though once free-living, these symbionts were shown to be highly integrated within certain strains of *Paramecium:* those that possessed a certain nuclear gene K.

The relationships between kappa, *Paramecium*, and environmental conditions were also striking. The reproduction rate of *Paramecium* could be controlled by such environmental factors as nutrition and heat. When *Paramecium* is kept in a medium where reproduction is rapid, they tend to multiply more rapidly than kappa, until finally the organisms lose kappa and cease to be killers. The inherited transformation in response to the environmental agent is adaptive. The killer trait, not needed when conditions for reproduction are optimal, is accordingly lost. The inherited change is irreversible.

The same sort of mechanism operated in some insects, as exemplified in the 1940s by the infectious viruslike plasmagene called sigma, inherited through the egg cytoplasm of *Drosophila.* L'Héritier demonstrated that sigma could migrate from body cells to the germ cells, which then pass it on to later generations through the egg cytoplasm.[22] Plasmagenes that could migrate from soma to germ cells provided a possible mechanism for the inheritance of acquired characteristics in animals. As Sonneborn commented in 1950, "The work on kappa in *Paramecium* and other plasmagenes shows that acquired characteristics can be inherited if the characters fall in a certain sub-division of the non-Mendelian category."[23] Kappa and sigma were "migratory or infectious plasmagenes," but there was also evidence for environmentally directed adaptive hereditary changes in *Paramecium* due not to any genetic particle as such, but rather to the perpetuation of interlocking metabolic pathways.[24]

The behavior of kappa in *Paramecium* was analogous to the behavior of chloroplasts in the unicellular green algae, *Euglena mesnili*. When this alga is cultured in the dark, the rate of divisions of its chloroplasts is slowed down, a process that leads to a decrease in the average number of chloroplasts per organism. Eventually organisms arise that are devoid of chloroplasts. Chloroplasts, which are not needed in the dark, are lost, and the change is adaptive and hereditary. As J. B. S. Haldane saw it in 1954, this was "a perfect Lamarckian example of irreversible heritable loss of a function through disuse."[25] He believed that such directed changes might be possible in plants and animals as well. Haldane was one of several socialist biologists who quit the Communist Party because of Lysenkoism. But he thought it would be difficult to get a fair hearing for the inheritance of acquired characteristics in the context of the cold war. As he commented in 1954, "It is apparently vain to hope that the existence of such a series of organisms will prevent dogmatic assertions both as to the non-existence of this phenomenon and as to its universality."[26]

The experience of the Belgian embryologist Jean Brachet, who broke with the Communist Party after a meeting with Lysenko, offers testimony of Lysenko's own attitude about academic debates over cytoplasmic inheritance.[27] In 1949, Brachet was sent by the Party in Belgium to meet with Lysenko to try to evaluate the situation in the Soviet Union. Brachet suggested to Lysenko that plasmagenes were viruslike genetic particles (like the RNA-based microsomes he investigated) that might be able to invade the flowers from the somatic cells and thus could explain Lysenkoists' reports on the inheritance of acquired characteristics by grafting. He also offered an experimental test for the hypothesis: repeating the grafting experiments and inserting a membrane between the two parts of the graft combination, which would prevent viruses from getting through but would allow smaller nutrients to pass. In Lysenko's view, Brachet's suggestion showed that he came from a capitalist country, and it was out of the question for two reasons: there were no plant viruses in the Soviet Union, and such an experiment would be due to the pure curiosity of scientists who were not working for the people, and so it was a waste of time. It did not matter how it worked. Lysenkoists in the Soviet Union were interested in cytoplasmic inheritance only insofar as it supported their extreme claim that all of Morganist genetics was wrong. As I. I. Prezent put it in 1948:

> The cytogenetic system is crumbling. No wonder the Morganists are thinking up, in addition to the gene, all sorts of "plasmagenes," "plastidogenes," and similar terms designed to veil the complete theoretical and factual discomifture of Morganism. . . . Mendelism-Morganism has already betrayed its utter hollowness: it is also rotting from within, and nothing can save it.[28]

In the United States, the inheritance of acquired characteristics became a heated political issue, with Communism on the Left and McCarthyism on the extreme Right. As one journalist put it in 1950:

> The argument has long since ceased to be a scientific one, however. Not to accept Lysenko, lock, stock and gene, in the U.S.S.R. is close to heresy. To admit in the U.S. today to the possibility of some basis of fact in the Lysenko approach is tantamount to having subversive thoughts. And, unfortunately, science can never flourish without completely free inquiry, so that a proper evaluation of Lysenko will have to wait for the expiration of the Cold War.[29]

As president of the American Genetics Society in 1950, Sonneborn found himself in an awkward position. He had no sympathy for Marxism, and less for Lysenkoism, but ever since his days as a graduate student with H. S. Jennings at Johns Hopkins, he had had "a closer than distant relation with Lamarckism."[30] He also understood that there were biologists early in the century who had made experimental claims about the inheritance of acquired characteristics who were, like Lysenko and his followers, either frauds or incompetent, and some had come to a bad end.

The experiments of the Austrian Paul Kammerer represented the most notorious case before Lysenko.[31] Just before the First World War, Kammerer reported that he had induced special organs on the midwife toad, *Alytes obstetricans*. Most species of toads mate in the water, and the males use special pads on the forelimbs to grasp the female. The midwife toad was an exception. It had adapted to living on land and no longer possessed those nuptial pads. But Kammerer claimed to have produced males with pads by raising and breeding the midwife toad in moist conditions. He preserved them in a jar. After the war, it was found that the pads in the specimens were faked: they had been marked with India ink that someone had injected under the skin. Kammerer insisted that he was innocent and that an assistant must have made the injections in order to preserve the original induced marks after they had begun to fade. He subsequently shot himself, leaving the scientific community with the impression he was indeed guilty.

Genetic studies on plasmagenes did provide evidence for the inheritance of acquired characteristics, but Sonneborn, Ephrussi, and others emphasized that they did not support Lysenkoism; they did not refute the existence of Mendelian genes.[32] Nonetheless, other leading geneticists of the 1950s, including H. J. Muller and George Beadle, rejected cytoplasmically inherited particles as part of any general theory of heredity and evolution. Some apparent examples of cytoplasmic inheritance could be given alternative Mendelian interpretations; other examples, they argued, were symbionts or parasites, which they considered to be outside of genetics and of little significance to heredity and evolution.[33]

University Politics

During the 1950s, belief in the inheritance of acquired characteristics was targeted by American geneticists' initiatives to counteract communist propaganda and the threat to genetics and professional academic science.[34] This effort culminated in 1951 with the text *Genetics in the Twentieth Century*, a compilation of papers by twenty-six leading geneticists.[35] The authors were asked to emphasize the positive achievements of genetics, both theoretical and practical, and to "speak for the cause of the freedom of science" from political intervention. To further contradict Lysenkoists' assertions about the lack of practical value of Mendelism, geneticists pointed to dramatic increases in yields due to hybrid corn, a billion-dollar industry in the United States by the middle of the century.[36]

The controversy over Lysenkoism was especially complicated in France. In the country of Lamarck, the inheritance of acquired characteristics was a favored theory. Neo-Lamarckians, who were often associated with the political Right in France, also had high hopes for Lysenkoism insofar as it tended to show that Lamarck was right. On the political left, some leading biologists, such as Georges Tessier, director of the Centre National de la Recherche Scientifique (CNRS), were Communists, and some were also heroes of the resistance movement during the German occupation of France. In the midst of this was the Russian-born Boris Ephrussi (once an anarchist) and his collaborators, who worked on cytoplasmic inheritance of respiratory mutations in yeast (later associated with mitochondria).[37] Ephrussi had no sympathies for Lysenkoism. But as the leader of French genetics, which was focused on problems of genetic regulation and cytoplasmic heredity he found himself in an awkward situation during the Lysenko controversy. He felt obliged to make his views on chromosomal genes unambiguous in the preface of his little book on cytoplasmic inheritance of 1953: "I do not deny the general occurrence and significance of nuclear (Mendelian) heredity."[38] Nonetheless, he suggested, as had embryologists before him, that cytoplasmic inheritance might be involved in the fundamental aspects of heredity, whereas the nuclear genes might be largely concerned with relatively trivial traits.[39]

The real threat to genetics in France, as Ephrussi saw it, was not Lyskenkoism, but neo-Lamarckism, which had persisted in France since the 1880s, when Darwinian theory became identified solely with natural selection.[40] Many French biologists placed the inheritance of acquired characteristics at the center of an anti-Darwinian view of evolution. There were several reasons for this. First, the religious beliefs of many biologists and philosophers created an intellectual environment uncongenial to Darwinism and mechanistic biology. The vitalist philosopher Henri Bergson's book *Creative Evolution* was widely read by French biologists prior to the Second World War.[41] Spiritual overtones resounded in the writings of several leading French neo-Lamarckians throughout the twentieth century.

Neo-Lamarckism in France also had a nationalistic component. The defeat of France in the 1870 war with Bismarck's Prussia brought with it a patriotic tendency of French biologists to encourage support of their own theories rather than to accept those of other countries, especially those of Germany. Leading French biologists of the 1920s and 1930s trivialized the significance of genes, insisting with many embryologists that Mendelism did not deal with the most fundamental properties of the organism but rather with superficial traits.[42]

The neo-Lamarckian opposition to genetics was coupled with a more pervasive problem in Europe: the hierarchy of bureaucratic authority within universities and the centralized structure of the university systems, which greatly impeded the development of new specialties. Historians and sociologists have long compared the European university structure with that of the United States.[43] Their analysis usually begins with Germany. Whereas in the typical American university department, professors had more or less equal bureaucratic authority and research autonomy to compete for funds and students, the institute of the typical German university was effectively the feudal domain of a single chair professor, who possessed the power to allocate facilities and funds to junior faculty and thus dictate the direction of research in his institute.[44] The conservative interests of the German professoriate were strengthened by the ministries of culture of the German states, which were reluctant to allocate the additional funds required for a new chair in a new specialty.

The first chair of genetics in a German university was not founded until 1946.[45] Genetics was generally taught in institutes of botany or zoology, where it often remained subordinate to the broader aims of those traditional fields. The Kaiser Wilhelm Institutes (later renamed the Max Planck Institutes), founded in 1913, compensated for the conservative effects of the university. In 1914, Carl Correns was appointed director of the Kaiser Wilhelm Institute for Biology in Berlin-Dahlem, which was structured not around specialties per se, but around four men, each with his own division: Correns (botany and genetics), Richard Goldschmidt (zoology and physiological genetics), Hans Spemann (embryology), and Max Hartmann (protozoology and general biology).

In no country were the conservative effects of the centralized university system more obtrusive than in France. French science, which had flourished in the eighteenth century, had become stifled by the bureaucratic and centralized structure of the university system constructed by the Napoleonic administration. Terrance Clark described the university system: "The Napoleonic structure was rigidly hierarchical. It was a mixture of an ecclesiastical control of ideas, government bureaucracy, and the military style of the emperor."[46] The procedure for university innovation in France was to convey ideas to the government Ministry of Education, which legislated on the examinations, hours of each class, and method of teaching throughout the state-run university system. Any piecemeal change was formally resisted. Paris was at the center, and Sorbonne professors advised the ministry about exams and promotions in their fields. Thus, the control of scientific research was left in the hands of a few individual chair holders who fostered the development of their own interests and thus induced intellectual conformity. It was not until the student protests of 1968 that the power of the professoriate in the French university was diminished somewhat.

The first chairs of genetics in France were finally established in 1945 for Boris Ephrussi and Georges Tessier. But even ten years after it was established, Ephrussi's chair of genetics possessed neither laboratories nor workroom nor library.[47] Genetics was not the only underdeveloped discipline. By 1957, there were no chairs of experimental embryology, nor of microbiology in the Faculty of Science, and though there was a chair of biochemistry, the laboratory of biochemistry was located at the Pasteur Institute. Genetic research also developed outside the university. André Lwoff, Jacques Monod, and François Jacob worked on bacteria and its phages at the Pasteur Institute in Paris. Ephrussi and his team worked about twenty kilometers outside of Paris at newly founded Institute for Genetics at Gif-sur-Yvette, which was under the auspices of the CNRS.[48]

Morgan's Smile

For those who investigated non-Mendelian heredity, its chief theoretical importance was not for the inheritance of acquired characteristics, but for resolving the paradox of cellular differentiation. How could cells become biochemically and morphologically different and yet possess the same chromosomal genes? The extent to which this issue had polarized geneticists and embryologists is illustrated by Ephrussi's own recollection about a conversation with T. H. Morgan in the summer of 1934 at Woods Hole. Morgan's book *Embryology and Genetics* had just come off the press, and he gave a copy to Ephrussi with the request that he offer his frank opinion about it.[49] Ephrussi returned a few days later and said that he found the book very interesting, but the title was misleading because he did not try to bridge the gap between embryology and genetics as he had promised in the title. As Ephrussi remembered, "Morgan looked at me with a smile and said, 'You think the title is misleading! What is the title?' 'Embryology and Genetics,' I said. 'Well,' he said, 'is not there some embryology and some genetics?'"[50]

Embryologists had shown that differentiation was dependent upon differences in the cytoplasm, as well as upon the relationship between cytoplasm and the environment. Studies in tissue culture had shown that many of the directed changes in cells during the course of ontogeny were hereditary: epithelial cells give rise to epithelial cells, fibroblast cells to fibroblast cells. As Ephrussi put it 1951: "*Unless development involves a rather unlikely process of orderly and directed gene mutation, the differential must have its seat in the cytoplasm.*"[51] Cytoplasmic inheritance offered a bridge between genetics and embryology. It could account for cell heredity—the fact that somatic cells of a given type would give rise, when they divide, to more cells of the same type. Cytoplasmic inheritance in yeast and *Paramecium* was comparable to inherited somatic cell changes in the individual animal or plant, which affect its phenotype and in some cases might also be inherited sexually.

By being differentially transmitted during cell division under the influence of the environment, plasmagenes offered one plausible mechanism for cell heredity. But cytoplasmic geneticists were more eclectic; they offered several mechanisms and emphasized that nucleus and cytoplasm also affect each other's activity. Some cases of cytoplasmic heredity might be due to self-perpetuating metabolic states—a system of cross-reacting and inhibiting chemical reactions leading to hereditary steady states.[52] One of the functions of the cytoplasmic part of the genetic system, then, would be to regulate which nuclear genes come to expression. The inverse relationship was also evident: petite mutations affecting respiration in yeast showed that the nucleus could activate cytoplasmic genetic factors.[53]

Plasmagene theorists also proposed a third kind of heredity based on the structural properties of cells. The cell is not "a bag of enzymes." It is organized. And this orderliness, they argued, was not determined or built up anew in each cell generation by genes and/or plasmagenes. It was likely that another genetic role of the cytoplasm was to provide a self-perpetuating structural organization or "molecular pattern." This inherited structural organization would exist at two levels: one for structures such as mitochondria and chloroplasts, and another for the structural organization of the cell as a whole, which seemed to involve "the fibrous ground substance of the cytoplasm." In 1951, Sonneborn phrased the problem in the following terms:

> Perhaps it will be objected that there are some self-duplicating cytoplasmic elements which the nucleus cannot make. Then suppose these too can be cultivated *in vitro*. Is anyone willing to believe that, if all such self-duplicating components of the cell were thrown together in a test tube in the proper proportions with adequate food for their multiplication, a *Chilomonas* cell or any cell at all would result? Although the whole picture is admittedly imaginary, it makes the nature of the problem sharp and clear. If cells cannot be reconstituted in the way suggested, then it seems to me we are forced to admit that the molecular and particulate arrangement of the cellular materials, their organization into a working system, is itself a part of the genetic system of the cell.[54]

Ephrussi framed the same problem in traditional embryological terms: cytoplasmic mechanisms based on plasmagenes and their relationships with genes were only instruments of cellular differentiation: *"Development is an orderly process: it follows a 'plan'* which dictates when and where the instruments of differentiation come into action."[55] Experimental embryologists had revealed the rough outline of this plan in the cytoplasm of the undivided egg (see chapter 11). As Ephrussi saw it, "*the fundamental anisotropy of the egg cytoplasm itself has a genetic basis.*"[56]

The organization of the cell was a genetic problem for the future (see chapter 17), but some observations pertaining to this problem had already been assembled in 1950 by André Lwoff (1902–1994) in a small book titled *Problems of Morphogenesis in Ciliates: The Kinetosome in Development, Reproduction, and Evolution*. Lwoff's aim was to show how studies of kinetosomes in ciliates visibly confirmed two theoretical conclusions of embryologists: that cellular differentiation involves cytoplasmic genetic particles, and that the fundamental control of differentiation operates under the influence of a "morphogenetic field."[57] "Cytoplasm is not just a collection of enzymes or a plastic and complaisant receptor passively submitted to the dictatorship of genes,"[58] he wrote; "it is a very differentiated system with cortex, mitochondria, kinetosomes, and chloroplasts, each endowed with genetic continuity."[59]

Kinetosomes are microscopically visible granules located at the base of each cilium of ciliated protozoa and of the ciliated cells of animals. Kinetosomes were always formed from preexisting kinetosomes, Lwoff argued, and they multiplied and were organized into orderly chains or rows of kinetosomes (kineties) along the cell surface or cortex. As he saw it, "the morphogenesis of a ciliate is essentially the multiplication, distribution, and organization of populations of kinetosomes and of the organelles which are the result of their activity."[60] Like the killer particles of *Paramecium* and the viruses of plants, he emphasized that kinetosomes were sensitive to the environment of the "host" in which they lived. The metabolism of the cell, light, and temperature could cause variations in the relative multiplication rates of kinetosomes in a way similar to the behavior of chloroplasts in *Euglena*.

Kinetosomes were also endowed with another special property, which embryologists called prospective potencies: they were pluripotent and turned out different structures and systems according to their position in the cell and the phase of the cell cycle. They produced various products: cilia for cell motility, trichocysts (cylindrical rods that elongate toward the inside of some cells), and centrioles, which were concerned with cell division.[61] Thus Lwoff argued that the kinetosome was a model for visible plasmagenes.[62]

Each kinetosome formed part of a complex called a ciliary unit—a sophisticated structure that also included a cilium, a variety of subcortical fibers, and specialized membranes. These ciliary units were arranged along the cortex in a repeating pattern that was remarkably constant and was reproduced faithfully through a regular sequence of events during growth and fission.[63] What was responsible for this orderly pattern? According to Lwoff, this power of organization did not stem from any inherent properties of kinetosomes themselves. Kinetosomes did not control their own destinies; they did not "command," they "obeyed." Like the cells of multicellular organisms, the fate of kinetosomes in ciliates was controlled by "some mysterious and powerful field of forces."[64]

In Lwoff's view, the cell cortex had the properties of a morphogenetic field.[65] It provided the spatial conditions for the organization of dispersed molecules in the cell. And it was entirely reasonable to compare morphogenesis in ciliates with animals.[66] The importance of cell surface had been stressed by several embryologists. The famed African American biologist Ernest Everett Just went so far as to consider that "in the entire animal kingdom, with the exception of mammals, the embryo arises from the egg surface."[67] Systematic genetic investigations of cell structure were begun in the 1960s (see chapter 17).

The main aim of microbial genetics of the 1950s was not on cell structure, morphogenesis, or hereditary symbiosis, but rather on the nature of the gene and its relations with proteins. While nucleocentric geneticists emphasized that the gene may be protein, some cytoplasmic geneticists, such as Ephrussi, emphasized that it was possible that nuclear genes were responsible only for activating proteins and did not play a role in determining their primary structure.[68] Jean Brachet suggested that RNA-based structures in the cytoplasm (microsomes, later renamed ribosomes) associated with protein synthesis were viruslike plasmagenes. Studies of plant viruses indicated that the hereditary material could be either DNA or RNA; and RNA-based plasmagenes fit well with embryologists' preference for the cytoplasm as the basis of cellular differentiation.[69] By the end of the decade, all these molecular conceptions had been refuted as molecular biologists reordered the relationship between the gene, RNA, and protein.

IV

MOLECULAR BIOLOGY AND ORGANISMIC COMPLEXITY

16

Conceiving a Master Molecule

> To do molecular biology, it was no longer enough to use one technique, to investigate all the parameters of one particular phenomenon. It became necessary to exploit all the means available in order to define the architecture of the compounds involved and the nature of the relations. . . . The organization of a macromolecule, the "message" formed by the arrangement of chemical patterns along a polymer chain, held the memory of heredity.
>
> —François Jacob, 1976

THE DISCOVERY OF DNA as the basis of the gene, how it is reproduced, and how it affects protein synthesis was the pinnacle of the life sciences in the twentieth century. It is said to have deep philosophical implications affecting our innermost understanding of ourselves and the nature of all of life, and it has spawned promises of medical panaceas and agricultural advances. Gene splicing, recombinant DNA, transgenic organisms (or genetically modified plants and animals), and the patenting of genes from microbes to humans all have roots in fundamental discoveries of molecular biology of the 1950s and 1960s. The realities and prospects of biotechnology entail ethical, legal, social, and political issues no citizen can afford to ignore.

The molecular approach to the gene involved a merger of microbiology and genetics with three techniques from physics and chemistry: radioisotopes were used to help identify DNA as the basis of the gene; X-ray crystallography was used to reveal the three-dimensional structure of proteins and of DNA; and chromatography was used to analyze the composition of DNA and proteins. Molecular biology rested on the premise that the functions of the cell are based on the configuration of its macromolecules, and it involved new analogies from communication technology and a whole new set of doctrines.

DNA or Protein?

That DNA was the basis of the gene was not obvious.[1] Retrospectively, the trail begins with investigations in the 1860s and 1870s of the chemical constitution of the cell nucleus by the Swiss chemist Johann Friedrich Miescher. Using fresh pus cells obtained from the dressings of surgical wounds, supplied daily from the Tübingen surgical clinic, he found a substance that contained nitrogen and was rich in phosphorus. He called it nuclein, and his student Richard Altmann renamed it nucleic acid (see chapter 8). In 1870, Miescher returned to his native Basel and in the waters of the Rhine found a more pleasant source of nuclear material in the sperm of salmon for which the river was then famous. "A knowledge of the relationships between nuclear materials, proteins, and their immediate products of metabolism," he wrote prophetically in 1871, "will gradually help to raise the curtain which at present so completely veils the inner process of cellular growth."[2]

By the 1920s, chemists led by Phoebus Levene and Walter Jacobs at the Rockefeller Institute showed that there were two fundamentally different kinds of nucleic acids: one, now called ribonucleic acid, or RNA, found in abundance in the cell cytoplasm, and the other, desoxyribonucleic acid, or DNA, found in abundance in the cell nucleus. Both comprise three main constituents: phosphoric acid, a five-carbon sugar (in RNA this is ribose, whereas in DNA deoxyribose is the five-carbon sugar), and four nitrogenous bases (RNA contains adenine, guanine, cytosine, and uracil; DNA does not contain uracil, but contains thymine instead). The base, sugar, and phosphate, are linked together to form a unit called a nucleotide, later identified by the letters for the bases: AGCU for RNA, and AGCT for DNA. By the 1930s, biochemists showed that nucleic acid molecules contained many of such nucleotides linked together.

The substance of the gene would have to possess a high degree of complexity to accommodate the enormous diversity of life. The available chemical data ruled out DNA in this regard. To Levene and his colleagues, it seemed to be a monotonously uniform macromolecule, which, like other monotonous polymers, such as starch, did not vary much between organisms. They championed what was called the tetranucleotide hypothesis, according to which DNA was made up of repeated sets of the four nucleotides, and following one another in a fixed and orderly manner, for example: AGCT AGCT AGCT. . . . Thus, on the basis of the best evidence available, it was believed that equal amounts of AGCT were present in the DNA of each organism.[3]

DNA was in the nucleus, but proteins were also components of chromosomes. They were the more likely candidates for the gene. Complex and diverse, they were the most important molecules in the cell. Biochemists of the early 1940s had shown that each protein was made up of chains of about twenty essential amino acids, from a few hundred to over a thousand. Proteins were also known to differ in their amounts of various amino acids—a particular protein, such as human hemoglobin, for example, has a well-defined sequence of amino acids. When the theoretical physicist Erwin Schrödinger wrote *What Is Life?* in 1944, enticing physicists into the field, it was generally accepted that genes were a special kind of protein.

Transformation and Transduction

The first experimental evidence that DNA was actually the stuff of genes did not come from chemical studies, but rather from studies of transformations in bacteria and from medical investigations of disease. They began with the now-legendary research of the microbiologist Frederick Griffith (1881–1941) in London on a certain type of pneumonia-causing bacterium (*Diplococcus pneumoniae*), commonly called pneumococcus.

Griffith was interested in the effects of different strains of pneumococcus on humans. In 1928, he reported that an extract of one pneumococcus strain could actually alter the character of a different strain.[4] One strain grew a smooth, sugarlike pellicle over itself; he called it the S (smooth) form. The other strain produced no smooth pellicle; its colonies were rough. He called it the R form. When Griffith injected the S form bacteria into mice, it caused pneumonia, whereas the R form caused no ill effects on mice. To find out why, he killed the virulent S form of bacteria by boiling them, then he mixed them with the live R form. When this combination of live and dead forms was injected into mice, the mice still contracted pneumonia. He then collected and analyzed blood samples from the mice; they contained the live S form of bacteria that had caused the disease. He concluded that the R form had actually been transformed into the virulent S. In other words, the R form had acquired something from the dead S form that transformed it into the virulent type.

$$\text{Dead S} + \text{Living R} \rightarrow \text{Living S}$$

What was the transforming factor the R form acquired to make it virulent?

Griffith was killed in his London laboratory during a German air attack in 1941. That year his transformation experiments were taken up by Oswald Avery (1877–1955), a research physician and bacteriologist at the Rockefeller Institute for Medical Research commissioned to study the biochemistry of pneumonia.[5] Avery's attempt to identify the transforming substance in the S strain culminated in 1944 with a now-famous paper he wrote with his collaborators, Colin MacLeod and Maclyn McCarty.[6] They exposed the R strain to purified protein, RNA, sugar, and DNA extracted from the S strain. Only the bacteria exposed to DNA were transformed. To ensure that the transforming principle was indeed purified DNA, they eliminated protein, RNA, and sugar by subjecting them to specific enzymes that chopped them up. When these substances were chopped apart, the R form was still transformed into the S form. However, when an enzyme that split DNA apart was introduced into the chemical extract, the transformation no longer occurred. They reasoned, therefore, that the active material was "a highly polymerized and viscous form of sodium desoxyribonucleate," that is, a form of DNA. As Avery wrote in his famous letter to his brother Roy on 13 May 1943:

> If we are right and of course that is not yet proven, then it means that nucleic acids are not merely structurally important but functionally active substances in determining the biochemical activities and specific characters of cells and that by means of a known chemical substance it is possible to induce predictable and hereditary changes in cells. This is something that has long been the dream of geneticists. . . .
>
> Sounds like a virus—may be a gene.[7]

Their results were inconclusive. Other interpretations were possible and indeed seemed more likely, given the chemists' conception of DNA as a monotonous molecule, a tetranucleotide that lacked diversity. Some leading geneticists suspected that the transformation of the R form to the virulent S form was a case of environmentally directed genetic modification by some kind of specific mutagen—not by the acquisition of DNA. Theodosius Dobzhansky commented that "we are dealing with authentic cases of induction of specific mutations by specific treatments—a feat which geneticists have vainly tried to accomplish in higher organisms."[8] George Beadle also referred to the phenomenon as a "first success in transmuting genes in predetermined ways."[9]

There was still another reason why Avery, MacLeod, and McCarty's conclusions about DNA as the basis of the gene could not be taken as proven. When they published their paper in 1944, it was still not certain that bacteria actually had genes, as did plants and animals. The previous year, Salvador Luria and Max Delbrück had conducted experiments that supported the notion that at least some bacterial mutations seemed to behave like changes in genes.[10] The results of Avery and his colleagues in New York were made known to Luria and Delbrück immediately through the above-mentioned letter to Avery's brother Roy.[11] The pneumococcus results induced Joshua Lederberg to take the next logical step to show genetic recombination—that bacteria did indeed have genes like plants and animals (see chapter 14).[12] For Lederberg, the pneumococcus transformations were formally analogous to, but not the same as, bacteria viruses (or phages), which were able to induce similar changes by transporting DNA from one bacterium to another. In 1952, he and Norton Zinder coined the term "transduction" for this phenomenon when it resulted from viral infection.[13] The only difference between the hereditary material of viral infection and the pneumococcus transformations was that the former is encapsulated in a protein coat, the other, naked DNA: the gene (see chapter 19).

The next step in demonstrating that DNA, not protein, was the basis of the gene came with the use of radioactive isotopes. A radioactive element emits radiation that can be detected and followed wherever it occurs in an organism. The use of such isotopes made it possible to follow the tangled process of "intermediary metabolism"—to follow the step-by-step elaboration of small molecules and how they are strung together into larger ones. In 1952, Alfred Hershey and Martha Chase reported their celebrated experiments with radioactive bacteriophages, which pointed to DNA.[14] They used a T2 phage—a type of virus that infects the common bacterium *Escherichia coli*, a normal inhabitant of the intestines of animals.

First, Hershey and Chase grew bacteria on culture media containing radioactive isotopes of either phosphorus ^{32}P or sulfur ^{35}S. DNA contains phosphorus but not sulfur, and most proteins contain sulfur but usually not phosphorus. Thus, bacteria grown on these media would incorporate the respective radioactive isotopes in their DNA and proteins. Then they infected the bacteria with phages. Those phages that were propagated from the bacteria grown in ^{32}P contained radioactive DNA. Those phages propagated from bacteria labeled ^{35}S contained radioactive proteins. The two types of radioactive phages were then used to infect unlabeled bacteria. The ^{32}P- labeled DNA was found inside the infected bacteria. The ^{35}S-labeled protein was left outside on the bacteria's surface. These results convinced many geneticists that the gene is DNA, not protein. And by that time there was also new supporting data from chemistry itself, from chromatography, which showed that DNA was a complex and variable enough molecule after all.

Chromatography

Before chromatography was used by chemists, it had been developed by botanists of the early twentieth century, who discovered a way of purifying and isolating various plant pigments. They poured plant extracts through long columns of calcium carbonate, which they then washed with various solvents. The pigments remained on the columns in different zones and could be studied separately. Later they used sheets of filter paper onto which the chemical compounds could be poured out, each moving at its own characteristic velocity, and thus be separated and analyzed. This technique allowed researchers to distinguish qualitatively and quantitatively between very closely related compounds. The simplicity and efficiency of this technique transformed the investigation of the constituents of macromolecules.

Chromatography allowed biochemists to investigate the actual order of the amino acids in a protein—the way they were arranged along a giant molecule. By using specific enzymes that split a protein molecule at different points, they were able to collect fragments containing several amino acids. They could then cut each fragment into smaller pieces and investigate the composition of each. Reconstructing the order of the amino acids was then a matter of marking the relative position of the fragments and pieces, and fitting them together to form the original design as one would a jigsaw puzzle.

The success of chromatography for the analysis of proteins led the German emigré Erwin Chargaff (1905–2002) and his co-workers at Columbia University to apply it to nucleic acids in the late 1940s. Their comparisons of nucleic acids in a number of organisms from bacteria to humans indicated that the nucleotide base composition (AGCT) of DNA differed from one kind of organism to another. In 1950, Chargaff discovered another important aspect of the structure of DNA: there was a characteristic symmetry in the base composition of all DNAs.[15] The amount of adenine is always nearly equal to the amount of thymine (A = T), and the amount of guanine closely followed the amount of cytosine (G = C). These regularities would provide crucial clues for unraveling the three-dimensional structure of DNA.

Chromatography gave a one-dimensional picture of the structure of a macromolecule. It was the technique for determining the primary structure of a macromolecule—the sequence of amino acids along the polypeptide chain of a protein, for example. An enzyme's three-dimensional structure (the way it is folded up) is actually responsible for its specific catalytic action. And to study that structure, X-ray diffraction was required.

X-Ray Crystallography

When a beam of X-rays is directed at a substance whose atoms are arranged in an orderly fashion, the X-rays are diverted off the regular spacings of atoms to form regular patterns. A beam of X-rays passing through a crystal, for example, will form a beautifully symmetrical pattern of dots radiating from a central spot. By measuring the distances between these groups of dots and the angles they make, crystallographers of the early twentieth century were able to calculate the relative position of the atoms within the crystal. The same technique was subsequently applied to macromolecules. However, the X-ray photographs of macromolecules are not quite as orderly as those of a crystal: the diffraction pattern is fuzzier and harder to interpret. The technique was difficult and required considerable expertise.

The use of X-ray diffraction patterns to discern molecular structure was invented in 1912 by two Cambridge scientists, W. H. Bragg (1862–1942) and W. L. Bragg (1890–1971), father and son, who were awarded the Nobel Prize in 1915. It was applied first to simple inorganic salts and then to ever more complex organic molecules. X-ray analyses of proteins got underway in the 1930s through the efforts of two students of the Braggs, W. T. Astbury and J. D. Bernal. In the 1940s, Astbury also took some X-ray diffraction photographs of DNA, but they were of insufficient quality to reveal much detail of fine structure.

During the early 1950s, three teams were interested in the X-ray crystallographic analysis of DNA: Linus Pauling and his group at the California Institute of Technology, Maurice Wilkins and Rosalind Franklin at Kings College London, and James Watson and Francis Crick at Cambridge. Pauling's work touched on many aspects of chemistry and medicine. He wrote one of the most influential chemistry books of the century, *The Nature of the Chemical Bond* (1939). His interest in the structure of proteins led him to suggest in the 1940s that sickle-cell anemia was due to abnormality of the hemoglobin molecule, which he and his co-workers demonstrated in 1948, after three years of experimentation, in their classic paper "Sickle-Cell Anemia: A Molecular Disease."[16] Pauling's biggest biological success came from his proposal about the general structure of protein molecules. In 1951, he and his co-workers demonstrated that polypeptide chains in such proteins as collagen are arranged in helixes held together by hydrogen bonds. He and his longtime collaborator Robert Corey recognized the helical shape to be a basic configuration of protein molecules in many life-forms. Pauling called it the alpha helix.[17] For this and for a great deal of other work on the bonds between atoms, Pauling was awarded the Nobel Prize in Chemistry in 1954. He was also well known for his opposition to nuclear weapons and his struggles for peace, for which he was awarded a Nobel Peace Prize in 1962.[18]

Before Pauling's work on the three-dimensional study of proteins, it was generally assumed that proteins were too large and too complex to study. Pauling's success was due in part to his novel approach. He used guesswork and model building in addition to the analytic procedures of more conventional crystallographers. During the 1940s, he surmised that it ought to be possible to deduce the structure of polypeptide chains from knowledge of the exact spatial conformation of the peptide bond. He therefore concentrated his X-ray crystallographic analysis on determining the lengths and angles of bonds that link the backbone atoms of amino acids. Using these data, he constructed a model made of balls of light plastic of different colors to illustrate the configuration of protein molecules, and he announced his discovery to a dazzled audience in Copenhagen in 1951. His phenomenal success with protein encouraged him to apply the same techniques to DNA. In 1953, he and Corey proposed a three-dimensional model of DNA that was perceived to be wrong the moment it was published.[19]

In the early 1950s, Rosalind Franklin (1920–1958) accomplished the major technical feat of obtaining an X-ray diffraction photograph of DNA that showed a wealth of detail. In the winter of 1952–53, this picture was shown to Watson (b. 1928), a postdoctoral fellow at Cambridge, and Crick (b. 1916), a Ph.D. student. Crick, fascinated by the line between the living and the nonliving, migrated from physics into chemistry and biology. Watson, twelve years his junior, had graduated from the University of Chicago when he was nineteen years old, and at twenty-two he completed his doctorate from Indiana University, where he studied genetics and bacteriophage. At the recommendation of his thesis director, Salvador Luria, Watson went to Copenhagen in 1950 to learn biochemistry and study the fate of the DNA of infecting viruses. He saw the first vague, ghostly image of a DNA molecule rendered by X-ray crystallography at a conference in Naples in May 1951, where he met Maurice Wilkins. Subsequently, Luria and John Kendrew arranged for Watson to work at the Cavendish Laboratory at Cambridge University in early October 1952, where he met Crick and discovered their common interest in discerning the structure of DNA. They reported the correct structure of DNA as a double helix in the spring of 1953.

The Watson and Crick collaboration is one of the most famous of its kind; that between Wilkins (b.1916) and Franklin is one of the most infamous. Franklin's strained relationships with her colleagues in a sexist workplace have been the subject of many commentaries and studies.[20] She died of cancer in 1958 at age thirty-seven. In 1962, the Nobel Prize was awarded to Watson, Crick, and Wilkins. If Franklin had lived, it would surely have been awarded to her instead of Wilkins.

Watson and Crick had already considered a variety of possible helical structures, but because of the poor quality of the X-ray photographs from which they had been working, they were unable to reach definitive conclusions—that is, until Wilkins showed Watson one of Franklin's best photos.

From Franklin's photograph Watson and Crick were able to conclude that the DNA polynucleotide chain had the form of a regular helix, that the helix has a diameter of 20 angstroms (Å), and that the helix makes one complete turn every 34 Å along its length, and therefore, since the internucleotide distance is 3.4 Å, it must contain a stack of ten nucleotides (bases) per turn. Considering the known density of the DNA molecule, they concluded that the helix most likely contained two polynucleotide chains. But the X-rays in themselves did not directly point to the double helix; as Crick later commented, they "provided only half the required data. For this reason a good model is worth its weight in gold."[21]

To discern the double helix, Watson and Crick used Pauling's unorthodox approach of model building. They toyed with large plastic models and cardboard cut-out representations of the four bases, A, T, G, and C. Watson fit the final piece into place when he realized that an adenine-thymine pair held together by two hydrogen bonds was identical in shape to a guanine-cytosine pair. These pairs of bases could therefore serve as the rungs on the twisted ladder of DNA. The complementarity between A and T and between C and G would account for Chargaff's equivalence rule: A = T and G = C. That insight of base pairing led them to understand that the whole DNA molecule is self-complementary. Every DNA molecule would carry two complete sets of information, albeit written in a complementary notation. So Watson and Crick built a regular helix composed of two nucleotide chains that could contain any arbitrary sequence of nucleotide basis, so as to be diverse enough to account for hereditary differences. The two chains would be held together by hydrogen bonds between each of the pairs of bases.

A mechanism for copying DNA could be immediately envisaged from this complementarity. Cleave or unzip the hydrogen bond between base pairs that hold strains together, and each strand would act as a template for the formation of a complementary strand. A single strand of the genetic alphabet, say CTA, is paired rung by rung with a complementary strand, GAT. When the helix unzips, the complementary strand becomes a template; its GAT bases attract bases that amount to a carbon copy of the original strand, CTA. Imitating the rhetorical style of Pauling, Watson and Crick concluded their 1953 article in *Nature* with one of the most famous understatements in the literature of science: "it has not escaped our notice that the specific pairing we have postulated immediately suggests a possible copying mechanism for the genetic material."[22]

Digital DNA

How did DNA specify the structure of a protein, the most important cellular entity, if that is what it did? Biochemical genetic research in *Neurospora* and other microbes since the 1940s had shown a relationship between gene mutation and enzymatic protein function. However, those experiments did not show how changes in a gene could result in changes in a protein: they did not and could not show a physical link connecting gene and protein.

The implications of the double helix were the main topic of interest at the Cold Spring Harbor Symposium of 1953. As Gunther Stent recalls, "No one who listened to Watson's lecture at that meeting needed much imagination to realize that with the discovery of the double helix, the understanding of the gene was about to reach a new plane. A new era was obviously dawning for genetics."[23] New metaphors were introduced for understanding the relationship of DNA and protein. Discussions focused on the idea of some sort of code in a linear order of "letters."[24] It was assumed that the sequences of nucleotide bases in the DNA polypeptide chains represented information of the gene. The nucleotide chain was thought of as a tape on which information is inscribed in a language or a code that used a four-letter alphabet—A, G, C, and T—that somehow specifies the amino acid sequence of a particular protein. This information was inscribed twice, in each of the complementary strands of every gene. The fundamental question was whether the code was based on units of two, three, or four nucleotides.

There are twenty amino acids. It was obvious that the correspondence would not be one-to-one between each DNA base and each amino acid, because taken one at a time that code would make only four amino acids. If two adjacent pairs of bases specified an amino acid, it would still not be enough, since this would mean that no more than $4 \times 4 = 16$ nucleotide combinations, four short of the required 20. Therefore, at least three were required to specify one amino acid. If the code was triplet, then there would be $4 \times 4 \times 4 = 64$ different trinucleotide combinations. More than enough.

The physicist and cosmologist George Gamow, champion of the big bang theory of the universe, proposed the first formal scheme for a genetic code in 1954.[25] It was based on the notion that the code was triplet, but overlapping. This would cut down the number of triplets to about twenty. Every other nucleotide base forms part of two triplets. For example, the last U in a sequence GCU would also be the first of UCA. Then A would be the first in the next triplet—say, AGA. This model had the merit of comprising about as many triplets as there were amino acids. But a few years later it was realized that this model did not match empirical data. There would always be some forbidden sequences of amino acids in any overlapping code. For example, an amino acid whose code would be, say, ATT could not be the neighbor of an amino acid whose code was ACC. However, no such neighbor restrictions existed. Surveys of amino acid sequences published in 1957 by Sydney Brenner (b. 1927), who coined the word "codon" for each triplet, ruled out an overlapping code.[26]

The problem of the code was intensely studied by several researchers who proposed various models.[27] By 1961, Crick and his colleagues had provided experimental evidence for a nonoverlapping triplet code.[28] They used point mutations that added (or deleted) bases to DNA in T_4 phage so as to make a specific protein nonfunctional. When they added one nucleotide base to the DNA, the infected bacteria could not make the protein functional again. They then added a second nucleotide, but it still did not work. But when they added a third nucleotide, the proper function of the sequence was restored: the protein could be synthesized. They fully restored the triplet reading frame. If the code was triplet, then there would be sixty-four possibilities to code for twenty amino acids. Two interpretations were possible. One could assume either that many of the triplets were nonsense words, or that more than one triplet may stand for the same amino acid. Any code in which two or more symbol combinations all stand for the same thing is referred to as "degenerate." In the early 1960s, Marshall Nirenberg and J. Heinrich Matthaei at the National Institutes of Health led the race to crack the code. By 1966, the genetic code was fully deciphered: for example, UUU codes for the amino acid phenylalanine. Nirenberg shared the Nobel Prize in 1968 with Robert Holley and Har Gobind Khorana. The code was universal: it held for all organisms from elephant to virus. A new era in biology had arrived. The implications for genetic engineering were staggering.[29] Today, many writers refer to DNA as a digitized program, "uncannily computer-like."[30]

Transcription and Translation

During the early 1960s, molecular biologists not only resolved what they referred to as DNA's ability "to reproduce itself *autocatalytically*" based upon complementary strands as templates for copying, they also resolved what they called the gene's "heterocatalytic" ability: the process that leads to the corresponding amino acid sequence of the protein.[31] The protein subunits, amino acids, were not assembled directly on the gene, but on small RNA-protein particles in the cytoplasm called ribosomes. That ribosomes were the sites of protein synthesis had been indicated by biochemical studies of the 1940s led by Torbjörn Caspersson in Stockholm and by Jean Brachet in Belgium.

The structure of ribosomes was actually made visible through another technological improvement by physicists: the electron microscope, commonly employed since the mid-1950s. Substituting a beam of electrons for visible light increased resolving power a thousand-fold. Thus, it was possible to observe the fine details of cell organelles such as ribosomes, mitochondria, and chloroplasts, and even to distinguish the shape of certain very large molecules. A typical bacterium contains about 20,000 ribosomes, a human cell about 100,000.

It was not long before biologists deduced how the information encoded in DNA is made available to the ribosomes during protein synthesis. DNA serves as the template for the synthesis of a complementary RNA nucleotide. That is, the information from each gene (DNA) is first transcribed into RNA. These "messenger RNA" molecules then migrate to the cytoplasm and enter into temporary combination with ribosomes. The nucleotide sequences of the messenger are then translated into the amino acid chains that make up the primary structure of proteins.[32] Protein synthesis thus involved two stages, first transcription of the message and then translation of the message into protein.

The ribosome in the cell body or cytoplasm was considered the workshop in which proteins were made.[33] Ribosomes were found to have a grooved structure that enables them to provide a temporary abode for the messenger RNA (mRNA) as its message is being deciphered during protein synthesis. Individual amino acids are brought to ribosomes by specific transfer RNA molecules. These transfer RNA "adapter" molecules must recognize the information in the messenger RNA triplet nucleotide (codon) so that the appropriate amino acid is added to the chain at the correct location in the linear sequence. Each transfer RNA recognizes one amino acid and one specific codon. Carrying suitable adapters, the ribosomal particles move from end to end of the messenger RNA like the reading head of a tape recorder passing over the tape. The protein chain is thus synthesized in a stepwise fashion from one end to the other.

Turning Genes On and Off

The concept of genes as reservoirs of information represented only one half of the new molecular meaning of heredity. By the early 1960s, it had been shown that the information transfer from genes to proteins could be turned off and on. The genome, it was argued, not only contained a series of blueprints for the amino acid sequences of proteins, but it was capable of regulation. The most influential study supporting this conception was carried out by François Jacob, Jacques Monod and their collaborators at the Institut Pasteur in Paris.[34] It involved the regulation of the enzyme ß-galactosidase in bacteria. *E. coli* had ß-galactosidase activity only when it was grown in a medium containing the sugar lactose as a carbon source: the presence of lactose "induced" the formation of the enzyme. However, Jacob and Monod did not view this as a case of environmentally directed adaptive gene mutations. Instead, they argued that it was a matter of regulating the expression of existing genes. To understand this regulation, they postulated the existence of elements that control gene expression.

Jacob and Monod's work culminated in 1961 with a model they called the operon.[35] It postulated that structural genes that specified enzymes were under the influence of an operator: genetic elements that stimulate genes under their control to produce messenger RNA. The operator in turn, was under the control of a regulatory gene, which produced a repressor molecule that prevented the action of the operator. However, the repressor itself could be blocked by certain substances called inducers (e.g., the presence of lactose in the cytoplasm). So the operator could exist in either of two states: opened or closed. It is open when it is free of repressor, and closed as soon as it has combined with the repressor.

The demonstration that genes could be switched on and off in bacteria represented what was seen as a revolution in developmental biology. When genes were endowed with the dual functions of regulating and specifying proteins, it was clear how cellular differences could be perpetuated even among cells with the same sets of genes. As Jacob and Monod put it, "The biochemical differentiation (reversible or not) of cells carrying an identical genome does not constitute a 'paradox' as it appeared to do for many years to both embryologists and geneticists."[36]

Cellular differentiation ultimately depended on specific cytoplasmic substances that activate or repress the genes that make the differentiating proteins. If, for example, when an egg divides, a nucleus enters into one region of cytoplasm containing substance A, certain genes will be put into action by that substance, while others will be repressed. Another cytoplasmic environment, B, may have another effect. This reorientation guided genetic research on cellular differentiation. In as much as the operon model provided a way for understanding how genes could participate in cellular differentiation, it offered a unified perspective and brought chromosomal genetics and embryology together.[37] Jacob, Monod, and André Lwoff were awarded the Nobel Prize in physiology and medicine in 1965.

Analogies from electronics and cybernetics and an information-based cell society were added to the old metaphors. Models of cellular regulation in terms of conflict and cooperation among plasmagenes that responded differentially to environmental influences were replaced by communication networks, circuitry, feedback loops, and information. Jacob and Monod commented that "the cell must be visualized as a society of macromolecules, bound together by a complex system of communications regulating both their synthesis and their activity."[38] Jacob and Monod repeatedly compared their regulatory entities to "the basic elements of electronic engineering, which could be organized into a variety of circuits fulfilling a variety of purposes."[39] As their analogy pertained to the operon model, enzyme synthesis would be regulated by "circuits of transmitters" (regulator genes) and "receivers" (operators) of cytoplasmic "signals" (repressors) that controlled the rate of messages sent from the nucleus.

Classical Doctrines of Molecular Biology

By the end of the 1960s, molecular biology had transformed the gene concept in several fundamental and interconnected ways that, as far as I know, have not been brought together explicitly.

The Concept of the Gene

The classical geneticists' concept of genes as consisting of individual structures arranged like a string of beads was radically transformed. Molecular biologists described the gene in terms of chemical symbols, as a chemical message. Jacob described the new concept in his famed book *The Logic of Life*, first published in 1970: "Just as a sentence represents a segment of text, so a gene corresponds to a segment of nucleic acid. In both cases, an isolated symbol means nothing;

only a combination of symbols has any 'sense.' In both cases, a given sequence, sentence or gene, begins and ends with special 'punctuation' marks."[40]

Relationship between Gene and Protein

The relationship between gene and protein—the one-gene: one-enzyme hypothesis of biochemical genetics—was redefined. The gene of biochemical genetics was still the abstract hereditary unit and, for promoters of the one-gene: one-enzyme hypothesis, could have been a protein rather than DNA. The gene of molecular biology entailed further refinement of the distinction between genotype and phenotype by the separation of gene and protein, but it still maintained its one-to-one correspondence. As Gunther Stent phrased the new view in his *Molecular Genetics* in 1971, "the gene directs the assembly of amino acids into a polypeptide chain of a given primary structure."[41] Not all proteins were enzymes; they played other structural roles as well. But all enzymes were thought to be proteins, and each catalyzed one particular reaction. It did so by transforming a particular metabolite, because on the surface of its three-dimensional curled-up structure there is a specific kind of crevice into which only one molecular species fits exactly. Once a substrate is lodged on its site, the forces holding some of its atoms are perturbed; the bond is broken, the substrate no longer fits the protein, and the protein is free to bind another molecule of the substrate. All of this happens in a fraction of a second.

Hereditary Information

One and only one kind of hereditary information existed: that contained in DNA. This entailed an exclusive one-way flow of information from gene to protein. "Genes give the orders, proteins execute them." The nucleic acid sequence determines the order of the protein subunits, the amino acids. And once this sequence is arranged, the chain folds back on itself in a complicated and unique pattern conferring its particular properties; the twisted three-dimensional shape gives the protein its specificity. How this one-dimensional amino acid structure is transformed into a three-dimensional one was not understood in detail, but it was assumed to occur spontaneously and to be fully determined by the amino acid sequence.

Thus, the "text" of DNA fully determined protein specificity. And this information always flowed from nucleic acid to protein, never in the other direction. Proteins do not determine the sequence of nucleic acids. And proteins do not reproduce.[42] In a famed paper on protein synthesis of 1958, Francis Crick gave molecular biology what he called its "central dogma": "This states that once 'information' has passed into protein it cannot get out again. In more detail, the transfer of information from nucleic acid to nucleic acid, or from nucleic acid to protein is possible, but transfer from protein to protein, or from protein to nucleic acid is impossible."[43]

The Genetic Program

The demonstration that the genome contained information for its own system regulation gave rise to the idea that not only did it contain blueprints for proteins, but it also contained a computerlike program for generating the whole organism. Again we can turn to Jacob's *Logic of Life* for an explicit statement: "The whole plan of growth, the whole series of operations to be carried out, the order and the site of synthesis and their co-ordination are all written down in the nucleic acid message."[44] Earlier in the century, embryologists and physiologists such as Jacques Loeb had argued that order in the organism was due to cell structure, which acted as an internal mold. But Jacob emphasized that biologists at that time had worked within the conceptual confines of colloid chemistry. Colloids were substances that did not manifest specific characteristic structures but formed structurally undefined gels. The principles of molecular biology contradicted this. Proteins did have shapes, and it was a central tenet of molecular biology that the properties of life could be understood in terms of the structure of molecules.

According to the doctrine of self-assembly that was formed in the 1960s and 1970s, the transformation of disorder into order in the cell—the construction of cell membranes and organelles—was due to the physicochemical properties of gene-determined proteins and their random collisions in a cell "soup." This argument was fortified with the assembly of the shell of small viruses containing enough nucleic acid information for three or four proteins. Their shells, shaped like rods or spheres, are built of several proteins of the same species. Molecular biologists had shown that this architecture of the virus could indeed be formed spontaneously from the aggregate of proteins in a process analogous to crystallization. How more complex structures, including the cell membrane, were built was unknown, but the same principles were held to apply. As Jacob put it, "The order of the living organism therefore is based on the structure of a large molecule."[45] The Weismannian conception of the organism as a product of a "self-replicating" genome reemerged in new guise—as articulated by many neo-Darwinians of the late twentieth century.

The doctrines of the first generation of molecular biologists were clear, precise, and revolutionary. But before the end of the century, virtually every tenet they erected was either challenged, refuted, or made much more complex, including the notion of the gene itself as a specific unbroken sequence or "sentence" of the DNA "text." Hereditary mechanisms in addition to DNA in chromosomes were considered and/or reconsidered: proteins as hereditary units, the inheritance of complex cell structures and of cell structure as a whole, and "epigenetic" mechanisms of cell heredity. The mechanisms of gene regulation for humans and other animals turned out to be far more complex than those for bacteria (see chapter 17). Studies of developmental and hereditary symbiosis further refuted the Weismannism concept of one-germ plasm: one-organism, and they contradicted several tenets of the neo-Darwinian synthesis (see chapter 19).

17

Beyond the Genome

It was naturally assumed that this bacterial gene structure was universal. It followed that if the gene structure was the same, then the mechanisms of regulation were probably very similar, and thus what was true of a bacterium would be true of an elephant.

—Phillip Sharp, 1994

A cell has a history; its structure is inherited, it grows, divides, and as in the embryo of higher animals, the products of division differentiate on complex lines. Living cells, moreover, transmit all that is involved in their complex lines. I am far from maintaining that these fundamental properties may not depend upon organization at levels above any chemical level; to understand them may even call for different methods of thought; I do not pretend to know. But, if there be a hierarchy of levels, we must recognize each one.

—F. G. Hopkins, 1932

THE DISCOVERY OF DNA has been trumpeted globally: it has been called the "blueprint of life," "the book of life," and the "autobiography of the species." The story is well known. The DNA molecule is a code that contains all the information required to specify the heritable characteristics of the organism. The information is translated into protein structure by a process in which DNA determines the specificity of proteins. Once the information has been so translated, all of the chemical reactions of the cell—wholly determined by protein enzymes—have also been specified. Thus we are told that the basis of heredity, "the secret of life," is solved. All that remains is to apply that knowledge to agriculture, and to medicine to correct inborn "errors" of metabolism.

In 1990, the United States began the Human Genome Project. A fifteen-year effort coordinated by the Department of Energy and the National Institutes of Health, its aims were to identify all the genes in human DNA, determine the sequence of its 3 billion nucleotides, store this information in databases, improve tools for data analysis, transfer related technologies to the private sector, and address the ethical, legal, and social issues that may arise from the project.

The human genome was labeled the "book of man" and the "holy grail." Knowledge of it promised to reduce human suffering caused by genetic diseases due to single-gene defects, among them cystic fibrosis and sickle-cell anemia. The ability to screen for genes and analyze them directly would open up a new medical era of gene therapy, and products derived from genome research would boost the international pharmaceutical business. By licensing technologies to private companies and awarding grants for innovative research, the human genome project would catalyze the multibillion-dollar U.S. biotechnology industry.

In June 2000, a working draft of the entire human genome sequence was announced, with analyses published in February 2001. The hyperbole about self-knowledge was boundless.[1] The sequencing of the genome was likened to landing on the moon, splitting the atom, and inventing the wheel. The code, it was said, will tell us what distinguishes us from other species, what makes us human; it has been considered to be "the secular equivalent of the soul, our inner stable true nature throughout our individual lives."[2] The most profound social, ethical, and legal issues permeate the "brave new world" with old questions about eugenics reconsidered and new ones over human gene patents, genetic screening, and concern that knowledge of an individual's genome be kept from employers and insurance companies as well as governments, no matter how apparently benign or benevolent.[3]

Reports and speculations about genes for violence, depression, impulsiveness, novelty seeking, alcoholism, and homosexuality have permeated public consciousness. No sooner has one group of researchers tied a gene to a behavior than along comes the next study asserting that the link is false or even that the gene in question has exactly the opposite effect.[4] Debates have often become polarized between genetic determinism on the one hand and cultural determinism on the other. Untangling the effects of genes from their social context is hazardous.

Critics of genetic determinism have probed various social aspects of "the DNA mystique" in an effort to understand the cultural fixation with genes.[5] When emphasizing the roles of environment and culture, however, they commonly assume that our heredity is solely a matter of our genome. But should we accept the premise that the biological basis of heredity is essentially solved? Protests against the gene as the sole basis of heredity and evolution have paralleled the development of genetics from its inception (chapters 9, 12, and 15). They have continued into the twenty-first century against the centrality of the genome, and against the classical tenets of molecular biology.[6]

Complexity and the Human Genome

That organismic complexity cannot be reduced to the information in the genome was suggested by one of the most striking and unexpected results of the human genome project: the human genome is made up of very few genes. At the outset of the project, it was estimated that there would be about 100,000–140,000 genes. However, by 2001, the Human Genome Sequencing Consortium had spotted 22,000 protein-coding genes and estimated that our genomes contain a total of about 30,000 different genes.[7] This is only twice as many genes as a fruit fly or a flatworm has, and not as many as rice, which has about 40,000.[8] In terms of gene number we are on a par with *Arabidopsis*, a mustard-like weed (which has 26,000 genes). The 30,000 genes in our genome also contradicted the principal scientific premise on which the human genome project was based: that there would be a one-to-one, linear relationship between genes, proteins, and genetic disease, allowing one to detect defective genes from the DNA code. As one of the pioneers of the human genome project, Craig Venter, recalled in 2002, "People so wanted it to be the beginning of the end of everything. . . . One gene, one disease, one treatment."[9]

Decoding the human genome has turned out to be much more complicated than had been expected. Proponents of genomics-based medicine have begun to recognize how erroneous is the view that one gene leads to one enzyme and one disease (see also chapter 14). This was the basis of previous estimates of gene numbers. Classical studies of the molecular biology of the gene were based on the genetics of bacteria. It was assumed that the linear structure of genes and the relationships between DNA, RNA, and protein in bacteria (prokaryotes), which do not have a nucleus, would be the same as those for animals whose cells have a nucleus (eukaryotes). However, by the 1980s, molecular biologists were recognizing more and more that gene structure and gene expression in eukaryotes was more complex than the molecular biology of bacteria.[10]

Long before the Human Genome Project, it had been known that the DNA content of animal sex cells varied significantly between phyla, without an apparent variation in the number of genes. The human genome was 200 times larger (in the sense of the number of nucleotide sequences) than baker's yeast, but 200 times smaller than some amoebae. It had been recognized since the early 1970s that eukaryote genomes contained huge regions of nucleotide sequences that do not code for proteins or RNA—so-called junk DNA.[11] Today, it is estimated that more than 98 percent of human DNA is made up of such noncoding sequences—although the adaptive value of junk DNA is still debated.[12]

Whether or not gene structure in eukaryotes was the same as that in bacteria was not really questioned until the discovery of additional noncoding regions within chromosomal genes. In the late 1970s, Phillip Sharp and his collaborators at MIT, and Richard Roberts at New England Biolabs, showed that, unlike bacteria, eukaryotic genes have long intervening sequences, called introns, between coding regions, called exons.[13] Before these so-called split genes can be expressed, the RNA has to be "edited," that is, the intervening sequences have to be removed by special enzymes and then spliced back together again. After a DNA sequence is transcribed to messenger RNA, often a small body called a spliceosome (or splicing body), a protein-RNA complex, assembles at sites along the mRNA, where it cuts out various segments. When coding regions are spliced together, certain exons are excluded from some transcripts and included in others. The remaining pieces of RNA may be spliced together into a number of alternative combinations, and these are used as messages (mRNA). Each new nucleotide sequence is different from that coded from the original gene.

Alternatively spliced RNAs code for different proteins.[14] It is comparable to a game of Scrabble in which, for example, the seven letters STARTLE can be alternatively spliced, in the order in which they appear, to give the words "start," "star," "stale," "sale," "tar," "tart," "tale," "art," "at," "are," and "ale." The original informational unit STARTLE, the "gene," is important, but so too is the information in the spliceosome that arranges the message. Many human diseases are caused by mutations that interfere with RNA splicing.[15] In 1993 Sharp and Richards shared the Nobel Prize in physiology and medicine for their discovery of split genes.

Alternative RNA splicing can have an enormous impact on the gene-to-protein ratio. One gene can give rise to dozens of proteins.[16] At first thought to be an exceptional occurrence in humans, by the end of the century alternative splicing was estimated to occur in about 40 percent of human genes, and this, genomic researchers argue, partially explains the earlier inflated gene number predictions.[17] Alternative RNA splicing is recognized as one of several important forms of genetic regulation during cellular differentiation. Different portions of the transcribed RNA are excised in specific tissue cells. In addition to the alternate proteins arising from the same sequence of DNA (gene), those individual proteins that are produced often become modified after they are made (post-translational modifications), and these modifications alter the function of the protein.

Because of such mechanisms for regulating and modifying protein synthesis and other post-translational modifications, there may be 300,000 proteins in the human body—ten times the number of genes. Not only can one gene have many effects, but the reverse is also true: many genes could affect one hereditary trait, including genetically based maladies in humans.[18] A gene's effect on inheritance cannot be completely predicted from its nucleotide sequence. And one cannot search for gene-based diseases solely in the primary sequence itself. The champions of Human Genome Project, including James Watson and Craig Venter, have acknowledged that the promise of therapeutic breakthroughs by the map of the genome in fact may not come for decades—and then we might suggest using the map only as a guide.[19]

The complexity of genetic information and its modifications with RNA editing greatly complicated the definition of the gene. The gene was first defined as simply as a unit of inheritance, then as a locus on a chromosome, and then as a specific nucleotide sequence of DNA, but the "gene" has become increasingly abstract. As Sharp noted in his Nobel lecture of 1994, what exactly the gene is has become somewhat unclear.[20] Molecular biologists have generally returned to the old concept of the gene popular in the early twentieth century—that genes are operational entities (see chapter 12). Raphael Falk commented that "living systems are essentially complex and integrative systems. It is meaningless to identify entities of such systems on an ontological basis. The gene is a generic term. This is the pragmatic approach adopted by many practicing molecular biologists."[21]

Alternative splicing and post-translational modifications are not the only discoveries over the past four decades that complicate molecular biology's classical concept of DNA as "the master molecule." Although their views hardly ever reach the wider public, many cell biologists point to different levels of hereditary organization, from molecules to cells, each with its own fundamental properties.

A Genetic Plan?

Molecular biologists of the 1970s often spoke of a "genetic program," asserting that the blueprints for proteins and organisms were encoded in DNA. The famed bacterial geneticist François Jacob expressed this idea in 1970: "The whole plan of growth, the whole series of operations to be carried out, the order and the site of synthesis and their co-ordination are all written down in the nucleic acid message."[22] This same idea has been perpetuated outside of science. The author of *The Evolving Self* writes confidently: "Elephants are only a by-product of the genetic information contained in elephant chromosomes. Theoretically one could build elephants provided one had the blueprint of their genes."[23]

That DNA is a genetic program theoretically capable of creating an organism, even when operating within the egg of a different species, is the premise for Michael Crichton's best-selling novel *Jurassic Park*, later made into a blockbuster movie by Steven Spielberg. The computer program is no longer just a metaphor for the genome. We can read that our genetic information, "the essence of our being," can be stored in computers and reemerge to reproduce us. Marveling over the biotechnological possibilities of the human genome project, the author of a scientifically acclaimed paperback, *The Genome*, introduces his book by asserting that "you can now download from the Internet the near-complete instructions for how to build and run a human body."[24] Venter and his colleagues also allude to a genetic plan when they emphasize that one of the main tasks of human biology is to find out how "genes orchestrate the construction and maintenance of the miraculous mechanism of our bodies."[25]

Is DNA a genetic plan, the blueprint, or the recipe or director of the organism? Does such information really exist in DNA? To give meaning to such a proposition, one must describe explicit mechanisms by which linear, one-dimensional nucleotide sequences determine living forms in three-dimensional space and time. This is morphogenesis, one of the main problems of heredity before genetics took over early in the twentieth century: how like reproduces like. During the last years of the twentieth century, interest in morphogenesis and cell organization began to revive somewhat, and those concerned with the genesis and reproduction of cell structure and organization insisted that the popular notion of a genetic program or DNA-based blueprint for an organism made no sense. They argued that morphogenesis is not orchestrated by DNA; it defies reductionism.[26] Instead, it is the manifestation of higher levels of order, often corresponding to the cellular scale of size and organization. The American cell biologist Franklin Harold commented in 1995:

> Offspring resemble their parents in form as well as function: roses and rabbits, yeast, and *Escherichia coli* display the same forms, generation after generation, within a narrow range of variations. How does that come about? The answer is not known, not even in principle, for the quest reaches deep into the abiding mysteries of organized complexity.[27]

Cell biologists concerned with the genesis of cell structure insist that the overwhelming focus on genes and molecules detracts from the study of how molecular constituents come together in space and time. They pointed to a huge disparity in funding to account for the fallacy that the genome is the major repository for developmental information.[28] While molecular genetics and gene-based remedies have attracted extremely generous funding from pharmaceutical companies, agricultural industries, government agencies and financial speculators, researchers investigating fundamental problems of cell structure above the molecular level complain of the difficulty of obtaining suitable financial support in the fast-paced and commercial world of science.[29] To appreciate their criticisms of reductionism, let us begin with the central tenets of classical molecular genetics, often expressed as:

DNA → RNA → Protein → Cell → Organism[30]

Confronting Old Dogmas

Molecular biologists of the 1960s and 1970s often referred to DNA as "self-replicating" or as an "autocatalytic "molecule. Yet, critics have argued that DNA is, in fact, one of the most inert molecules known to science.[31] DNA by itself can actually do nothing. Multiple enzymes are generally required to reproduce DNA, and many others are required to transcribe DNA into RNA and still others to edit the messenger RNA. A number of other enzymes, ribosomes, and the rest of the protein synthesis machinery are required to actually transcribe RNA to the amino-acid sequence of a protein. As the discoveries about split genes, RNA editing, and spliceosomes show, the causality between DNA and protein is far from linear.

The first generation of molecular biologists had maintained that the shape and function of a protein (its tertiary structure) was determined by its amino acid sequence (primary structure), and this in turn by the information in DNA. Crick referred to this as the "central dogma": that "the transfer [of information] from protein to protein, or from protein to nucleic acid is impossible."[32] However, this now appears to be a misstatement—in light not only of protein-based spliceosomes, but of another family of proteins called chaperonins. It is the three-dimensional form of the protein—how it folds—that allows it to function properly. The molecular biologists who cracked the code had assumed that the folding of the protein was determined by its amino acid sequence, which, in turn, was determined by DNA (see chapter 16). However, flaws appeared in this conceptual edifice beginning in the 1980s, when it was recognized that most large proteins in eukaryotic cells will not fold properly if not guided by "chaperone proteins."[33]

More problems with the idea that DNA determines protein folding are indicated by particles called prions, the infectious agents of mad cow disease (bovine spongiform encephalopathy) or Creutzfeldt-Jakob disease, which afflicts cattle and humans.[34] Although it was once dismissed as an impossibility, prions are now widely recognized to contain no DNA or RNA.[35] In 1997, Stanley Prusiner of the University of California, San Francisco, was awarded the Nobel Prize in physiology and medicine for his work on the infectious nature of prions and for showing that these infective agents are pure proteins. When the prion invades the brain, it refolds a normal brain protein to match its own infectious three-dimensional shape. Thus no information is conferred from nucleic acid to the protein; only its conformation—its three-dimensional shape—changes. The newly folded protein then becomes infectious and acts on other normal proteins, setting up a chain of reactions that propagates the disease. Thus, according to contemporary prion theory, information is passed from protein to protein in contradiction to the central dogma of classical molecular biology.

The prion represents a totally new model for agents of infectious disease, and its implications are far-reaching.[36] But, if some proteins can be said to reproduce themselves, what about larger structures in the cell? Embryologists and cytoplasmic geneticists have long argued that more than genes are inherited (chapters 10 and 15). In fact, an egg, before fertilization, contains a complete apparatus of reproduction, as evidenced in parthenogenesis.

Cell Architecture and Spatial Information

Morphogenesis is usually likened to architecture. The fundamental question at the cellular level is, how do organisms impose order on the confusion at the construction site? How are proteins organized into cell structures, such as endoplasmic reticulum, Golgi bodies, lysosomes, secretary granules, centrioles, mitochondria, and chloroplasts? And how is the structural organization of the cell cytoplasm determined? Since the nineteenth century, cytologists have referred to the physical continuity of such organelles as chloroplasts, mitochondria, and perhaps centrioles (see chapter 8). This in itself suggested that they cannot be formed *de novo* by the cell. Beginning in the 1960s, geneticists showed that mitochondria and chloroplasts possess their own genomes of about 50 to 200 genes. However, neither mitochondrion nor chloroplast is formed anew from proteins: each is formed only from a preexisting organelle. If these organelles were removed from the cell, they could not be formed again, because their reproduction requires a preexisting supramolecular structure.[37]

Preexisting structures provide crucial spatial information; they are sites at which proteins made elsewhere assemble. Any cell structure passed on from one cell to its daughters without being totally deconstructed to its component molecular parts could, in principle, act as a seed for the propagation of similar structures. Centrioles lack DNA, but, in many organisms, they still need to be inherited.[38] As revealed by electron microscopy, centrioles are cylindrical, comprised of nine groups of triplet microtubules. It was shown that, as had long been suspected, centrioles are homologous to, and indeed the same as, kinetosomes, or basal bodies, which lie close to the cell surface of many organisms. How centrioles reproduce is unclear. In the cells of some species, they seem to arise *de novo*, and in the cells of others their reproduction requires cytologically visible preformed centrioles. In the latter case, new cartwheel structures develop from amorphous material in a stepwise orderly sequence, tubule by tubule, usually at right angles to, but not in direct contact with, preexisting centrioles.

Recall that at the outset of genetics, embryologists had insisted that development was essentially a cytoplasmic phenomenon, and that genes played a minor role in development. The force of such claims was considerably weakened in the early 1960s by evidence that many developmental changes could be explained in terms of regulatory genes and selective control of gene action (see chapter 16). Certainly, what cells can do is constrained by the proteins within them. The hemoglobin in red blood cells enables them to carry oxygen, and the presence of actin, myosin, and tropomysin and other proteins enables muscle cells to contract. However, while changes in gene activity do occur during cellular differentiation, developmental biologists continue to insist that differentiation is not simply a matter of gene regulation. As Lewis Wolpert observed in 1998, "To think only in terms of genes is to ignore crucial aspects of cell biology, such as change in cell shape, that may be initiated at several stages removed from gene activity. . . . The route leading from the genes to a structure such as the five-fingered hand may be tortuous."[39]

Those who study morphogenesis emphasize that the form of a single cell or a protist is not related in a simple fashion to the shape of its molecular constituents. Cell form emerges as a collective deployment of the activities of hundreds of gene products. For this reason, Guenter Albrecht-Buehler has remarked, in "defense of nonmolecular cell biology," that so many different molecular interactions are involved as to often make the expression "molecular analysis" meaningless.[40]

The first generation of experimental embryologists had postulated a primordial structural organization, a rough body plan, in the cytoplasm of the egg that would provide a guide for cell products and determine the early stages of embryogenesis. Developmental processes up to gastrulation (during which the gut is formed and the main body plan emerges) all occur without much influence from the nuclear genome of the fertilizing sperm. How such patterns are formed—how cells become organized—has remained one of the great unsolved problems of biology. Embryologists and physiologists spoke about this organizing principle as a spatial property that brought parts together so they would combine in the right way and at the proper time. They located it in the egg cortex (see chapters 10 and 14). To believe that genes alone could build a cell required that one abandon one of biology's few universal laws, as formulated in the mid-nineteenth century: cells, Rudolf Virchow proclaimed, never arise spontaneously but invariably descend from a preexisting cell.

Virchow was right. Cells never arise by aggregation, but by the growth of preexisting cells; they model themselves upon themselves. As a cell grows and makes itself, the macromolecules specified by the genes are released into a context that is already spatially structured. The fact that nonrandom cell structure persists throughout growth and division immediately implied to biologists of the 1960s that preexisting structure plays a decisive role that may not be explicable by mere random self-assembly of gene products. The doctrine of self-assembly was strengthened by the startling demonstration that a linear code could be transformed into three-dimensional organization of a virus. But a virus is not a cell and does not divide like a cell, which preserves its structure through growth and division. Viral geneticists themselves were particularly aware of this. Salvador Luria, for example, commented in 1966 that macromolecules did not have all the "knowledge written inside themselves" to make a cell any more than individual humans had all the information to make a complex culture. Information also had to be transmitted by cultural means. As Luria put it: "You have to have cultural information in order to continue to make a complex product whose blueprint is not written out."[41]

But again the metaphor must be made flesh. If not the genome, then what orchestrates the hubbub of filaments, membranes, and polymers into recurring patterns recognizable as cells? Those biologists who exclude cell form from being "hardwired" into the genome really do not know, and they have searched for new metaphors and concepts to understand cell organization and hereditary information.[42] In 1970, the phage geneticist Alfred Hershey made similar comments to Luria when discussing problems of cellular organization that seemed to lie beyond the reach of current molecular principles:

> If cells draw on an extragenic source of information, a second abstraction must be invoked, another vital principle superimposed on the genotype. A likely candidate already exists in which is usually called cell polarity, which tradition places in a rigid ectoplasm for good reason—it's a spatial principle and as such requires mystical language. Seemingly independent of the visible structures that respond to it, polarity pervades the cell much as a magnetic field pervades space without the iron filings that bring it to light. Biological fields are species-specific, as seen in the various patterns and symmetries of growing things.[43]

Embryologists and cytologists had used the term "polarity" to refer to the visible directionality of cellular processes. Cell polarity is also especially noticeable in tip-growing organisms such as fungal hyphae, pollen tubes, and growing nerve cells. These cells orient their activities toward a unique site, called the apex. Building blocks of the cell, made elsewhere, are translocated to the apex, where they are assembled into new plasma membrane, new cell wall, new cell cortex, and so on. Cellular organelles also become oriented toward the apex and migrate to it along tracks supplied by the microtubules and microfilaments. The cytoskeleton, a network of microtubules and microfilaments in the cytoplasm of cells, gives the cell its characteristic shape.[44]

In no organisms has the morphogenetic role of preexisting cell structure in guiding the elaboration of new cell structures been more studied than in ciliates.[45] The complex patterns that make up the surface (the cortex) of ciliates are their most impressive and readily observable structural features. The cortex is composed of linear arrays of a large number of fundamentally similar ciliary units, complex structures that include kinetosome (centriole), cilium, a variety of subcortical fibers, and specialized membranes. These ciliary units are arranged in a precise repeating pattern. It was difficult to imagine how this organization could arise *de novo* by purely random collisions of proteins. And on cytological observation alone, the cortical pattern appeared to reproduce faithfully through a regular sequence of events during growth and fission.

In his book *Problems of Morphogenesis in Ciliates* (1950), André Lwoff linked his observations about the morphogenetic role of the ciliate cortex with those of embryologists who emphasized the importance of the cell surface of eggs of multicellular organisms (see chapter 15).[46] In 1961, the American protozoologist Vance Tartar also suggested that studies of a large trumpet-shaped ciliate called the stentor offered solutions to the "great unsolved problem of organic form." As he commented, "A cytoarchitecture which has repeatedly been postulated as necessary to explain the orderly development of eggs is visibly displayed in stentors and does in fact play a cardinal role in their morphogenesis."[47]

Strictly speaking, the observations of protozoologists said nothing about the origin of the guidance mechanism. Did it represent a principle of inheritance that was relatively independent of genes? Did it also represent a basis upon which evolutionary change could occur? In other words, if the structure or organization were changed, would the change be inherited? In principle, cellular organization could be recreated every time an egg was created or every time a ciliate emerged from a cyst. If the continuity of spatial organization is so crucial in the propagation of form, it might be expected that a structural modification could be passed from one generation to the next without alteration in genes.

Beginning in the 1960s, Tracy Sonneborn and his colleagues demonstrated that gross differences in cortical organization bred true to type through both sexual and asexual reproduction, independently of DNA. In 1963, Sonneborn coined the term "cytotaxis" to designate "the ordering and arrangement of new cell structure under the influence of pre-existing cell structure." Without cytotaxis, he argued, "an isolated nucleus could not make a cell even if it had all the precursors, tools and machinery for making DNA and RNA, and the cytoplasmic machinery for making polypeptides."[48]

In classical grafting experiments in 1965, Janine Beisson and Sonneborn inverted a small patch of ciliary units and showed that the progeny inherited the inverted row or rows for hundreds of generations independently of genes. "Our observations on the role of existing structural patterns in the determination of new ones in the cortex of *P. aurelia*," they wrote, "should at least focus attention on the information potential of existing structures and stimulate explorations, at every level, of the developmental and genetic roles of cytoplasmic organization."[49] Subsequent studies with *Paramecium*, *Tetrahymena*, and other ciliates in the 1970s and 1980s produced many examples of such structural inheritance showing how preexisting cell structure constrains and molds the structural arrangement of cell progeny.[50]

Sonneborn favored the possibility of "a parallel, independent, and selectively correlated evolution of genome and cortex."[51] In 1968, one of his former students, David Nanney, expressed the relationship of genes to preexisting structures in the following architectural terms:

> In an extreme polar interpretation, one might postulate that nucleic acids specify only proteins, which must be appropriate for cellular design, but not decisive. In this case the cellular architects (that is, preexisting structures) might be required to determine whether the eventual edifice constructed of the building blocks would be a railroad or a cathedral. I doubt the value of this extreme analogy, but some intermediate position may be more consonant with the larger biological realities than either extreme.[52]

Field Heredity

So, what does a cell pass on to its daughter over and above the information controlled in its genome? Do biologists have to invoke nonmolecular principles to account for the propagation of cell structure? In the 1960s, Sonneborn did not think so. He accounted for cortical inheritance within the conceptual framework of molecular biology. It was easy to imagine that preexisting structures would play a role in providing a scaffolding or template through which building blocks were assembled in jigsaw-puzzle fashion from smaller to larger structures. Cell form would be determined by the structure of parts, but one could not go all the way back to DNA and proteins. However, limitations were soon found with this model. Some inherited phenomena seemed to lie beyond the reach of molecular principles. In addition to local preexisting cell structure—that is, the ciliary unit—ciliatologists concluded that another mysterious invisible spatial property pervades the cell as a whole and is responsible for the perpetuation of morphogenetic patterns.

The most spectacular illustration of this unknown spatial property is the propagation of experimentally produced double monsters: ciliates that have not completely divided. They are fused back to back by aborting cell division or by surgical intervention. Doublet cells can propagate indefinitely as doublets, even though no gene has mutated. Beginning in the 1980s, ciliatologists concluded that what maintains the doublet organization is something other than the visible ciliature itself. Evidence for this comes from experiments with the ciliate *Oxytricha fallax*. When *Oxytricha* is starved, it forms cysts in which there appear to be no visible preformed cortical structures.[53] Neither cilia, nor kinetosomes, nor other microtubular structures can be seen in cysts by electron microscopy. Nevertheless, doublets emerge from their cysts as doublets, and singlets emerge as singlets—that is, the emerging cells "remember" their original architectural pattern. Thus, ciliatologists conclude that some other aspect of structural inheritance or cortical organization persists through the cyst stage.[54]

In other words, at least two kinds of mechanisms operate to maintain cell structure.[55] One is a local constraint involving microscopically visible cell structures acting as a scaffold—ciliary units, for example, which are specified by local interactions at the molecular level. The other mechanism is a global one that pertains to cell structure as a whole. For example, the placement of the oral apparatus and contractile vacuole pores, or the ciliates' overall handedness, appears to be established in relation to the cell as a whole by reference to some sort of global cellular grid that specifies the placement of organelles independent of their molecular architecture. These patterns are superimposed upon local structural guidance, and they exhibit properties that have traditionally been associated with embryonic fields.

The idea of a morphogenetic field was prominent in embryology of the 1920s and 1930s (see chapters 10 and 15) and was revitalized somewhat in the 1980s. Embryologists defined the field as a "group of cells whose position and fate are specified with respect to the same set of boundaries. A particular field of cells will give rise to a particular organ when transplanted to a different part of the embryo, and the cells of the field can regulate their fates to make up for the missing cells in the field."[56] While the reality of fields was not questioned, their nature and their evolution were targets of heated debate among developmental biologists. Did embryological fields have properties not determined by genes? Did such structures put constraints on evolution?

Beginning in the 1980s, evolutionary developmental biology reemerged under the nickname "evo-devo."[57] Many in this field took as their starting point that the gene is the unit of heredity. They aimed to complete the evolutionary synthesis, by emphasizing that development is hierarchical and characterized by emergent properties whose features cannot be predicted from properties at the lower level in the hierarchy. They promised that an analysis of the evolution of developmental stages, processes, and mechanisms would enhance understanding of how organisms, organs, tissues, cells, and genes evolve. Others wanted to do more than complete the neo-Darwinian synthesis, they aimed to replace it with a developmental evolutionary biology.[58] A movement of "biological structuralism" led by Brian Goodwin and Gary Webster sought noncontingent physical laws underlying evolutionary change.[59] In their view, the genome "is no more the 'directing center' of organismic structure than a lexicon or a dictionary is the directing center of a sentence or a text."[60]

Debates over the nature of the field and its relationship to genes had begun in England. On the one hand, the distinguished embryologist Lewis Wolpert proposed that fields could be understood in terms of the diffusion of a gene-produced chemical he called a morphogen. Cells would behave according to their position in the organism and local genomic activation. Such positional information would place no constraints on Darwinian evolutionary change. Wolpert's theory had great heuristic value and continues to be used to explain various developmental processes.[61] On the other hand, biological structuralists such as Goodwin and Webster maintained that morphogenetic fields are not specified by genes—they exist independent of the entities or the final gradients that bring them to light—and they argued that there were internal structural constraints on evolutionary change. For them, the structure, the field, and the changes that occur in them were governed by "generative principles" embodied in the processes that set them up. Morphogenetic fields, they asserted, have properties that would result in changes of the phenotype and might lead to evolutionary transformations.[62]

For structuralists, evolution is not simply a matter of chance and necessity, adaptations, and historical contingency. They allied their approach with the early-twentieth-century ideas of D'Arcy Thompson (1860–1948), as expressed in his famed text *On Growth and Form*.[63] They saw themselves following in the tradition of Geoffroy St. Hilaire (see chapter 1); they praised the tradition of comparative morphology and of comparative embryologists who emphasized the "essential" properties of the type and the concept of a structural plan; and they argued that adaptations to external transformations were secondary only to the internal organization of organisms and lawful transformations.[64] Evolution was not merely a matter of tinkering by natural selection. Not all shapes are possible—contrary to Richard Dawkins's quip that "it may be that the only reason pigs have no wings is that selection has never favored their evolution."[65] Complexity theorist Stuart Kauffman, at the Sante Fe Institute in New Mexico, searched for "ahistorical universal laws" underlying evolution, "principles of order like physics."[66]

In a volume boldly entitled *Beyond Neo-Darwinism: An Introduction to the New Evolutionary Paradigm*, Goodwin challenged what he called genocentric biology, which claims "that all aspects of organismic form are determined by the hereditary particulars encoded in DNA."[67] To refute this claim, he pointed to the work on ciliated protozoa, which showed that specific patterns of cortical morphology are inherited by a mechanism that is relatively independent of DNA. In other words, the specific causes of morphology in these organisms do not come solely from genes. "Reproduction in this large category of organisms," he wrote, "does not conform to Weismann's scheme . . . nor to any of its subsequent modifications."[68]

Ciliates are analogous both to a single cell of a plant or animal, and to the organism as a whole. The ciliatologist Joseph Frankel at the University of Iowa emphasized that a morphogenetic field operates both within the organism and within the cell. But what this field is exactly remains unclear:

> The problem still stands where the problem of inheritance stood before the rediscovery of Mendel. . . . What is needed is a more explicit and mechanistic field theory, or a wholly new concept that provides a superior understanding of the same underlying phenomena, much as the concept of oxidation replaced the phlogiston theory in accounting for the facts of chemical combustion.[69]

The perpetuation of global cell patterns (cellular organization and form) thus presents not a puzzle, but a mystery. That cellular organization is a mode of heredity distinct from that encoded in genes continues to disturb the conceptual foundations of the modern consensus—perhaps, as some have suggested, as an outrageous assault on established truth, and tinged with vitalism to boot.[70] Several biologists, including André Lwoff, have interpreted cortical inheritance as experimental proof of the inheritance of acquired characteristics.[71] Ciliatologists also insist that the highly organized cytoplasm of a metazoan egg cannot be solely the consequence of direct gene activity.[72] The inheritance of a particular structural arrangement in the cell cytoskeleton can have far-reaching consequences for morphogenesis, such as the shaping of an organism's adult body plan.

Neo-Darwinian evolutionists have occasionally admitted their concern about structural inheritance. In 1983, when discussing the inheritance of acquired characteristics and why he himself upholds a Weismannian and gene-centered view of evolution, the distinguished evolutionist John Maynard Smith commented:

> There are a few well-established exceptions, of which the phenomenon of "cortical inheritance" in ciliates is perhaps the most important. Neo-Darwinists should not be allowed to forget these cases because they constitute the only significant threat to our views.[73]

One of the reasons structural inheritance is considered to be an exception is that the crucial evidence comes entirely from ciliates. Those who believe in the generality of the phenomenon argue that it is reported primarily in ciliates because ciliates are such good organisms for detecting and investigating the phenomenon.[74] All organisms use preexisting cell structures for the reproduction of their cytoskeleton and cell surface.[75] The importance of the cortex in morphogenesis is clear, but its general evolutionary importance as a hereditary mechanism for evolution is not.

Epinucleic Inheritance

"Extranucleic inheritance" is pervasive—from prions, to centrioles, to cell cortex and cell structure as a whole.[76] Today these forms of inheritance are often discussed together with other nongenic or "epinucleic" mechanisms.[77] Some of them were considered earlier in the twentieth century in relation to cell heredity (see chapter 15), and advances in molecular biology have led to the recognition of others. One involves changes in chromatin structure due to methylation.[78] In many eukaryotes, some of the cytosines in DNA can be modified by the enzymatic addition of a methyl group. The methyl group does not change the coding properties of the base, but it does influence gene expression—that that is, highly methylated DNA is usually transcriptionally inactive. DNA methylation patterns are inherited from one cell generation to the next, and evidence also suggests that the directed changes are transmitted through sexual reproduction.[79] Thus, methylation may also provide evidence of the inheritance of acquired characteristics.

The phenomenon known as genomic imprinting provides some of the best evidence that epigenetic changes can be transmitted from parent to offspring. Normally, it does not matter whether a gene or chromosome is transmitted from mother or father. However, during the 1980s, it was increasingly found that the transmission and expression of a gene, a part of a chromosome, a whole chromosome, or even a whole set of chromosomes depended on the sex of the parent from which it was inherited. The process that establishes the differences between paternally and maternally inherited chromosomal genes is known as imprinting. It was first discovered in insects. During the formation of sperm, all chromosomes derived from the paternal line are eliminated in some insects. Offspring inherit only maternally derived chromosomes.[80]

Imprinting is also widespread in mammals (and may be important for understanding a number of human diseases, including some cancers).[81] Homologous chromosomes or genes may be differentially methylated, depending on which parent they were derived from: in some tissues only the gene inherited from the mother is active; in others, only the gene inherited from the father. In fact, a substantial part of the mammalian genome carries imprints of its parental origin.[82] This suggests that some sort of "label" on a gene indicates its paternal or maternal donor. Exactly how parental imprints are established, modified, and expressed is not understood.[83] The importance of these epigenetic mechanisms for evolution is still a subject of debate.[84]

Epinucleic mechanisms of inheritance, from structural inheritance to steady states, are often lumped together and discussed under the rubric of "epigenetic inheritance" so as to distinguish them from the more well-known inheritance based on nucleic acids, a genetic system.[85] However, this bipartite labeling tends to obscure important differences. To refer to structural inheritance as epigenetic inheritance belies the fact that the term "epigenetic" was coined as an antonym for preformation, and yet the phenomenon of structural inheritance is an argument precisely for preformation.[86] We can avoid the contradiction by referring to structural inheritance as such, or as extranucleic inheritance, and distinguishing it from epigenetic inheritance and nucleic inheritance.

Molecular biology has profoundly transformed many aspects of the life sciences, not the least itself. Genes did not "command," nor did proteins execute their orders. The informational flow from DNA to proteins was not unidirectional as the central dogma of the 1960s would have had it. It is incorrect to say that the specificity of proteins is determined by nucleic acid sequence of the genome. In the form of spliceosomes, proteins edit the genomic sequence to determine the amino acid structure of proteins. In the form of chaperonins, proteins are often required to determine the tertiary structure of proteins. In the form of prions, proteins perpetuate their kind as infectious particles, by mechanisms not fully understood. Experiments on ciliates also show the importance of preexisting cell structure and the inheritance of global cellular fields.

A unified theory of heredity and evolution, one that encompasses nongenomic inheritance, would have to go further still. It would have to include the fundamental role of symbionts in morphogenesis and the inheritance of acquired bacteria and viruses. The importance of symbiosis in the genesis of new forms and symbiotic phenotypes have been proposed since the late nineteenth century (see chapter 19). Advances in molecular biology have been crucial for detecting and analyzing symbiosis and underscoring its importance in evolution. And, as we shall see in the following chapter, molecular biologists also developed new techniques for investigating the evolution and diversity of bacteria, and in so doing have recognized other non-Mendelian mechanisms for evolutionary change left unconsidered by classical Darwinian evolutionists.

18

Molecular Evolution and Microbial Phylogeny

> The technologies that permit the manipulation and sequencing of genetic material are revolutionizing biology. . . . New strains of agriculturally important plants and animals can now be engineered, as can organisms to help remove the vast quantities of pollutants our race inflicts on this planet. New, faster, and more accurate means of diagnosing diseases are being developed, and powerful, specific treatments for some incurable diseases seem in the offing.
>
> The culmination of this revolution is seen as that remarkable undertaking, the sequencing of the human genome in its entirety—sometimes referred to the Holy Grail of Biology. What medical miracles we could then perform; *Homo sapiens* could come to know himself to the very molecular essence of his being!
>
> However, this is not the revolution I see. The real revolution is a far quieter one. . . . The revolution we are witnessing is in evolutionary biology. . . . To know the sequence of every gene in the genomes of a number of prokaryotes is to possess wisdom of incredible antiquity and enormous value—something that was unimaginable to the biologist even as recently as the 1970s.
>
> —Carl R. Woese, 1991

SCIENTISTS AGREE TODAY that life on Earth originated between 3.5 and 4 billion years ago. To appreciate the long natural history that preceded our own evolution, biologists use the analogy of a twenty-four-hour geological time span. During the first few hours of Earth's existence, beginning 4.5 billion years ago it was blasted by an intermittent stream of huge, ocean-vaporizing meteorites, making life impossible. Around 4:00 A.M., life emerged, and at 5:30 A.M., the oldest bacterial fossils were entombed. Eukaryotic microbes, with nuclei and chromosomes, appeared around 2:00 P.M., and, at 6:00 P.M., planktonic organisms capable of reproducing by sex emerged. At 8:30 P.M., larger seaweeds arose, and a few minutes later so did early wormlike and jellyfish-like creatures (known as the Ediacaran biota). Subsequently, around 9:00 P.M., following a major ice age, shelled animals such as trilobites and snail-like creatures arose, marking the beginning of the Cambrian system of the oldest fossilized plants and animals. The evolution of the plants and animals with which we are familiar represented the last three hours of the twenty-four-hour geologic day. We humans arose about one minute before midnight.[1]

The evolutionary synthesis of the 1930s and 1940s dealt with the evolution of plants and animals over the last 560 million years. During the last two decades of the twentieth century, biologists developed methods to trace life back billions of years to investigate early microbial evolution with the aim of creating a universal phylogeny, something Darwin could only imagine. Once considered far outside the boundaries of respectable scientific discussion, the study of microbial evolution is one of the most vibrant fields of modern evolutionary biology. This revolution is due to the development of new techniques for comparing nucleic acids and proteins, with the genesis of what is called molecular evolution. These explorations of the first 75 percent of Earth's biological history have involved more than an extension of classical evolutionary theory. It has contradicted long-held assumptions about the course of evolution and the diversity of life on Earth, and it has brought forward evidence for the fundamental importance of non-Mendelian mechanisms in the evolution of microorganisms.

Precambrian Explosion

Ever since Darwin, evolutionists assumed that life evolved from microbial ancestors, but they lacked fossil evidence about what these few forms were and how they evolved.[2] Most of the fossil record comes from extinct plants and animals which burst onto the scene between 560 and 495 million years ago, the Cambrian period, named after the location where the fossil-containing rocks of this age were first formally described: Cambria, the Roman name for Wales. In 1909, Charles Doolittle Walcott (1850–1927), secretary of the Smithsonian Institution, discovered a fossil locality in the Burgess Pass in the Canadian Rocky Mountains on the eastern border of British Columbia. The Burgess shale locality preserved a wide diversity of fossil invertebrate animals from the Cambrian explosion: arthropods, the most common animals on Earth; trilobites, now extinct; lobsters, crabs and shrimp; spiders and scorpions; and insects. The Burgess shale not only contained all the invertebrate marine life extant today but also many more kinds of arthropods now extinct. The extent of the loss of life forms has been so great that some evolutionists consider the history of animal life to be one of massive removal.[3] Why the Cambrian explosion occurred is not fully understood.

Some of the first convincing Precambrian fossils were found in 1946 in the Ediacara Hills of Australia, and subsequently elsewhere. The Ediacaran biota were soft-bodied marine creatures. These unusual fossils were originally interpreted to be those of jellyfish, strange worms, and frondlike corals that lived 600–700 million years ago. Their relationship to Cambrian animals is still uncertain. They may be their ancestors or life forms that went completely extinct.[4]

Paleontologists considered the chances of finding microbial fossils to be exceedingly small. Not only do microbes lack shells, bones, or other hard parts resistant to decay, but, even if they were entombed in sediment, they would be almost always flattened beyond recognition. Then there are problems of finding ancient rocks that might contain them: the older the rock, the rarer it is. Finding microbial fossils was a matter of knowing precisely where to look and how to measure them. Fossil hunters had been greatly aided by geological surveys aimed at exploring and identify rock strata and at making maps of importance to a nation's economy. Determining the age of rocks is important for such surveys because certain types of economically important rocks were formed abundantly only during particular geological times. (For example, iron-rich rocks for making steel and certain types of uranium deposits for nuclear reactors are plentiful only in Precambrian strata older than 2 billion years, whereas coal and oil reserves are common in particular periods of the past 550 million years.)

Specific radioactive isotopes are needed to date ancient rock. Radioactive carbon—^{14}C—is used for dating fossil humans or human artifacts. However, radioactive carbon rapidly decays to a stable isotopic form of nitrogen (^{14}N), after 50,000 to 60,000 years. So it is useless for dating anything older than 60,000 years. To determine the age of ancient rocks uranium ^{238} is used; its rate of decay (to ^{206}Pb) is extremely slow: its half-life (that is, the time it takes for half of it to be converted to the new form) is about 4.5 billion years.[5] Paleontologists were able to push the fossil record back billions of years.

The oldest rock-bearing fossils are some 3.4 billion years old, and paleobiologists generally argue that microbes may well have existed much earlier.[6] Mat-building communities of microbes form large structures called stromatolites (Greek *stromatos*, "bed covering" and *lithos*, "rock") were ubiquitous in the Precambrian. The ancient microbial fossils would include cyanobacteria (previously known as blue-green algae). Cyanobacteria have essentially the same photosynthetic apparatus as plants. They use energy from the Sun to synthesize glucose from carbon dioxide and water, giving off oxygen ($CO_2 + H_2O \Rightarrow CHO_2 + O_2$). Also like plants and animals, all cyanobacteria can utilize oxygen (by aerobic respiration). These are complex ways to live. Simpler microbes, paleobiologists argue, which would have lived in the absence of oxygen (that is, anaerobically), must have arisen far earlier than ever had been imagined.[7] The subsequent emergence of the oxygen-producing cyanobacteria would have changed the earth dramatically. Oxygen is toxic to anaerobic bacteria, and the pumping of oxygen into the atmosphere by cyanobacteria would have resulted in the first mass extinction of life on Earth (the last one was caused by an asteroid that wiped out the dinosaurs about 65 million years ago). Anaerobes then would have retreated to oxygen-free habitats, which are quite common on Earth.[8]

The existence of microbes at such an early date would entail a radical shift in biologists' thinking about how life would have been formed. It has been generally assumed that life was a low-probability event that occurred by chance collision of molecules over billions of years.[9] During its first 500 million years, the earth was a steaming, belching system bombarded by meteorites. Evidence that microbes had emerged some 4 billion years ago, very soon after the surface of the earth cooled enough for water to condense and oceans to form, would greatly curtail the role of chance.

Molecular Clocks

Ever since Pasteur, the study of bacteria had developed outside an evolutionary framework. There was no natural genealogy for them, no universal "tree of life." Bacteria were important for pathology and for biochemists, and one bacterium in particular, *E. coli*, had proven to be a most productive tool for molecular geneticists. However, bacteria hold a special place in the global ecology. Organisms do not simply adapt to their surroundings; they modify them. And nothing on Earth has done that in a more dramatic fashion than bacteria. They are the source of our oxygen atmosphere, the basis of our food chains, the main mechanisms for recycling organic material, the agents of mineral deposition, and the basis of fertile soil. And, as will be discussed in the following chapter, bacteria have played the main role in our own evolution. They are at the roots of our very being: our bodies are filled with them, and the cells that make up our bodies contain vital vestiges of bacterial symbionts. Yet, until recently, biologists knew little more about the natural history of bacteria than had Pasteur.

Throughout most of the twentieth century, bacteria were classified on the basis of their morphology and physiology, including their disease-causing traits, a taxonomy of considerable importance for pathologists. But attempts at constructing a natural classification—a genealogy—for bacteria were unsuccessful. Evolutionists had constructed the phylogenetic relationships among plants and animals by comparing the similarities and differences in phenotypes. But bacteria possessed few morphological traits. In their influential book *The Microbial World* (1963), Roger Stanier (1916–1982) and his colleagues Michael Douderoff and Edward Adelberg concluded that "classification of bacteria into natural relationships became the purest of speculation, completely unsupported by any sort of evidence." They lamented that "the ultimate scientific goal of biological classification cannot be achieved in the case of bacteria."[10] An attempt at a natural classification of bacteria began again in the next decade with new techniques from molecular biology.

In a landmark paper of 1965, Emile Zuckerkandl and Linus Pauling proposed that phylogenies could be reconstructed from comparisons of molecular structures.[11] Instead of examining only anatomy and physiology, they reasoned, one could base family trees on the order or sequence of the building blocks in selected genes or proteins. The approach was eminently logical. Individual genes composed of unique sequences of the nucleotides adenine, guanine, cytosine, and thymine (AGCT) would typically serve as the blueprints for the primary structure of proteins, which consist of strings of specifically ordered amino acids. As genes mutate, the simplest change would be to replace one base for another, for example, the G in the sequence ". . . AAG . . ." to C, yielding ". . . AAC . . . ," which in turn could change the amino acid glutamic acid (AAG) to aspartic acid (AAC) at a certain position in some protein. Some of these small changes would have little effect, but others could drastically change the encoded protein: small changes can have large effects. (Consider, for example, that DNA sequences between humans differ by only 0.1 percent, and between humans and chimps by only 2 percent.)

Genetic mutations that either have no effect or that improve protein function would accumulate over time. As two species diverge from an ancestor, the sequences of the genes they share also diverge. And as time advances, the genetic divergence will increase. One could therefore reconstruct the evolutionary past of species and make phylogenetic trees by assessing the sequence divergence of genes or of proteins isolated from those organisms.[12] Pauling and Zuckerkandl thus introduced what they called "the molecular clock." They assumed a rate constancy at the molecular level: that is, changes in the amino acid sequence of a protein from different species should be "approximately proportional in number to evolutionary time."[13] In other words, molecules may not evolve in the same (irregular) way that morphological features of an organism did. This was not the first announcement of molecular evolution, nor of the idea behind a molecular clock. In 1963, Emmanuel Margoliash (b. 1920) and his colleagues compared similarities and differences in amino acid sequences of cytochrome c molecules from horses, humans, pigs, rabbits, chickens, tuna, and baker's yeast to infer phylogenetic relationships.[14] Zuckerkandl and Pauling had also pioneered the use of amino acid sequence comparisons to infer evolutionary relationships in primate phylogeny with data from hemoglobin sequences. But their paper of 1965 crystallized the idea of molecular evolution for many who entered this field.

Techniques for sequencing RNA and DNA dramatically improved in the 1970s and 1980s.[15] In order to compare the nucleotide sequences of genes, one needed the means to clone DNA, that is, to make many copies of sequences from minute samples. The invention of the polymerase chain reaction (PCR) solved this problem.[16] PCR revolutionized many aspects of biology. In criminal investigations, DNA "fingerprints" could be prepared from cells in a tiny speck of dried blood or at the base of a single human hair. Gene amplification was also crucial to the development of the Human Genome Project and its promised new era of gene therapy. Sequencing to understand microbial evolutionary history developed in parallel, but at a comparatively modest scale.

The Origin of the Code

The new era in the study of bacterial evolution was led by Carl Woese (b. 1928) at the University of Illinois (Urbana). He developed an empirical framework for a natural classification of microbes based on comparisons of ribosomal RNA sequences. In doing so, he offered a radical revision of the history of life on Earth. Woese had two complementary interests: comparing genetic systems in order to unravel microbial genealogies, and using genealogies to understand the evolution of the genetic system itself. Educated in biophysics at Yale, Woese was already recognized for his influential book of 1967, *The Genetic Code*.[17] His subsequent aim was to understand how the complex mechanism for translating nucleic acids into the amino acid sequences of proteins had evolved. To appreciate the problem, we must first step back and consider how amino acids may have arisen in an abiotic soup billions of years ago.

In the 1920s, the Russian biochemist Aleksandr Ivanovich Oparin (1894–1980) suggested that if the prelife atmosphere on Earth lacked free oxygen, diverse organic compounds could be synthesized from atmospheric gases containing carbon, hydrogen, oxygen, and nitrogen (CHON) using energy from ultraviolet light, from volcanic heat or from lightning.[18] J. B. S. Haldane proposed the same idea in 1929.[19] The absence of free oxygen was an important assumption, because if there were free oxygen, all organic compounds would be rapidly oxidized (burned) to CO_2 and H_2O. But if the early environment lacked oxygen, then simple organic compounds could accumulate, giving rise to the organic constituents of the first cells, which would feed on other ingredients of the primordial soup from which they emerged.

These ideas were tested in 1953 by the American chemist Stanley Miller, a student of the Nobel laureate Harold C. Urey at the University of California, Berkeley. Urey, the world's expert on the early solar system, suggested that the Earth's primordial atmosphere would have been favorable to the formation of simple organic molecules. Carbon would have existed there as methane, nitrogen as ammonia, and there would be plenty of hydrogen. Under Urey's direction, Miller passed an electric current (simulating lightning) through a chamber containing hydrogen, water, methane (CH_4) and ammonia (NH_3) (simulating the earth's atmosphere). After a few months, he had brewed seven amino acids, including three (glycine, alanine, and aspartic acid) that are components of proteins today.[20] Using similar techniques, scientists subsequently produced a wide range of organic compounds including not only amino acids, but also sugars, purines, and pyrimidines, components of RNA and DNA. Although doubts remain about the underlying assumptions about the state of the earth's early atmosphere, these simulation experiments at least showed how the ingredients of proteins and genes could originate abiotically.[21]

Research on the origins of life increased dramatically when the National Aeronautics and Space Administration (NASA) became a major sponsor in the 1960s. Its original charter included a mission for the agency to search for extraterrestrial life, or to at least understand how life might have arisen on Earth. Its funding was also crucial for many of the major advances in microbial evolutionary biology.

During the 1980s molecular biologists also shed light on how the first genelike entities might have evolved. This might appear to be a nonproblem because we have heard so often that genes are "self-replicating." Actually, many protein enzymes are involved in DNA reproduction (see chapter 17). This has led to a longstanding paradox: DNA needs proteins in order to function, and yet proteins cannot be made without DNA. Sidney Altman of Yale University and Thomas Cech of the University of Colorado provided one possible solution. They were awarded a Nobel Prize in 1989 for showing that RNAs can function as enzymes, called "ribozymes." The first "replicator" may not have been DNA, but rather RNA, which could act as both template and agent of its own reproduction. Molecular biologists have envisioned an "RNA world" of self-reproducing entities, the forerunners of the gene. Certainly one could envisage alternatives. Some theorists have suggested the existence of self-reproducing polypeptides, others have argued that a membrane-bound metabolic system would have emerged first, and perhaps replicating molecules developed within it.[22] Still others suggest that there may have been a pre-RNA world based on a different chemistry, perhaps of clays or of other mineral surfaces.[23]

The origin of a genelike entity was only one problem. The amino acids of proteins and the nucleotide sequences of DNA and RNA, which specify those sequences, are strung together in a precise order. How did these sequences evolve and come together in the appropriate way to make a functional protein? What about the genetic code—the translation of the nucleic acid into the amino acid sequence of protein? How did the codons evolve for specific amino acids, and how did they come to be arranged in such a manner as to make the primary structure of a protein? Here was another paradox: making proteins from DNA requires translation machinery, itself partly made up of protein. A cell employs ribosomes to read the RNA copy of the gene, and ribosomes are made up of a complex of about fifty different proteins with several different forms of RNA bound within it. So how could the ribosomal proteins necessary for translation have originated without translation? This problem, the evolution of the genetic code, interested Woese.

Francis Crick had proposed that the evolution of the code could have been some sort of historical quirk, a "frozen accident." In other words, it evolved merely because the cell needed to have a relationship between nucleic acid and protein structure, not because relationships existed between nucleic acids and amino acids that forced the genetic code to evolve.[24] But Woese suspected that the translation machinery had to have evolved in steps, with selection acting on the mechanism's speed and accuracy.[25] And this is what led him to microbial phylogeny. He supposed that one might be able to follow the translation machinery's early evolution, before cells reached their present sophisticated complexity. To do so meant that one had to construct a universal evolutionary framework, a deep phylogeny—a universal tree that would in effect encompass all organisms.[26] The universal tree would therefore be far more than the ultimate ordering of life on Earth; it held the secret to its existence as well.[27] Thus, he set out with the hope of tracing cell life back to a universal ancestor, the "progenote," which might not possess a modern translation machinery.

A Code for Classification

Woese's first step in tracing life's origins was to construct a natural classification of bacteria, one that ordered them in terms of their evolutionary history or genealogy. Classifying bacteria according to their shape and metabolism without a phylogenetic understanding confined microbiology to the Dark Ages. As Woese commented years later, "It was as if you went to a zoo and had no way of telling the lions from the elephants from the orangutans—or any of these from the trees."[28] Not surprisingly, he focused on ribosomes, more particularly ribosomal RNAs (rRNAs). All cells need rRNAs to construct proteins, and therefore their similarities and differences could be used to track every lineage of life from bacteria to elephants. Ribosomal RNA is also abundant in cells, so it was easy to extract.

Ribosomes served at the core of an organism, and since they interacted with at least a hundred proteins, Woese suspected that their molecular sequences would change so slowly that they would hardly differ between species. They would be among the most "conserved" elements in all organisms and therefore make excellent recorders of life's long evolutionary past. Ribosomes are composed of two pieces or subunits, a smaller one slightly cupped inside a larger one. Woese chose to compare sequences of the small subunit ribosomal RNA (SSU rRNA). He believed that SSU rRNA sequences (or, more precisely, the genes encoding them) would change so slowly over evolutionary time that they would retain traces of ancestral patterns laid down of billions of years ago at the deep roots of the phylogenetic tree. SSU rRNA would thus serve as a universal molecular chronometer.

Woese's methods were at first indirect and tedious.[29] But by the mid 1970s, he and his collaborators had sequenced the SSU rRNA from about sixty kinds of bacteria and arranged them by genetic similarity.[30] Their results contradicted the standard classification based on morphological similarities of bacteria. For example, taxonomic reference texts on bacteria distinguished the gliding bacteria, the sheathed bacteria, the appendaged bacteria, the spiral and curved bacteria, the rickettsias, *Flavobacterium*, and *Pseudomonas*. But Woese argued that these groups had no biological or evolutionary meaning; they were really paraphyletic, that is, they were made up of parts of many different genealogical groups. Assigning an individual to these categories said very little about its actual identity. It would be almost like placing worms and snakes in the same group because both are long, thin, and slithery.

By the late 1980s, the study of microbial phylogeny by rRNA sequences had attracted many biologists, who classified several thousands of bacterial species so as to sketch an outline of a universal tree of life. With a universal evolutionary tree, biologists could begin to understand bacteria as they do the rest of life, as organisms with histories and evolutionary relationships to one another and to all other organisms. Ribosomal RNA sequencing held the promise of tracing relationships back to the common ancestor of all existing life, the mother of all cells, the progenote.[31] Under the critical eye of classical evolutionists, molecular phylogenies based on ribosomal RNAs were also applied to animals and to protists.[32] But no studies and interpretations caused more controversy and interest than those of Woese and his colleagues.

A Trilogy of Life

The most spectacular claim came in 1977 when Woese and George Fox announced that they had discovered a new form of life: a group of bacteria-like organisms that was genetically and historically very different. They named them the archaebacteria, to distinguish them from true bacteria or eubacteria.[33] In 1990, Woese and collaborators shortened these organisms' name to Archaea to further emphasize their fundamental difference from eubacteria.[34] In their view, Archaea were more different from eubacteria than humans were from plants. They were morphologically as diverse as the bacteria—rods, spirals, marble-like cells—but they all completely lacked the signature SSU rRNA sequences that Woese had come to recognize as bacteria. The first organisms they assigned to the archaebacteria were methanogens, and this suggested their antiquity. As Woese and Fox wrote in 1977:

> The apparent antiquity of the methanogenic phenotype plus the fact that it seems well suited to the type of environment presumed to exist on Earth 3–4 billion years ago lead us tentatively to name this urkingdom the archaebacteria. Whether or not other biochemically distinct phenotypes exist in this kingdom is clearly an important question upon which may turn our concept of the nature and ancestry of the first prokaryotes.[35]

The archaebacteria had other traits in common. They lived in extreme environments.[36] There were the salt-loving halophiles, found in brines five times as salty as the ocean, and thermophiles, found in geothermal environments that would cook other organisms and in anaerobic habitats where even trace amounts of oxygen would prove lethal to them. Archaeabacteria are "extremophiles" par excellence. Still other phenotypic features corroborated the conclusion from the RNA data about the uniqueness of the archaebacteria. Their cell membranes are made up of unique lipids (fatty substances) with quite distinctive physical properties, and the structures of the proteins responsible for several crucial cellular processes such as transcription and translation are different from those of their counterparts in eubacteria.

At Woese's suggestion, in November 1977, the National Aeronautics and Space Administration (NASA) issued a press bulletin announcing the discovery of a new form of life. It was splashed across the front pages of the *New York Times,* carried in other major newspapers and magazines, and featured on evening television news programs. For the press and the public, the discovery of the archaebacteria was a momentous event; it touched on the age-old concern of where we came from. But as the first organisms, this universal ancestor contradicted biochemists' assumptions about the conditions under which life first emerged on Earth. The leading theory had long held that life began when lightning activated molecules in the atmosphere, which then reacted chemically with one another, and then that atmosphere deposited those compounds in the oceans where they continued to react to produce warm soup of organic molecules. Darwin had said little about the origins of life, except for famous note in a letter to Joseph Hooker in 1871: "But if (and oh! what a big if!) we could conceive in some warm little pond with all sorts of ammonia and phosphoric salts, light, heat, electricity and etc., present that a protein compound was chemically formed, ready to undergo still more complex changes."[37] Woese's tree seemed to point to a strikingly different geochemical context. If the first organisms were Archaea-like creatures, instead of evolving in a mild soup of organic molecules, they may have been born in boiling, sulfurous pools, or hot, mineral-laden, deep-sea volcanic vents.

Comparative studies of RNA sequences also offered a novel conceptual scheme for the first billion years of life on Earth. Formerly, it had been supposed that life's history was straightforward and progressive: from simple bacteria, to more complex bacteria, to cells with nuclei, from which came plants and animals. However, the rRNA studies indicated that life divided into major independent lineages much sooner than biologists had ever imagined. The stunning implication of this branching was that the prokaryotic world was richly diversified, beyond what anyone had imagined. Plants and animals were but recent twigs on what amounted to a great microbial tree of life. Over the previous hundred years, there had been several spectacular discoveries in the world of biodiversity, such as okapi in the Congo forest, the only living relative of the giraffe; and *Latimeria* in the Indian Ocean, a living coelacanth, a fish believed to have been extinct for 60 million years. But these were tiny spots on the world map of biodiversity. Woese's discovery of the archaebacteria was compared to the discovery of a new continent.[38]

The postulation of the Archaea also contradicted the established belief in a basic dichotomy in the living world. Since the 1960s, with the development of the electron microscopy, cell biologists had come to agree that all organisms could be grouped into two fundamental forms, the prokaryotes and the eukaryotes, as Edouard Chatton had named them in 1925.[39] Eukaryotes had a membrane-bound nucleus, a cytoskeleton, an intricate system of internal membranes, mitochondria that perform respiration, and, in the case of plants, chloroplasts. Prokaryotes were smaller and lacked all of these structures. In *The Microbial World*, Stanier, Douderoff, and Adelberg declared that, "In fact, this basic divergence in cellular structure, which separates the bacteria and blue-green algae from all other cellular organisms, represents the greatest single evolutionary discontinuity to be found in the present-day world."[40]

While paleontologists referred to "the Cambrian explosion," those who studied cellular organization insisted that the real "big bang" of biology occurred at least 1.8 billion years earlier when the eukaryote arose. With its membrane-bound nucleus and all the associated features, such as mitosis, meiosis, and multiple chromosomes to package up to tens of thousands of genes per cell, it provided the organismic conditions for the differentiation of tissues, organs, and organ systems of plants and animals. Bacteria had only an unpackaged single strand of DNA, holding some four thousand genes. By the 1970s, biologists led by R. H. Wittaker and Lynn Margulis generally agreed that eukaryotes embraced four kingdoms: Planta, Animalia, Protista, and Fungi. Prokaryotes (the bacteria) were classified as a fifth kingdom.[41] Woese's model of three domains of life—Archaea, Eubacteria, and Eukarya—contradicted that order of things. As Woese saw it, the prokaryote-eukaryote dichotomy was an organizational distinction only. It had no phylogenetic meaning, and it was a hindrance to understanding microbial evolution.

The rRNA approach to microbial phylogeny and the three-domain proposal were developed by many biologists, including Otto Kandler and Wolfram Zillig in Germany; Mitchel Sogin, Gary Olsen, James Lake, and Norman Pace in the United States; and Ford Doolittle and Michael Gray and associates in Canada. This work led to an upheaval in bacterial systematics and a major revision of texts in regard to the universal genealogical tree. Some of those who worked on the evolution and phylogeny of Eubacteria and Archaea suggested that these microbes collectively possessed greater biological diversity than plants and animals combined.[42]

Microbes were largely ignored by most biologists and virtually unknown to the public except in contexts of disease and rot, on the one hand, and the making of bread, cheese, beer, and wine, on the other. Biology textbooks still teach biodiversity almost exclusively in terms of animals and plants; insects usually top the count of species, with about a half-million described to date. In terms of diversity, plants and animals obviously showed far greater and more elaborate morphological differences, but bacteria were far more diverse biochemically. As Pace remarked, within one insect species one could find hundreds or thousands of distinct microbial "species."[43] A handful of soil contains billions of them, so many different types that accurate numbers remained unknown. Most life in the ocean is microbial. Prokaryotes can live in an incredible variety of conditions from well below freezing to above the normal boiling temperature of water. Extreme halophiles thrive in brines so saturated that they would pickle other life. Other microbes live in the deep border of the trench at the bottom of the Red Sea in hot saline loaded with toxic heavy metal ions. They are also found growing in oil deposits deep underground. Microbial researchers argued that they had barely scratched the surface of microbial diversity. The entire surface of this planet down to a depth of at least several kilometers may be a habitat for eubacteria and archaebacteria. Prokaryotes, it was announced in 1998, constituted the greatest biomass on Earth.[44]

In the mid-1990s, the U.S. Department of Energy instituted a Microbial Genome Initiative, an offshoot of the Human Genome Project it had initiated with the National Institutes of Health five years earlier. The Human Genome Project was rationalized in terms of its medical benefits, and at first microbial genomics was justified similarly, each microbe for a specific practical purpose (medical, agricultural, or industrial).[45] But Woese and his colleagues also saw a much deeper and more fundamental rationale. Humans were stressing the biosphere, and soon a day would come when a deep knowledge of the biosphere and its capacity to adapt would be critical. Bacteria are largely responsible for the overall state of the biosphere: our oxygen atmosphere exists (directly or indirectly) because of them, and they are vital to the regulation of the planet's surface temperature through their roles in carbon dioxide turnover and methane production and utilization. Microbial genomics was needed to explore microbial diversity, to understand the interaction between microorganisms and their environments, and to reveal their evolutionary dynamics.

Studying bacterial diversity entailed certain technical problems as well as financial ones in the sense of obtaining funds. Technically, knowledge of microorganisms and their niches depends mainly on studies of pure cultures in the laboratory. However, those who studied microbial diversity estimated that more than 99 percent of organisms seen microscopically cannot so far be cultivated by routine techniques.[46] Beginning in the 1980s, Pace and his collaborators at the University of Colorado developed means to get around these limitations.[47] He reasoned that an inventory of microbes in a niche could be taken by sequencing rRNA genes obtained from DNA isolated directly from the habitat itself. Arguing that biologists' understanding of the makeup of the microbial world was rudimentary, in 1998 Pace called for a representative survey of the Earth's microbiodiversity with the use of automated sequencing technology.[48]

Dissension and Disaffection

By the late 1990s, just as the three domain proposal and the outlines of a universal phylogenetic tree were becoming well established, the microbial order based on rRNAs was challenged. Problems began with data from complete genome analysis of microbes begun in the mid-1990s. Phylogenies based on genes other than those for rRNA often indicated a different view of microbial genealogies. Moreover, the new genomic data showed that archaebacteria and eubacteria had many genes in common. Perhaps the Archaea were not that different after all. Ernst Mayr led the attack in 1998, and several disaffected molecular phylogenicists soon followed.

That one could do molecular phylogenies, whether of bacteria or primates, based on comparisons of one or a few molecules had long been a simmering issue. From the outset some had argued that one could not classify microbes on the basis of rRNA phylogenies.[49] And as early as 1990, Mayr had sent a note to the journal *Nature* protesting that separating the Archaea from bacteria and claiming they formed a "domain," "superkingdom," or "empire" was grossly misleading.[50] Indeed, classical evolutionists such Mayr, Theodosius Dobzhansky, and George Gaylord Simpson were opposed to the whole field of molecular evolution and any molecular approach to classification from the very beginning.[51]

There were several aspects to their resistance. At the most general level were basic institutional issues between molecular biology on the one hand, well funded and rapidly growing, and traditional evolutionary studies on the other. The swift rise of molecular biology was perceived to be in direct competition with the aims and interests of evolutionary biologists.[52] In 1963, Dobzhansky felt it necessary to remind readers of *Science* that there was also stimulating research going on in "organismic as well as molecular genetics."[53] His note was followed by a commentary from Mayr in the next issue arguing for "more financial and moral support for the classical areas."[54] When the ideas of molecular evolution and a molecular clock emerged, the architects of the synthesis not only resented the intrusion into their domain, but also argued that evolution was an "affair of phenotypes," and one simply could not reduce it to comparative molecular morphology. Not only did classical evolutionists reject molecular methods for studying evolution, they also rejected the molecular evolutionists' answers.

Attempts to stop the molecular clock came to a head when molecular evolutionists argued that the amino acid sequences of proteins did not evolve by adaptation and natural selection. Many changes simply had no effect on the protein structure and therefore no adaptive value. The evolution of proteins without selection was dubbed the neutral theory by Mooto Kimura and non-Darwinian evolution by Jack King and Thomas Jukes in the late 1960s.[55] Mayr and Simpson found it incredible that molecular and morphological evolution could be different in mechanism and rate. In any case, they argued, the only evolution that mattered operated by natural selection.[56] Changes at the molecular level that did not affect the phenotype were really of "no interest for organismal biologists as they are not involved in the evolution of whole organisms."[57] Though the debate between panselectionists and neutralists has fizzled in recent years, the scope and significance of the neutral theory in molecular evolution remain unsettled to the present day.

Mayr's 1998 offensive against rRNA phylogenies came just as the three-domain proposal was getting into the biology textbooks and microbial evolutionists were reporting that bacteria contained a diversity that rivaled and indeed surpassed that of all the macrobiological world. As Mayr saw it, Woese's three-domain proposal was absurd: prokaryotes did not possess a degree of diversity even remotely comparable to that of the eukaryotic world. It was simply preposterous to compare the molecular genetic differences between bacteria and Archaea to the huge morphological differences between eukaryotes and prokaryotes. Echoing comments from previous confrontations with molecular evolutionists, he reasserted that evolution was "an affair of phenotypes," and, on this basis, he insisted that "all archaebacteria are nearly indistinguishable"; even if one took prokaryotes as a whole, he argued, the group "does not reach anywhere the size and diversity of eukaryotes."[58] Microbial phylogenists had so far described only about 200 archaebacterial species and only about 10,000 eubacterial species, whereas Mayr suspected that within eukaryotes there were more than 30 million species. There were 10,000 species of birds alone, and of course hundreds of thousands of species of insects.[59] He remarked that

> the eukaryote genome is larger than the prokaryote genome by several orders of magnitude. And it is precisely this part of the eukaryote genome that is most characteristic for the eukaryotes. This includes not only the genetic program for the nucleus and mitosis, but the capacity for sexual reproduction, meiosis, and the ability to produce the wonderful organic diversity represented by jellyfish, butterflies, di-

> nosaurs, hummingbirds, yeasts, giant kelp, and giant sequoias. To sweep all this under the rug and claim that difference between the two kinds of bacteria is of the same weight as the difference between the prokaryotes and the extraordinary world of the eukaryotes strikes me as incomprehensible."[60]

Woese responded in detail. As he saw it, the differences between himself and Mayr were not simply a matter of a molecular versus an organismic approach to biology and evolution. Mayr looked at evolution from the top down, from the present to the past, observing the great phenotypic diversity of plants and animals that had evolved over the previous 560 million years. Woese looked from the bottom up: his concern was understanding evolutionary processes over the first two billion years of evolution, based on observing differences in molecules and genes. From his perspective, prokaryotes could not be defined negatively and in opposition to eukaryotes or in terms of the kingdoms some of them later gave rise to. Prokaryotes had to be understood in their own terms and from a historical perspective. As he commented, "The science of biology is very different from these two perspectives and its future even more so."[61] But as he also noted, Mayr's critique had come at an opportune time: the three-domain proposal seemed to be under considerable strain from molecular studies of whole genomes over the previous three years.

Defection from within the ranks of molecular evolutionists grew during the late 1990s. Several leading microbial phylogenicists saw in Mayr's critique much that they considered to be true, as central features of the Archaeal story were challenged. First, the hypothesis that the Archaea were so very different from bacteria because the two groups diverged when life was quite new was complicated when genes other than those for rRNA were compared. Analyses of whole genomes (seventy had been sequenced by 2002) showed that Archaebacteria and Eubacteria had numerous genes in common, and they shared a rich biochemical complexity. Therefore, some suggested that the Archaea actually may not be very old, and others suggested it was not a single domain. Comparisons of genes for other functions also seemed to contradict the phylogenetic lineages deduced from rRNA sequences. For example, although comparisons of SSU rRNA placed microsporidia low on the phylogenetic tree, comparisons of the gene for the enzyme RNA polymerase placed the microsporidia higher on the tree with the fungi. Thus, some insisted, as critics had from the outset, that one simply could not use the SSU rRNA trees to trace microbial life.[62]

Lateral Gene Transfer

Still another fundamental issue remained. Not only did the phylogenies from the new genomic studies disagree with the standard rRNA-based phylogenies, but the new genome data also conflicted among themselves. Comparisons of individual gene phylogenies (other than those concerned with the translation machinery) often indicated different organismic genealogies. Phylogenicists suspected that the mix-up was due to evolutionary mechanisms whose scope and significance they had underestimated: gene transfer between unrelated groups. In addition to the "normal" gene transfer from parent to offspring, known as vertical transfer, genes are also transmitted between distinct evolutionary lineages. This is known as horizontal or lateral gene transfer. Bacteria possess several mechanisms for transmitting genes between unrelated groups—through transformations, viral transduction, and conjugation (see chapters 16 and 19). Thus, a bacterium of one strain may have acquired one or several genes from a completely unrelated organism.[63] For example, if organism type A and organism type B carry the same gene for a protein, it may be not because they both belong to the same taxonomic group, but because one of them acquired that gene from a third type of organism C that is not ancestral to them. By the end of the twentieth century, all microbial phylogenicists had come to recognize the fundamental importance of lateral gene transfer in the evolution of bacteria.

Recognition of the pervasiveness of lateral gene transfer entailed modifications to two seminal views about microbial evolution and the course of early evolution: microbes' treelike branching genealogies and the genealogy's hierarchical nature. In the Darwinian order of things, you sort plants and animals into species. Then you sort species resembling one another into genera and label each species with a two-part Latinized name. Similar species are grouped in a genus, genera are grouped into families, families into orders, and so on. The ordering is essentially like a military organization: individuals into squads, squads into platoons, platoons into companies. Each species belongs to one and only one genus, each genus to one and only one family, and so on. This phylogenetic order is based on the idea common descent from an ever-decreasing number of ancestors leading back to the origin of life, the common ancestor of all living things. This was fine as long as there were barriers to gene transfer between species (and if all cells had one common ancestor). However, the evidence for gene transfer between bacterial groups contradicted this notion of species and this hierarchical view of evolution.

Bacteria are composites; they have acquired and integrated genes from diverse taxa. Thus, instead of the branching genealogical tree that Darwin imagined for plants and animals, the pattern of early evolution (at least) is reticulated. Lateral gene transfers between taxa make it resemble a web rather than a tree. Accordingly, the resulting evolutionary order may be nonhierarchical.[64] Lateral gene transfer blurs the boundaries between "species." The ease with which genes are interchanged among bacteria has led many microbiologists to suggest that the biological species concept does not apply.[65] Some have suggested further that the entire bacterial world should be thought of as a superorganism.[66]

Speculations about the nature and intensity of horizontal gene transfer led some microbial phylogeneticists to fear that the whole enterprise of classification may be insolvent, that a natural phylogeny of prokaryotes may be impossible. The problem was not one of reductionism, of believing one molecule or several could be used to discern deep phylogeny. It was that bacteria did not comply with the rules of evolution for eukaryotes. While eukaryotes do have inscribed in their organization the largest amount of their own history, horizontal gene transfer may exclude bacteria (sensu lato). Early evolution, the deep roots of phylogeny, may lie beyond the chronicles of history. This was the message that some had gleaned from comparisons of complete genomes since the mid-1990s.

Biologists who had once been Woese's chief advocates, such as Ford Doolittle in Halifax, departed from his three-domain proposal as he became skeptical of ever constructing microbial phylogenies.[67] At the same time, others offered alternative models. Each emphasized different data. For example, James Lake and his collaborators at UCLA distinguished, within the Archaea, a group he called the eocytes ("early cells") that, he argued, was more closely related to eukaryotes than to Archaea.[68] Radhey Gupta and his co-workers at McMaster's University in Ontario argued that the bacterial world exhibited a fundamental dichotomy between what he called monoderms (bacteria possessing a single membrane) and diderms (those with a double membrane).[69]

Yet, others, including Karl-Hans Schleifer in Germany, Charles Kurland in Sweden, and Woese and his collaborator Gary Olsen, upheld microbial phylogenies based on rRNA comparisons and the three-domain hypothesis. Indeed, they saw an irony in the phylogenetic turbulence, because Woese had anticipated these problems from the outset. The reason he picked rRNA in the first place was precisely that it would be less likely to evolve at the same rate as other genes. Because the SSU rRNA gene is at the core of the cell's most complex machinery, they argued, it would be unlikely to be transferable between phylogenetic groups without disrupting core cellular systems. Therefore, rRNA comparisons would be the only reliable means for tracking bacterial lineages. The central question for all microbial phylogenicists was whether genes for rRNA are exchanged frequently between groups.[70] Microbial evolutionists called for further studies in comparative genomics to establish the principles governing horizontal gene transfer across the bacterial phylogenetic system.[71]

In the meantime, Woese maintained the view he had begun with decades earlier: that one cannot track genealogies using metabolic genes, which might vary frequently, but only genes governing crucial complex tasks. Synthesizing the evidence in 1998, he offered a new conceptual model, pieces of which he had been putting together for two decades.[72] He interpreted the new genomic evidence indicating so many shared genes between bacteria, eubacteria, and eukarya in terms of his long-sought transitional stages in the evolution of the translation apparatus. Ever since Darwin, biologists had assumed that all life on Earth arose from a single ancestral cell.[73] But Woese disagreed with the canon of a single ancestral mother of all cells. He speculated that instead of the expected first cell, the progenote, was a population of precellular entities with underdeveloped and error-prone replication and translation machinery. Before the development of the modern translation apparatus, evolution would be driven by a different mode and tempo. At this time, no individual organisms could be distinguished as such because there had been so much gene mutation and intense horizontal gene transfer. These processes would generate enormous diversity very quickly. Primitive systems would be modular and exchange parts freely. But as the translation machinery evolved, becoming refractory to horizontal gene transfer, so too did definable lineages. This was the great Darwinian divide. The three domains—Archaea, Bacteria, and Eukarya—emerged out of the fuzz as the translation machinery became well defined.

Woese likened the emergence of the three domains to physical annealing: there would first be a period of intense genetic "heat" (high mutation rates and intense gene transfer between lineages that would have short histories) when cellular entities were simple and information systems inaccurate. It would be impossible to discern organismic genealogies. This would be followed by genetic "cooling" with the development of the modern cell with its sophisticated translation apparatus, resulting in the emergence of genealogically recognized domains and taxa. Thus, as Woese concluded:

> the universal ancestor is not a discrete entity. it is, rather, a diverse community of cells that survives and evolves as a biological unit. This communal ancestor has a physical history but not a genealogical one. Over time, this ancestor refined into a smaller number of increasingly complex cell types with the ancestors of the primary groupings of organisms arising as a result.[74]

By the end of the twentieth century, molecular studies of microbial phylogeny had dramatically transformed thinking about the first 2 billion years of life on Earth. Debates centered not only on the importance of lateral gene transfer in the genesis and evolution of bacteria, but also in the genesis of the eukaryotic cell. Indeed, while horizontal gene transfer was recognized to be rampant among bacteria, the acquisition of genes and whole genomes was also known to occur in eukaryotes. As we shall see in the next chapter, the study of symbiosis in development, heredity, and evolution also offered a dramatically different image of evolution, and of the organism, from that offered by classical neo-Darwinian theory.

19

Symbiomics

> The time has come to supplement the century-old philosophy of the germ theory of disease with another chapter concerned with the germ theory of morphogenesis and differentiation.
>
> —René Dubos, 1961

ANCIENT GREEK MYTHOLOGY TELLS of a chimera, a fierce fire-breathing monster, part lion, part goat, and part dragon. The creature caused devastation until it was slain by the young warrior Bellerophon on his winged horse, Pegasus. Today, the word "chimera" is generally used to denote a fantastic idea or figment of the imagination. Until recently, evolutionists and geneticists insisted that no chimera ever existed in nature: crossing the species barrier was alien to nature's laws. Others, though, had long insisted that all organisms are chimeric—symbiotic complexes evolved from genes and genomes of different species. But their views were either discredited or ignored by mainstream biology throughout most the twentieth century.[1]

Since the early 1970s, biotechnologists have learned to move genes across the phylogenetic spectrum from plants and animals to bacteria and back again. Most of the insulin used to treat diabetes is now obtained from bacteria that contain a human insulin gene. In agricultural applications, cultivated plants and animals are genetically engineered to resist pests and herbicides and to grow bigger and faster, all using genes from other species. The creation of such transgenic organisms has been, and continues to be, a subject of intense controversy involving sociopolitical and conservation issues.[2]

Early opponents of this biotechnology argued that the creation of such chimeras is in violation of natural law because genes are not naturally transferred between species. They maintained that evolutionary change always occurs in the most gradual manner. Appealing to what is natural to determine public policy is a precarious business. The transfer of genes across the phylogenetic spectrum is now known to occur naturally. Many bacterial "species" exchange genes, and many protists, plants, and animals harbor symbiotic bacteria that are transmitted hereditarily from one generation to the next. Biologists also agree today that all eukaryotes emerged from mergers between different kinds of bacteria. We humans and all other animals and plants are already chimeras. According to the contemporary conceptual consensus, the mitochondria teeming in the cytoplasm of all eukaryotic cells and the chloroplasts of plants and protists were once free-living bacteria that became incorporated in a primitive host cell some 1.8 billion years ago.[3]

Mitochondria are the organelles in which cellular respiration take place. Without them we cannot breathe. Their primary function is the combustion of foodstuffs using oxygen to assemble the energy-rich molecule adenosine triphosphate (ATP), the main source of energy in virtually all oxygen-dependent (aerobic) organisms. Chloroplasts are the organelles in plants in which photosynthesis takes place; they collect the electromagnetic energy of sunlight to produce organic compounds from water and carbon dioxide, and without them plants do not grow. Mitochondria are thought to have been first acquired as food by predatory microbes. They resisted digestion and proved to be of benefit to their host because the primitive atmosphere increased in oxygen, which otherwise would be toxic to their hosts; the engulfed mitochondria detoxified the oxygen. Chloroplasts originated subsequently when a microbe engulfed cyanobacteria. Their selective advantage to the host is obvious: protists that once needed a constant food supply henceforth could thrive on nothing more than light together with air, water, and a few dissolved minerals. Mitochondria and chloroplasts are inherited from one generation to the next through the cell cytoplasm. In the course of evolution, many of their original bacterial genes were lost, and some were transferred to the cell nucleus.[4]

Developmental Symbiosis

Every eukaryote is a superorganism, a symbiome composed of chromosomal genes, organellar genes, and often other bacterial symbionts as well as viruses.[5] The symbiome, the limit of the multicellular organism, extends beyond the activities of its own cells. All plants and animals involve complex ecological communities of microbes, some of which function as commensals, some as mutualists, and others as parasites, depending on their nature and context. It is estimated that there are about 10^{14} cells in our bodies: 10 percent of them are our own eukaryotic cells, and 90 percent are bacteria comprising about 400 to 500 kinds.[6] They form a sheath on our skin, and they cover the insides of our nose, throat, and gut. Between 20 and 50 percent of the human colon is thought to be occupied by bacteria. Our mouths contain about eighty kinds of bacteria, and our stomach typically contains several hundred kinds. Although little is known about their interactions, if we did not have them in the proper relationship, we would not be able to function. Biologists are just beginning to explore the nature of gut bacteria. Different kinds of bacteria in the human colon are stratified into specific regions in the gut tube. They are picked up as soon as the amnion bursts from the mother's reproductive tract. How do these bacteria know where to go when they colonize the gut in the first hours of human life? How do they hold their own against the influx of new, potentially pathogenic bacteria? Why do these symbionts live peacefully for years and then sometimes turn deadly against us? Why is not the immune system engaged in a constant battle with these intestinal microbes?

Contemporary research indicates that our cells may be fashioned physiologically and morphologically by our bacterial community, which not only provides vitamins K and B_{12}, but may regulate many of our own genes, and may be crucial in warding off pathogens. The intestines of germ-free mice cannot complete differentiation. They require their gut bacteria for that. And bacteria have been shown to regulate transcription of genes involved in several intestinal functions.[7] There is evidence that even the harmless bacteria on our tongue help to protect us against harmful microbes taken in with our food.[8] Because of the immunological importance of bacteria, some researchers are reconsidering the centuries-old practice of probiotics—eating live bacteria to prevent gut infections and other intestinal problems.[9]

Although the bacteria living in our bodies have recently begun to attract some medical interest, for most of the twentieth century, symbiosis research had developed close to the margins of the life sciences and in virtual conflict with the aims and doctrines of the major biological disciplines, as well as with medical concepts of germs.[10] Equipped with many examples of symbiosis with diverse physiological and morphological effects, leading researchers again urge neo-Darwinian evolutionists to reconsider symbiosis in heredity and development and as a general mechanism, in addition to gene mutations and recombination, of evolutionary innovation.[11] Although symbiosis as a source of evolutionary change has been discussed since the nineteenth century, it continues to be strikingly absent from all the standard histories and treatises on evolutionary biology.[12] To gain a better understanding of this aspect of biology, in this chapter I briefly sketch some of the phenomena, issues, and debates that have shaped attitudes about its scope and significance.

Symbiosis Silhouette

The introduction of the term "symbiosis" (from the Greek word for "living together") into biology is attributed to the famed German botanist Anton de Bary, who used it in 1879 when discussing evidence that lichens, which are double organisms—combinations of algae and fungi.[13] Today, it is universally recognized that all of the 15,000 species of lichens are made of fungi and either algae or cyanobacteria. A whole new structure, the thallus, emerges out of their association. For de Bary, lichens and other examples of symbiosis offered actual proof of evolution, and they showed a means of macroevolutionary change in addition to Darwinian gradualism.

In 1885, the German botanist Albert Bernard Frank (who used the term *Symbiotismus* a year before de Bary) reported another important symbiosis between fungi and the roots of forests trees, an association he called mycorrhiza ("fungus root").[14] He suggested that the fungi were of considerable benefit to the plant. For this he was severely ridiculed by his contemporaries, who found it incredible that fungi could be anything but parasitic.[15] Today biologists recognize that Frank was correct. Mycorrhizas are known to occur in practically all terrestrial plants; they mediate entry of the great bulk of nutrients into plants and are thus essential for almost every plant community.[16] The plants benefit from receiving essential nutrients, the fungi by receiving organic compounds from the plant. This symbiosis is thought to have been instrumental in the colonization of land by ancient plants some 450 million years ago.[17] Mycorrhizas are also at the very foundation of the world's most complex and biodiverse ecological systems: the trees of tropical rainforests, in nutrient-poor soils, rely on these fungi to bring them precious nitrogen and minerals.

Nitrogen-fixing bacteria in the root nodules of legumes have also been central to discussions of symbiosis since the nineteenth century. Plants cannot metabolize (fix) the molecular nitrogen that is so abundant in the atmosphere: they need nitrogen in the form of ammonia or other nitrogenous compounds. To overcome this problem, legumes (the second-largest family of flowering plants, including peas, beans, clover, and alfalfa) have formed a symbiotic union with the nitrogen-fixing bacterium *Rhizobium*. The rhizobia live in special nodules produced by the "infection" in the root of the legume, where in return for the nitrogenous compounds they provide by their waste, they are fed sugars exuded by the plant. Symbiotic nitrogen fixation underpins the global nitrogen cycle. Molecular biologists have shown that during the production of the nodule, there is gene transfer between host and bacteria: some of the plant DNA is transmitted to the bacteria.[18] We do not know yet whether the reverse is also true.[19]

A somewhat analogous symbiosis exists between luminescent bacteria and several groups of fish and other marine animals in which they are found. About 1500 marine animal species are bioluminescent; some produce their own light, while others use light-emitting bacteria and have evolved special organs designed to hold them. Luminescent bacteria were first discovered in the late nineteenth century, and the study of their adaptive significance has benefited from the research of Margaret McFall Ngai and her collaborators on the bacterium *Vibrio fischeri* and the small Hawaiian squid *Euprymna scolopes*.[20] The luminescent bacteria protect the squid from predators when it forages near the water's surface at night. It turns out that the light the bacteria emit is of the same spectrum as the moonlight and starlight filtering through the water. When viewed from below, the squid is virtually invisible. The bacteria are acquired from the ocean by each newborn squid, which has a pouch lined with cilia to trap them. When the bacteria take up residence and multiply, a major morphological change occurs: the organ used to catch them is transformed into a light organ. The ocean water is full of bacteria, yet somehow only the light-emitting kind are captured and cultured in this pouch.

Sea anemones, hydra, giant clams, sponges, and the corals that build coral reefs acquire algae from the ocean and harbor them in their cells, where they are nourished by the algae's photosynthetically produced carbon compounds. Corals acquire up to 60 percent of their nutrition from their algal symbionts (symbiodinium), which in return obtain from the coral polyp nitrogenous compounds that are scarce in the crystal-clear tropical waters. The worldwide crisis tropical corals are now experiencing is indicative of what happens when this delicate balance is broken. A prolonged increase in sea-surface temperature, solar irradiation, sedimentation, and inorganic pollutants all cause coral bleaching: corals lose their algae, leaving their tissues so transparent that only the white calcium carbonate skeleton is apparent. Without the algae, the corals starve to death.

Microbial symbionts are especially prevalent in insects, especially sap-sucking, blood-sucking, and wood-eating insects, for which they provide enzymes, vitamins, energy, and sugars.[21] The hind-gut microbial communities of wood-eating insects teem with many kinds of protists and bacteria. Termites, leaf-eating insects, and ruminants would starve to death without the symbionts in their guts to break down cellulose. Even an insect as small as an aphid has some five million bacterial symbionts; this relationship developed 250 million years ago and is obligatory for both bacterium and insect. They cannot live apart.[22]

In many cases, the microbial infection is cyclical: bacterial symbionts are acquired anew in each generation. The symbionts may be harboured in special symbiotic cells or organs, with various morphological and physiological effects. In addition to developmental symbiosis, hereditary symbiosis is prevalent, especially among the insects. As will be discussed later, more than 16 percent of all insect species so far examined have bacteria of the genus *Wolbachia*, which are inherited through the egg, just like the mitochondria. The acquisition of these bacteria has resulted in novel organs and behaviors, and even species.[23]

Microbial symbionts carry out many chemical reactions impossible for their hosts; they can photosynthesize, fix nitrogen, metabolize sulfur, digest cellulose, synthesize amino acids, provide vitamins and growth factors, and ward off pathogens. We can see that, from the standpoint discussed above, every individual is a superorganism in which multiple species function to maintain the whole.

That microbial symbiosis is a fundamental aspect of life was first suggested in the late nineteenth century. The dual nature of lichens, nitrogen-fixing bacteria in the root nodules of legumes, fungi (mycorrhiza) in the roots of forest trees and orchids, photosynthetic algae living inside the bodies of protists, hydra, and the flatworm *Convoluta roscoffensis* all suggested a temporal continuum of dependency of microbe and host, from transient to permanent interdependence. When these phenomena were considered together with cytological evidence for "self-reproducing" bodies within the cells of plants and animals, they led several biologists of the late nineteenth century to conceive of the cell itself as a symbiotic community. By the early twentieth century, it had been proposed that the nucleus, cytoplasm, chloroplasts, mitochondria, and centrioles had all evolved in the remote past from distinct beings that came to live together.[24]

Symbiosis researchers have formed their own journals and international society only in recent years. Throughout most of the twentieth century, studies of symbiosis were carried out on an individual basis by biologists in several countries, separated by geography, language, and war. The idea that chloroplasts originated as symbionts, first suggested by Andreas Schimper in 1883, was developed most prominently by the Russian botanist Constantin Mereszhkowsky (1855–1921) at Kazan University, who in 1909 coined the word "symbiogenesis" for the synthesis of new organisms by symbiosis.[25] Mereszhkowsky developed this insight following his studies of single-celled algae: diatoms in California (1897–1902). In Geneva in 1918 he wrote an elaborate paper, which he considered his most important work, arguing that chloroplasts had originated from symbiotic blue-green algae (cyanobacteria) in the remote past. He also maintain that nucleus and cytoplasm had originated as a symbiosis of two different kinds of microbes.[26]

In his book *Les symbiotes* (1918), the French biologist Paul Portier (1866–1962) at the Institut Océanographique de Monaco developed an elaborate theory of symbiosis as a fundamental aspect of life. He argued that mitochondria originated as symbionts, and that they had been transformed over eons by their intracellular existence.[27] Thus, he declared, "All living beings, all animals from Amoeba to Man, all plants from Cryptogams to Dicotyledons are constituted by an association, the '*emboitement*' of two different beings. Each living cell contains in its cytoplasm formations which histologists call 'mitochondria.' These organelles are, for me, nothing other than symbiotic bacteria, which I call 'symbiotes.'"[28]

Portier's interest in symbiosis began with studies of bacteria in the gut of cellulose-eating insects such as termites. He suggested that symbiotic bacteria were necessary for digesting cellulose, supplied the host with essential vitamins, and played important roles in the development of their hosts. Bacteria in the root nodules of legumes represented a transitional stage in the evolution of mitochondria, which he called pro-mitochondria. Portier's speculations about mitochondria went too far when he postulated that bacteria, entering with food, fused with and rejuvenated the mitochondria. His contemporaries also went too far in their criticisms and ridicule, especially bacteriologists at the Institut Pasteur, who effectively pasteurized his symbiotes, rejecting all of his ideas.[29]

In the mid-1920s, the French-Canadian Félix d'Herelle (1873–1949) discussed the perpetuation of mixed cultures of bacteria and their viruses (he named them bacteriophages) in terms of symbiosis.[30] He referred to the bacteria that harbor viruses (lysogenic bacteria) as microlichens. The morphological and physiological changes resulting from symbiosis led him to assert in 1926 that "symbiosis is in large measure responsible for evolution."[31] The idea of virus-harboring bacteria was rejected as unbelievable for decades before it was revitalized by bacterial geneticists of the 1950s (see chapter 16).[32]

In the United States, during the 1920s, Ivan Wallin (1883–1969) at the University of Colorado argued that mitochondria originated as symbiotic bacteria. Like many other cytologists of his generation, he maintained that chloroplasts and centrioles were products of mitochondria. He also emphasized the importance of bacterial symbiosis as generator of new tissues and new organs. In his book *Symbionticism and the Origin of Species* (1927), he proposed that the inheritance of acquired bacteria was the source of new genes and the primary mechanism for the origin of species. He also claimed that he had cultured mitochondria to prove their actual bacterial nature.[33] Wallin's book met with virtual silence in the United States.[34]

The morphological and physiological effects of microbes transmitted through the eggs of many species of insects were investigated by Paul Buchner (1886–1978) in Germany in the 1920s and early 1930s, and later in Ischia, Italy, where he lived as an expatriate.[35] He brought his work together in several editions of his major treatise, *Endosymbiosis of Animals with Plant Microorganisms,* the fourth edition of which was published in English in 1965.[36] Throughout his career, Buchner divorced himself from the claims of Portier and Wallin, who claimed that mitochondria were symbiotic bacteria, and from those of others, such as Hugo von Schanderl, who claimed that he had "regenerated symbiotic bacteriods" from many sterilized plant parts and observed mitochondria transform into free-living bacteria.[37] Buchner saw such incredible claims as a liability to his own empirical work. They only hampered research because "theoretical misjudgments tended to obliterate the limits of what is understood as endosymbiosis today and thus to discredit the results achieved."[38] Buchner focused on a more basic struggle: changing the prevalent view that microbial symbiosis in animals was a rare phenomenon. As he observed, many biologists were still reluctant to accept that microbes, especially bacteria and fungi, played any beneficial role in the tissues of animals.

Why It Has Been Difficult to Imagine

Different people have used the term "symbiosis" to mean different things. For some, it has meant mutualism—two or more different organisms cooperating for a common good; for others, it has included parasitic relations as well.[39] But no matter how it was conceived, from the nineteenth to the late twentieth century, the evolutionary effects of interspecies integration due to microbial infections remained close to the margins of polite biological society. There were several reasons for this.[40]

1. Studies of symbiosis conflicted with the ever-increasing specialization that characterized the growth of the life sciences. Those investigating symbiosis necessarily crossed over institutional boundaries separating microbiology, cytology, zoology, botany, embryology, and genetics.
2. That bacteria played any beneficial role in the tissues of plants and animals conflicted with the basic tenets of the germ theory of disease. Bacteria had no natural history, and ever since Pasteur and Robert Koch, they had been defined largely as disease-causing germs and portrayed as the the enemy of humankind. The life-giving properties of bacteria and their fundamental role in the biosphere were overshadowed by the disease-causing aspects of some

of them. Bacteriologists only searched for an infectious microbe when tissue was diseased, not in healthy tissue, and for many, it was ridiculous to suggest that bacteria living in tissue could be part of the physiological well-being of animals.

Portier's work in *Les symbiotes* was framed by an opposition between symbiosis and germ theory. Rather than viewing microbes from "the window of medicine," he looked at "microbiology from the window of comparative physiology" and envisaged "a new form of bacteriology: physiological and symbiotic bacteriology."[41] This aspect of the debates was echoed in the United States by Wallin, who argued that only a small proportion of bacteria were pathogenic, but they were the ones most studied:

> It is a rather startling proposal that bacteria, the organisms which are popularly associated with disease, may represent the fundamental causative factor in the origin of species. Evidence of the constructive activities of bacteria has been at hand for many years, but popular conceptions of bacteria have been colored chiefly by their destructive activities as represented in disease.[42]

Indeed, the meager evidence of the beneficial effects of bacteria and viruses was no match for the evidence of their destructive effects. Consider the cataclysmic effects of microbial infections at the time *Les symbiotes* appeared. In 1918–19, an estimated 20 million people died from the influenza pandemic that followed the First World War. Consider too that some 1 billion people have died of tuberculosis over the past two centuries, and 300 million people have died of smallpox in the twentieth century alone.[43] The year after Wallin's book *Symbionticism and the Origin of Species* appeared, Alexander Fleming (1881–1955) discovered penicillin, which, when obtained in large amounts and purified by Howard Florey (1898–1968) and Ernst Chain (1906–1979), saved thousands of human lives during the Second World War and led to the ceaseless antibiotic war waged against bacterial infections. Fleming, Florey, and Chain shared the Nobel Prize in medicine in 1945.[44] The destructive image represented by microbial infections remains predominant today as we encounter emerging and reemerging diseases and threats of biological warfare.[45]

3. The importance of hereditary symbiosis in evolution also conflicted with Mendelian geneticists' insistence that chromosomal genes were the sole basis of heredity. Geneticists studied differences between individuals; symbiotic phenotypes common to individuals of a species would escape their notice most of the time. Their conceptions of evolution were based on gene mutations. The same year Wallin's book appeared, H. J. Muller reported that X-rays could dramatically increase the frequency of gene mutations in *Drosophila* by some 1500 times.[46] Bacterial symbiosis as the source of new genes was zapped by a wave of radiation genetics. Geneticists maintained that their artificially produced gene modifications were the same as those that occurred naturally.

Mendelian genetics was based on the Weismannian concept of one-germ plasm: one-organism. By definition, hereditary infections were considered contamination. As the Harvard corn geneticist E. M. East remarked in 1934, hereditary infections through the cytoplasm of eggs were a source of experimental error that geneticists needed to be wary of:

> There are several types of phenomena where there is direct transfer, from cell to cell, of alien matter capable of producing morphological changes. It is not to be supposed that modern biologists will cite such instances when recognized, as examples of heredity. But since an earlier generation of students used them, before their cause was discovered, to support arguments on the inheritance of acquired characteristics, it is well to be cautious in citing similar, though less obvious, cases as being illustrations of non-Mendelian heredity.[47]

As late as 1952, such leading geneticists as the Nobel Prize–winning H. J. Muller dismissed infectious hereditary particles from having any significance in genetics.[48]

4. The role of symbiosis as a source of evolutionary change also conflicted with central tenets of the evolutionary synthesis of the 1930s and 1940s. As discussed in chapter 13, that synthesis was built on the premise that natural selection acted on small hereditary differences between individuals in interbreeding populations. Populations evolved gradually because of changes in gene frequency within a population. Crucial microbial symbionts had been described in many protozoa, worms, sponges, coral, hydra, and molluscs, as well as insects and plants. However, they were regarded as exceptions, curiosities, "special aspects of life."[49] The evolutionary synthesis was essentially zoocentric. Microbes in evolution were not considered.
5. The creative effects of symbiosis were overshadowed by illustrations of conflict and competition—a view of nature that, it had long been argued, only reflected views of human social progress (see chapter 4). The evolution of cooperation between individuals of different species had been left largely unexplored.[50] In the middle of the twentieth century, the British botanist F. G. Gregory lamented that "the analysis of the relations between organisms has been dominated by the notion of 'competition' or 'struggle,' and the converse notion of 'cooperation' has in consequence been disregarded."[51]

The first international conference on symbiosis was held by the Society for General Microbiology in London in April 1963, six months after one of the most confrontational moments of the cold war: the Cuban missile crisis. As the editors of the resulting volume commented:

> The pressing problems of coexistence in world affairs may have influenced the Committee in their choice of subject for this year's Symposium. If so, it is to be hoped that the more bizarre examples of symbiosis illustrated in this volume will not be followed in the world at large; there are many other ways of escaping the Hobbesian predicament that without "commonwealth" life must be "nasty, brutish and short."[52]

6. Studies of the life histories of organisms and of the morphological and physiological effects of symbioses were also out of step with trends in ecology. Ecologists generally paid little attention to microbial symbiosis. Their textbooks usually contained only a few paragraphs on the standard cases: lichens, mycorrhiza, and nitrogen fixation. One of the major trends in ecology after the Second World War was understanding the ecosystem in terms of exchanges in energy and chemical substances.[53] Ecologists measured energy flow and circulation of water, carbon, and other essential elements, and they tended to avoid referring to species and species-specific phenomena as much as possible. This attitude persisted into the 1980s.[54] By that time, a major turning point for studies of symbiosis had already occurred within genetics as a few bacterial geneticists developed an understanding of symbiosis as a normal means of gene transfer and as a macromechanism of evolutionary change.

Toward a Unified Theory

The first generation of bacterial geneticists recognized that bacteria had various means for exchanging genes. In addition to the main circular DNA genome or genophore (erroneously called the chromosome), bacteria contain various other bits of DNA in the form of bacteriophage, as well as small circular pieces of DNA called plasmids. Whole plasmids and fragments of the genophore can be transferred between different kinds of bacteria by conjugation.[55] Bacterial genes can also be transferred by the uptake of DNA fragments from dead bacteria (transformations) and by viral infections (transductions).

In the 1950s, the diverse mechanisms of gene transfer among bacteria were discussed in relation to extranuclear mechanisms of heredity in protists, fungi, plants, and animals (eukaryotes). Consideration of infectious plasmagenes such as *kappa* in *Paramecium* and *sigma* in *Drosophila* in light of lysogeny in bacteria reinforced the idea that genes and genomes acquired by infection could become well integrated into, and form an essential part of, the genetic constitution of host organisms.[56] Cyril Darlington and Joshua Lederberg therefore called for an extension of the term heredity to embrace "infective heredity." As Darlington commented in 1951, "We are gradually being drawn to conclude that there is a wider range of cytoplasmic determinants of greater power than our predecessors had dared to suppose."[57] He prophesied that recognition of cytoplasmic genetic entities would enable geneticists "to see the relations of heredity, development and infection and thus be the means of establishing genetic principles as the central framework of biology."[58]

In 1952, Lederberg offered the word "plasmid" as a generic term for any extrachromosomal hereditary determinant regardless of its origin.[59] He saw lysogeny as "a stable symbiotic association" and suggested that the transport of genes from one bacterium to another by a virus (transduction) was, "functionally and perhaps phylogenetically, a special form of sexuality."[60] He argued that hereditary symbiosis in which phylogenetically distinct genomes were brought together was analogous to hybridization. He conceptualized a graded series of symbiosis, from cohabitants of a single chromosome through to plasmids and to extracellular ecological associations of variable stability and specificity. Such symbioses, he reasoned, obscured biological definition of the individual.

Lederberg adopted an operationalist approach, arguing that "the delineation of organic units, be they genes, plasmids, cells, organisms, genomes or colonies, is a tool of investigation and communication and not an absolute ideal."[61] What counts as an individual should be defined in terms of the practices of various specialities rather than as an absolute ontological entity:

> The cell or organism is not readily delimited in the presence of plasmids whose coordination may grade from the plasmagene to frank parasites. . . . The geneticist may well choose that entity whose reproduction is unified and hence functions as an individual in evolution by natural selection. The microbiologist will focus his interest on the smallest units he can separate and cultivate in controlled experiments, in test tubes, eggs, bacteria or experimental animals. Genetics, symbiotology and virology have a common meeting place within the cell. There is much to be gained by any communication between them which leads to the diffusion of their methodologies and the obliteration of semantic barriers.[62]

During the 1960s, the role of symbiosis and gene transfer between different species in evolution remained largely unexplored. As the French bacteriologist and environmentalist René Dubos at the Rockefeller Institute lamented in 1961, virologists and bacteriologists focused on diseases, for which research funds were available, thus maintaining themselves as "poor cousins in the mansion of pathology."[63] Dubos emphasized the creative role of microbial infections in bringing about new structures, functions, properties, and products. He pointed out that diphtheria was due to the infection of a bacterium with a toxogenic virus, that symbiotic phage brought about profound changes in the morphology of *Salmonella*, and that *lambda* phage carrying certain genes from a host cell could confer on a recipient the ability to produce enzymes for utilizing galactose.

He added these to the classic cases of the nitrogen-fixing bacteria of legumes, the dual nature of the lichen, and crown galls induced by inoculating certain plants with bacteria. Nonetheless, examples of the creative and evolutionary effects of microbial infections could not compare in funding or interest to the war between humankind and microbes. The beneficial effects of microbial symbiosis continued to conflict with the very history that had brought science and medicine into intimate association. Dubos prophesied, as others had decades before him, that there would "soon develop a new science of cellular organization, indeed perhaps a new biologic philosophy."[64]

Bacterial viruses and plasmids, first used by geneticists to study the nature of the gene, later became a tool for the development of genetic engineering. During the 1970s, molecular geneticists learned to artificially transmit targeted genes from the genome of one species to the genome of a distantly related species. The natural occurrence of such gene transfer became a stake in the technological debates and policy decisions.[65] Those who protested against recombinant DNA technology on ethical grounds or grounds of public health and safety, such as the famed biochemist Erwin Chargaff, referred to eukaryote-prokaryote hybrids as freakish forms of life that counteracted the evolutionary wisdom of millions of years of evolution.[66] The Nobel laureate George Wald at Harvard University insisted that it took millions of years for a single gene mutation to establish itself as a species norm.[67] The Caltech biologist Robert Sinsheimer argued that "nature has, by often complex means, carefully prevented genetic interactions between species. Genes, old and new, can only reassort within species."[68] Thus the classical understanding of evolution was invoked in opposition to biotechnology.

Symbiogenetic Renaissance

During the 1970s and 1980s, the venerable notion that eukaryotic cytoplasmic organelles had evolved as bacterial symbionts in the remote past was championed by Lynn Margulis (b. 1938). She had learned of symbiosis as an undergraduate at the University of Chicago. After completing her Ph.D. at the University of California, Berkeley, in 1965, on chloroplast genetics on the unicellular alga *Euglena*, she began to revitalize symbiosis theory with contemporary data from molecular biology and genetics.[69] By the late 1960s, the deployment of the electron microscope had revealed the structural similarities between mitochondria, chloroplasts, and bacteria. Furthermore, DNA had been discovered in these organelles; it was circular, as is bacterial DNA, and the organelles possessed ribosomes and a full protein synthesis apparatus.[70] During the 1960s, genetic research programs had emerged for dissecting chloroplasts and mitochondrial genomes, led by Ruth Sager, Nicholas Gillham, and others.[71] And the genetic studies of *kappa* in paramecium showed how chloroplasts and mitochondria could have originated as bacterial symbionts. The image of cytoplasmic organelles as quasi-independent organisms subsequently reemerged.[72] Margulis further placed the theory in a biogeological context to tell a story about how the eukaryotic cell may have originated eons ago by a series of events in which bacteria invaded a primitive "amoeboid microbe" and how mutual aggression eventually evolved to tolerance as the invading bacteria served useful functions for their host.

Margulis also extended the reach of symbiosis to account for the origin of centrioles and kinetosomes and therefore for the origin of cell motility and eukaryotic cell division. Since the nineteenth century, centrioles had been thought to divide by fission to spin out spindles and to play a crucial role in cell division by mitosis in many kinds of organisms (see chapter 8). C. O. Whitman's Japanese student Shôsaburô Watasē suggested at a Woods Hole lecture in 1893 that centrioles may have arisen as symbionts, and the idea was later mentioned by Wallin and others.[73] In the late 1960s, it was strengthened with evidence that centrioles might possess their own DNA. Margulis offered a far-reaching symbiosis theory to account for what was considered to be the greatest of all discontinuities in nature, that between bacteria and eukaryotes.[74]

The symbiotic theory of eukaryotic organelles attracted criticisms from many sides during the 1970s. Crucial evidence was lacking. One way to demonstrate that an organelle is a symbiont is to culture it outside the cell in a test tube or on a petri dish. Initially, Margulis imagined that biologists might indeed learn to culture chloroplasts, mitochondria, and centrioles. However, it soon became evident that these organelles were highly integrated into the nuclear genetic system: only a small fraction of the genes needed for mitochondrial and chloroplast functions was actually located in the organelles themselves.[75] This lent support to the alternative theory that these organelles had arisen by direct filiation, that is, they had evolved gradually from within the nucleated cell.

Leading cell biologists also argued that theories about eukaryotic cell origins were unscientific because they could not be proven. The influential cell biologist Roger Stanier spoke for many in 1970 when he commented that "evolutionary speculation constitutes a kind of metascience, which has the same fascination for some biologists that metaphysical speculation possessed for some medieval scholastics. It can be considered a relatively harmless habit, like eating peanuts, unless it assumes the form of an obsession; then it becomes a vice."[76]

By the end of the decade, the field of molecular evolution had emerged. Comparing nucleic acids of chloroplast, mitochondrion, and nucleus with each other and with different kinds of bacteria provided the rigor and closed the main controversy about the origin of mitochondria and chloroplasts.[77] Assembling the evidence in 1982, Michael Gray and Ford Doolittle considered it resolved that mitochondria and chloroplasts were of eubacterial origin (*alpha*-proteobacteria and cyanobacteria, respectively).[78]

No comparable data existed to test Margulis's most intriguing theory that centrioles and kinetosomes arose as symbionts. The evidence for DNA in centrioles had been on-again, off-again since the 1960s, but it was effectively refuted in the 1990s by evidence from electron microscopy and molecular hybridization, which indicated that genes affecting centriolar and flagellar function are located in the nucleus and that the linkage group is linear, not circular, as would be expected if they were of prokaryotic origin.[79] Although centrioles may not have arisen as symbionts in the same manner as mitochondria and chloroplasts, Margulis and colleagues subsequently suggested that the mitotic and motility apparatus of cells may have emerged through a symbiosis between two kinds of bacteria, which had resulted in the first nucleated cell.[80]

By the end of the century, the origin of the cell nucleus emerged as one of the most pressing problems of microbial evolution. Several researchers proposed that its genesis involved some kind of symbiosis (see chapter 18). The general idea is an old one.[81] That the nucleus may have evolved as a microbial symbiont living in a primitive host cell was suggested in the nineteenth century by Watasē in 1893 as part of the field he envisioned as cytogeny. Theodor Boveri discussed the same idea in 1903, and between 1905 and 1918, Mereszhkowsky developed it as part of his theory of symbiogenesis.[82]

Somewhat similar ideas about the nucleus reemerged in the late twentieth century from a number of different routes. [83] In 1974, Jeremy Pickett-Heaps supported the idea that the nucleus was a symbiont by noting that some dinoflagellates have two nuclei, "one characteristic of dinoflagellates, and the other a more typical eukaryotic nucleus."[84] In the 1980s, James Lake and Maria Rivera at UCLA argued that the nucleus evolved from an engulfed (eocyte) Archaebacterial symbiont of a Eubacterial host.[85] During the subsequent decade, Radhey Gupta at McMasters proposed that the nucleus originated not from an engulfed bacteria, but from a fusion of what he called monoderm and diderm bacteria.[86] Still other symbiotic scenarios were offered. William Martin at Düsseldorf suggested that the nucleus may have originated from a merger between an Archaebacterial organism and the Eubacterial ancestor of the mitochondria.[87] Margulis, Michael Dolan, and their collaborators at the University of Massachusetts, Amherst, suggest that the nucleus motility of features of protists developed from a symbiosis with a spirochete.[88] All these theories contradicted the traditional Darwinian conception of gradual evolution from within the cell.[89]

Macroevolutionary Change

Symbiosis is recognized to have played an important role in the evolution of eukaryotes some 1.8 billion years ago. But many who have studied symbiosis, including Lederberg, Margulis, Kwang Jeon, Werner Schwemmler, Mary Beth Saffo, Max Taylor, and Paul Nardon, have insisted that hereditary symbiosis has played a general and major role in macroevolutionary change.[90] Some have remarked that because microbial symbiosis draws genomes from the entire biosphere, the resulting changes are far greater than those that arose through gene mutation, hybridization, and the like.[91] Some interpret hereditary symbiosis as a neo-Lamarckian mechanism of evolutionary development.[92]

The inheritance of acquired bacteria is easy to imagine in the case of single-cell protists because symbionts are easily transmitted from one cell generation to the next. There are many illustrations of acquired bacterial genomes in protists, and there is also evidence that the acquisition of bacteria may occur rather quickly. The most remarkable study of such evolution in action was presented by Kwang Jeon at the University of Knoxville, Tennessee. In the late 1960s, one strain of the amoebae he studied (*Amoeba proteus*) had accidentally become infected, each with about 100,000 bacteria containing about 3000–5000 genes. Many of the infected amoebae were killed, but Jeon selected out the few amoebae that survived, and after a few years, he found amoebae that had become completely dependent on the bacteria.[93]

Symbiosis theorists from Portier to Dubos had often claimed that the relative scarcity of evidence for hereditary symbiosis was a reflection not of nature, but rather of the extent to which it was investigated. Research trends of the late twentieth century suggested they were right. Many new cases of bacterial symbiosis with profound effects on their hosts have come to light. In the late 1970s, it was discovered that deep in the oceans, hydrothermal vents are abundant oases of life. Ecological communities of tube worms, various mollusks, ciliates, and other marine creatures are formed around fissures in the rock where geothermal fluids containing sulfides, methane, ammonia, and hydrogen flow out from buried aquifers. It was immediately suspected that the abundant life surrounding these vents was due somehow to the energy contained in these geothermal fluids.

In the early 1980s, biologists found that many of these thermal-vent species contained sulfur-metabolizing symbiotic bacteria.[94] These symbiotic bacteria use chemical energy contained in reduced sulfur compounds to generate ATP and fix carbon for biosynthetic processes. Less than a decade later, biologists found symbioses between sulfur-oxidizing bacteria and various invertebrates in a wide variety of other sulfide-rich environments, including salt marshes, mangrove swamps, and sewage outfalls. In almost all of these cases, the animals rely on the bacteria for most or all of their carbon and energy requirements. These symbiotic sulfur bacteria have not been cultured in the laboratory, and it is unclear whether they can exist outside their host or are dependent "organelles." Biologists are now carrying out tests to see how symbionts are transmitted from one generation to the next and whether gene transfer has occurred between these bacteria and the host cells.[95]

Molecular techniques for screening nucleic acids and the use of polymerase chain reaction have dramatically increased the facility of detecting symbionts. Perhaps the most dramatic new discovery is the widespread occurrence of hereditary symbiosis in insects. Bacteria of the genus *Wolbachia* are maternally inherited through the cytoplasm of their hosts and are disseminated throughout the body cells of their hosts. Surveys based on molecular phylogenetic techniques for screening have so far found *Wolbachia* in more than 20 percent of all known insect species, including each of the major insect orders.[96] They are thought to be the most common hereditary infection on Earth, rampant throughout the invertebrate world; besides insects, hosts include shrimp, spiders, and parasitic worms.[97] Their complete distribution in arthropods and other phyla are yet to be determined. *Wolbachia* are *alpha*-proteobacteria, like mitochondria, and they appear to have evolved as specialists in manipulating reproduction and development of their hosts. They cause a number of profound reproductive alterations in insects, including cytoplasmic incompatibility between strains and related species, parthenogenesis induction, and feminization (that is they can convert genetic males into reproductive females and produce intersexes). Sometimes, as in the case of weevils (one of the most notorious pests of stored grain), *Wolbachia* are inherited together with other bacterial symbionts that allow the animal to adapt better to the environment by providing vitamins and energy and enhancing the insect's ability to fly.[98]

Wolbachia have attracted considerable evolutionary interest, especially as a mechanism for rapid speciation. All aphids carry bacteria of the genus *Buchnera* in their cells, and the symbionts are inherited through the host egg. Studies of hereditary symbionts of insects are well funded today not because of their evolutionary significance, but because of their potential for pest control in agriculture and as a mechanism for modifying arthropod vectors of human disease.[99]

Bacteria are certainly not the only microbes to profoundly modify their host animals. Viruses have also played a crucial role as a source of evolutionary change. They may also play a role in the acquisition of a larger symbiont, as in the case of chloroplasts in animals. Several phyla of animals are able to acquire chloroplasts within the cytoplasm of their cells, usually from a particular species of algae. One of the most remarkable examples is the sea slug *Elysia chlorotica*, found in salt marshes from Chesapeake Bay to Nova Scotia.[100] This sea slug feeds on a certain species of filamentous algae, *Vaucheria*, and in so doing acquires symbiotic chloroplasts that are sequestered by specialized cells in the epithelium. The maintenance of the chloroplasts in the slug's cells seems to be the result of a previous horizontal gene transfer of some of the algal genes to the sea slug's chromosomes, possibly by viruses the sea slug harbours in its genome.

The development of complex organs such as the eye has always been a problem for evolutionists. Walter Gehring at the University of Basel has recently proposed that the eye may have evolved from a microbial symbiont that became a chloroplast.[101] Evidence comes from a certain species of ocean algae (dinoflagellates) that has a chloroplast-derived eyespot capable of focusing light, a feature that may have been of benefit in avoiding ultraviolet light and searching for photosynthetic light. If the algae itself had become an acquired symbiont of an animal, then, as Gehring remarked, "It could be that eyes are coming from a symbiont within a symbiont." He is now screening specific algae for a gene known to be involved in eye development, and animals for traces of chloroplast genes.

In 2001, genome researchers suggested that there were 200 or so bacterial genes in the human genome. Although this has been refuted, the suggestion itself shows how thinking about horizontal gene transfer has changed in recent years.[102] However, the integrated retroviruses (and other processed RNA) in the human genome are indubitable examples of horizontal gene transfer from viruses to mammals. Endogenous retroviruses (RNA-based viruses), relics of ancient germ-cell infections, comprise 1 percent of the human genome.[103] In fact, these viruses may have been involved in key events leading to the evolution of all placental mammals from egg-laying ancestors.[104]

Hereditary symbiosis, taken together with evidence of horizontal gene transfer in bacteria, offers a dramatically different view of evolution from that offered by the neo-Darwinian synthesis (see Tables 1 and 2):

The gaps in the fossil record—the lack of fossil evidence for transitional forms between species—always had been a source of uneasiness to evolutionists (see chapter 6). Darwinian gradualists insisted that the fossil record itself was incomplete, like an old book with pages missing, but beginning in the 1970s, evolutionists reconsidered the idea that the gaps should be taken at face value. In 1972, Niles Eldredge and Stephen Jay Gould (1942–2002), two young invertebrate

TABLE 1. Tenets of the Modern Synthesis

1. Gene mutation and recombination are sources of evolutionary change.
2. Common descent: evolution is a hierarchical and a branching tree: one-species: one-genus; one- genus: one family; one family: one-order; etc.
3. Speciation generally occurs by geographic isolation.
4. Gradualism: evolution does not occur in jumps.
5. Mechanisms for macroevolution are the same as those for microevolution.

palaeontologists, insisted that the picture of evolution from the fossil record was not one of species gradually changing over time, but, rather, one of species emergence in a very short geological period of time followed by long periods of stasis. This was the basis of their theory of punctuated equilibria. Once established, a species actually remains relatively unchanged for the duration of its existence, before undergoing rather sudden transformation. Gould and Eldredge revitalized the idea that macroevolutionary events may occur rapidly, in spurts.[105] Their evidence came from studies of trilobites and snails, which, being preserved in large numbers, provided a reasonably intact record of their evolutionary past. They argued that typically, marine invertebrates species would endure in stasis for 5 to 10 or more million years, but speciation events occurred over a period ranging from 5000 to 50,000 years.[106]

Several explanations are compatible with this pattern. Speciation could be explained in terms of the processes of random genetic drift. That is, speciation begins with changes in peripheral populations, as Sewall Wright had proposed (by genetic drift), leading to reproduction. This would occur most rapidly in small populations accidentally isolated geographically from the main populations of the ancestral species. Accordingly, the gaps in the fossil record would be real evidence of a rapid process of speciation that involves only a fraction of the original population. However, the pattern of punctuated equilibria is also compatible with other mechanisms. Speciation could be due to novel kinds of genetic rearrangements, or it could result from the overthrow of developmental or genetic constraints that restrict organismic change to within the range of preestablished body plans. As the relations between development and evolution were explored during the last

TABLE 2. Recognition of Horizontal Gene Transfer, the Inheritance of Acquired Genomes, Wholes or Parts, Contradicts Several Tenets of the Classical Neo-Darwinian Synthesis

1. Transfer of genes across species and higher taxonomic groups occurred regularly.
2. Early evolution is not treelike; it is reticulated.
3. Speciation does not require geographic isolation.
4. Evolution may occur very rapidly by horizontal gene transfer, and the changes that result may be quite dramatic compared with those due to single gene changes.
5. By implication, the mechanisms for macroevolution are different from the mechanisms for microevolution.

decades of the twentieth century, mutations in those genes that regulate development also assumed great significance. Regulatory (homeobox) genes were reported that affected whole sets of genes, and some paleontologists have emphasized that mutations in such regulatory genes result in large discontinuous evolutionary changes (a reshuffling of parts) and may be a macromechanism of evolution.[107]

Gould and Eldredge did much to open up evolutionary biology from the tight hold of gradualism and selection to a reconsideration of random drift, the role of developmental constrains in evolution, and even macromutations. But symbiosis—the inheritance of acquired genomes, wholes or parts—was trivialized or ignored by paleontologists. Gould himself regarded the symbiotic origin of mitochondria and chloroplasts as "entering the quirky and incidental side" of evolution.[108] Leading neo-Darwinian theorists have also insisted that the inheritance of acquired bacteria is a rare exceptional phenomenon in plants and animals. John Maynard Smith and Eörs Szathmáry declared in 1999 that "transmission of symbionts through the host egg is unusual."[109] This view of life, which overlooks the many cases of hereditary symbiosis, was based on theoretical assumptions about the evolution of cooperation and the evolution of microbe-host relations, which I explore in the next chapter.

20

The Evolution of Relationships

> Most of the ideological influence from society that permeates science is a great deal more subtle. It comes in the form of basic assumptions of which scientists themselves are usually not aware yet which have profound effect on the forms of explanation and which, in turn, serve to reinforce the social attitudes that gave rise to those assumptions in the first place.
>
> —Richard Lewontin, 1991

THE EVOLUTION OF ASSOCIATIONS was not considered in the evolutionary synthesis of the 1930s and 1940s. Yet evolution is a process of integration as well as divergence; there is divergence in the production of new life forms, but there is integration when these entities unite so as to make new wholes. These unions are recognized to be at the basis of the major transformations in the evolution of life: the origin of cells, organelles, and multicellular organisms. How could natural selection lead to the cooperation among cells of a plant or animal, or cooperation within populations of deer, or birds? What about human cooperation? And what about cooperative relationships between species?

When these questions came to the fore of evolutionary thought in the 1970s and 1980s, they led to forceful debates among evolutionists over two fundamental issues. First, what precisely are the targets of selection?[1] Put differently, what are the units that are in struggle? The favored choices were genes, organisms, and populations. Second, how much could biology account for human social relationships, history, and culture? While some insisted that fundamental aspects of human cultures were due to human nature, others argued that cultural history was governed by forces beyond biology. Inversely, could one understand associations between animals, plants, and microbes in the same way as one could human social relationships? While some biologists used various human social categories to explain nonhuman associations, others cautioned against this, calling it anthropomorphism and circular reasoning. Still others argued that theories about the basis of many intimate biological associations were often untestable.

The Individual and the Group

Evolutionists have long noted that there is a tendency for individuals to risk their own lives for another. But how could natural selection have led to any trait that was detrimental to the survival of the individual? How could altruism have evolved by the struggle for existence between individuals? This paradox led Darwin, Spencer, and Kropotkin to underscore the necessity of a Lamarckian dimension to evolution (see chapter 6). Darwin described the dilemma in *The Descent of Man*:

> It is extremely doubtful whether the offspring of the more sympathetic and benevolent parents, or of those who were the most faithful to their comrades, would be reared in greater numbers than the children of selfish and treacherous parents belonging to the same tribe. He who was ready to sacrifice his life, as many a savage has been, rather than betray his comrades, would often leave no offspring to inherit his noble nature.[2]

Darwin suspected that human cooperation began with the "low motive" of being aided in return, which, over time, would become replaced by an innate feeling for sympathy and benevolent action, as the social habit of aiding his or her fellows would become hereditary over many generations.[3] But he also considered that an "even more powerful stimulus to the development of social virtues is afforded by the praise and blame of our fellow men." The instinct to "love the former and dread the latter," he argued, "was originally acquired, like all the other social instincts, through natural selection."[4] Once it arises in a population, an altruistic behavior such as bravery would be selected for in the context of warring tribes because they would be of benefit to the group: survival of the fittest *group*. He argued that a species, or a population within a species, whose individual members are prepared to sacrifice themselves for the welfare of the group may be less likely to go extinct than a rival group whose individual members place their own selfish interests first.[5] It would be possible to select for characteristics that are detrimental to the individual but of benefit to the group.[6]

Studies of the evolution of cooperation remained scarce, but during the First World War, Kropotkin, among others, lamented that there was too much focus on individual struggle and egoism, and not enough on group cooperation (and chapter 5). Those who portrayed nature as a Hobbesian war of each against all, he argued, were partly to blame for war, insofar as they justified it in the name of evolution. But for Kropotkin, no less than his Hobbesian adversaries, human social behavior was governed by nature of the individual. Only for him, individuals were by nature cooperative, not purely egoistic. The same view was advocated during the Second World War by the school of ecologists at the University of Chicago led by Warder Allee. They also protested that biologists played a major role in fostering war by promoting a view of nature and human nature as purely egoistic.[7] Although biologists provided a convenient, plausible explanation and justification for all the aggressive, selfish behavior of which humans were capable, Allee argued that the strong competitive egoist drives had to be kept "in their true place, somewhat subservient to the even more fundamental cooperative, altruistic forces of human nature."[8]

For Allee and his colleagues, the integrative forces of nature—as witnessed in the individual plant or animal, an ant colony, a flock of birds, a tribe, or the world economy—implied hope for humanity: progress and international cooperation were inevitable. It was futile for individuals to resist nature's force, and therefore they should submit for the good of the group. Cooperation and integration for the good of the group took on a different meaning when Hitler called upon pure-blooded Aryans to give their lives for the greater glory of the fatherland. The harsh lessons of the limits of group conformity were all too evident, as Donald Worster has commented: "The ideal of social integration, it became starkly clear, could harbor an unexpected danger: the possibility of a totalitarian police state based on the same appeals to self-sacrifice to the whole invoked by organicists."[9]

Natural selection for the good of the group took on a new social meaning and a new importance with the rise of environmentalism in the 1960s and 1970s. Group selection was developed by the distinguished British ethologist Vero C. Wynne-Edwards (1907–1997) in his influential book *Animal Dispersion in Relation to Social Behavior* (1962).[10] His arguments were targeted against Malthusian population principles as applied to ecology: that populations, if left unchecked would grow geometrically, while their food supply would increase only arithmetically. Malthusian implications shook the 1960s as issues about human overpopulation emerged.[11]

Do animals have mechanisms for regulating their own populations? If so, how do they do it? Wynne-Edwards emphasized that animals were not always striving to increase their numbers and competing for food; they collaborate for the benefit of the group, and they compete often for territory. Animals regulated their population growth by social displays, territorial behavior, and communal roosting, all of which, he thought, could be accounted for only by natural selection operating on the group. After all, if natural selection operated only at the level of the individual, would there not be a premium on higher and higher reproductive rates, and would this not lead a fixed population to overexploit its food resources and perish?

Selection acting for the good of the group would suppress individual interests and regulate reproduction in many species according to the needs of the population. One could see this in the evolution of sterile individuals, which are a large part of so many social insect societies. If selection acted only on the individual, Wynne-Edwards argued, then these sterile individuals would be quickly condemned to extinction. They "could only have evolved where selection had promoted the interests of the social group, as an evolutionary unit in its own right."[12] He assumed that humans shared many cultural forms with other animals and that they were predisposed through innate instinct or acquired conditioning to be governed by law and order, and to put group interests ahead of self-interest.[13] When discussing the evolution of social ethics, he argued against loyalty to international organizations, asserting that "on biological as well as traditional grounds it is to his sovereign state that the individual's first loyalty should continue to be given."[14] Wynne-Edwards's ideas were investigated by other ecologists of the 1960s and 1970s who searched for mechanisms of population control within species and who understood them in terms of group selection: evolution for the good of the group.[15]

Kin Selection

Group selection as an explanation for altruism and cooperative behavior was problematized in the 1970s when a new school of evolutionists called sociobiologists emerged, led by the writings of William Hamilton, John Maynard Smith, and Richard Dawkins in England, and by Robert Trivers, George C. Williams, and Edward O. Wilson in the United States. They agreed that animals possessed various characteristics that would limit their birth rates, but they argued that animal birth control was not a matter of individual altruism practiced for the good of the group, nor was it practiced for the good of the individual, but rather for the good of their genes.[16] Sociobiologists turned evolution into a sort of a game in which the object is to maximize an individual's genetic representation in the next generation—a struggle for existence among DNA molecules, or "selfish genes."

Altruism was an illusion, and cooperation for the benefit of the group, they argued, was a myth. Altruistic behavior was easy to account for when it occurred between family members: by helping a relative, an individual is propagating copies of its own genes. This idea had been expressed in 1955 by J. B. S. Haldane, who, after some quick mathematical calculations, and bearing in mind that brothers share on average only half their genes, quipped that he would sacrifice his life for (just more than) two brothers, four half brothers, or eight first cousins.[17]

In 1964 Maynard Smith labeled this kin selection.[18] The same year Hamilton independently gave it a mathematical formulation.[19] Although altruistic behavior may lead to individual death, it increased the probability of survival of the genes the altruist shares with its relatives in the population. This could explain sterility in insect colonies and the suicidal barbed sting of the worker honeybee. Similarly, if a human mother risked her life to save her child, it was because she was contributing to the survival of her own genes, which were invested in the child. Maynard Smith applied the same model to account for the cooperation of cells in the evolution of development.

The Organism as a Beehive

It was self-evident that a division of labor among cells such that different ones would specialize in mobility, protection and capture of prey, digestion, reproduction, and storage had many benefits.[20] But how did selection lead to that complexity? How did development evolve? During the last decades of the twentieth century increased attention was given to these questions. Some of the main issues were brought into focus by the Yale biologist Leo Buss in his book *The Evolution of Individuality* (1987).[21] He argued that the evolution of development resulted from the interplay of selection acting at two levels, the individual and the group—that is, "selection at the level of cell lineage, and selection at the level of the [multicellular] individual."[22]

As Buss saw it, much of development consisted in interactions in which one cell lineage limits the reproduction of another at specific stages while enhancing its own reproduction. This was prima facie evidence of intercellular competition. But if selection acted solely on individual cells, would this competition or self-interest not disrupt integration at the level of the multicellular whole? Buss assumed it would. Therefore, he reasoned that in the evolution of development, cell variants would have been selected for their effects on the whole multicellular individual and their ability to limit subsequent conflicts with other cells. Selection would therefore act on the group as a unit, as well as on the individual cell. In other words, there would be selection for the good of the group.[23] He also suggested that two other mechanisms may have helped to ensure that cells carry out their functions and to prevent a mutant rogue somatic cell from becoming a germ cell: maternal control of early development and the evolution of a specific cell lineage to become a sex cell early in development (for example, segregating the human germ line on the fifty-sixth day of the embryo's gestation).

Buss's theoretical treatment led to a renewed interest in the problem. But Maynard Smith rebutted his arguments about selection for the good of the group. Smith sketched out an alternative model, which he later elaborated in his acclaimed book with the Hungarian biologist Eörs Szathmáry, *The Major Transitions in Evolution* (1995).[24] They adopted a bottom-up explanation of how integrated wholes evolved despite selection between their components favoring selfish behavior. They asked, what cooperative mechanisms needed to be developed to prevent one kind of lower reproductive entity from getting out of hand? The aggregation of once-independent molecular "replicators" or "selfish genes" within a membrane-bound structure led to the first cell, and their further placement on chromosomes, they argued, was an evolutionary strategy designed by selection to prevent "cheating" among them. But they insisted that no additional mechanisms were needed to prevent cheating among cells in the evolution of development. In his discussion of group selection, Buss had failed to mention a crucial fact: all the cells of the individual are genetically identical. Just like sterile insects in a colony, cells would forfeit reproduction for identical sister cells and therefore behave in the interests of the organism as a whole (the group) because when one cell helps a genetically identical cell, it is propagating its own genes.[25]

Maynard Smith and Szathmáry also rejected the idea that maternal control over early development and germ-line segregation were crucial conditions for multicellularity. The emergence of the eukaryotic cell type between 1.5 and 2 billion years ago provided all the organismic conditions for the differentiation of tissues, organs, and organ systems of plants and animals. Mitosis was crucial: it ensured that chromosomes were distributed to daughter cells and that the plane of cleavage was at right angles to the spindle, and it provided the opportunity for unequal distribution of cytoplasmic components, the basis for cell divergence and differentiation. Cell adhesion is, of course, a phenomenon in many protists (as well as bacteria) that form colonies, and evidence indicated that cell-to-cell signaling is present among protists as well. So, it was argued, given these characteristics of the eukaryotic cell, little more was required to generate embryonic development. Differentiation and spatial patterning, crucial aspects of embryonic development, also appear well developed in protists, such as *Paramecium* or *Tetrahymena* (see chapter 17).[26] Maynard Smith and Szathmáry offered two alternative reasons for the evolution of a separate germ line.[27] A specialized gamete lineage may have been advantageous in keeping mutation rates lower. And because various epigenetic mechanisms are involved in somatic cell heredity when gametes are produced, all genes must be "reset" to a totipotent state.

The Lessons of Sociobiology

While gene selection could explain cooperation between closely related individuals, how could one explain cooperative behavior between individuals in which relatedness was comparatively low, such as grooming among birds and mammals or when one human saves another from drowning? To account for our kindness to strangers, sociobiologists employed the theory of reciprocal altruism, developed by Robert Trivers in 1971. He argued that natural selection operated on individuals in such a way as to ensure that acts of kindness can be recognized and reciprocated, with the result that the net fitness of both participants is increased. Again the idea was simple: "you scratch my back and I'll scratch yours." All altruistic acts are only apparent; they are actually carried out with the expectation of a payoff later on.[28] Trivers further suggested that many of our psychological characteristics, such as envy, guilt, gratitude, and sympathy, were genetically determined and shaped by natural selection for improved ability to cheat, to detect cheats, and to avoid being thought to be a cheat.

In 1975, the Harvard entomologist E. O. Wilson published *Sociobiology: The New Synthesis*, in the last chapter of which he stated that xenophobia, deceitfulness, aggressiveness, and many social organizations had deep biological roots.[29] His book was followed by several others that argued for a causal linear link from genes to society, none better known than of Richard Dawkins's blockbuster *The Selfish Gene* (1976). All of Dawkins's arguments were dedicated to the same reductionist point: that cooperation could be explained as a winning strategy through which an individual, "blindly programmed to preserve selfish genes," could best promote its own survival. He described humans as "giant lumbering robots" under the control of our genes, which "have created us, body and mind."[30] Any cooperative trait would be selected for if the trade-off value resulted in more benefit than harm to individuals with similar genes. As regards environmentalism, the issue of short-term individual self-interest versus cooperation for the common good was aptly phrased in the title of Garrett Hardin's famous essay of 1968, "The Tragedy of the Commons":

> Entities that pay the costs of furthering the well-being of the ecosystem as a whole will tend to reproduce themselves less successfully than rivals that exploit their public-spirited colleagues, and contribute nothing to the general welfare. Hardin (1968) summed the problem up in his memorable phrase "The tragedy of the commons" and more recently (Hardin 1978) in the aphorism, "Nice guys finish last."[31]

As T. H. Huxley had argued generations earlier, Dawkins maintained that there was a discontinuity between cosmic evolution and human morality. "Be warned that if you wish as I do, to build a society in which individuals cooperate generously and unselfishly towards a common good, you can expect little help from biological nature. Let us try to teach generosity and altruism, because we are born selfish."[32]

Ethics, aesthetics, politics, culture, war, and religion all fell within the scope of sociobiological inquiry. All were to be put on a firm universal biological basis. Sociobiologists sought more complex scenarios to explain risking one's life for the group, as in the case of war. And, following Hobbes, they argued that cooperation between egoistic individuals required a central government, payment of taxes, and other social forces that counteract our basic human nature.[33] But did ahistorical, universal human traits that were determined by a struggle between DNA for supremacy and dominance really exist? Are contemporary societies the inevitable result of our human nature?

About Just-So Stories

Sociobiological writings incited heated debates.[34] Coming at a time of movements for social change, their biological determinism—their advocacy of a chain of causality from genes to society—was taken by many biologists and philosophers as being designed to stifle a will to reform, just as had social Darwinian writings of a century earlier (see chapter 4). The Harvard evolutionists Stephen Jay Gould and Richard Lewontin, and the "science for the people" group in the United States, argued that the forms of human societies are only remotely associated with our genes, and that it is absurd to conceive of human cultural history as a by-product of our genes, an evolutionary product of natural selection. They protested that the biological determinists' conjectures and concepts transcended science and only reflected and reinforced a conservative establishment. At the surface of sociobiological theory, critics charged, was the obvious ideological commitment to modern entrepreneurial, competitive, hierarchical society, and deeper was the ideology of the priority of the individual over the collective. As Lewontin quipped, "Despite the name *socio*biology, we are dealing with a theory not of social causation but of individual causation."[35] Critics argued that sociobiologists' evolutionary reasoning was fundamentally flawed on three grounds:

1. The notion that individual genes were made visible to selection because of their more or less direct one-to-one relationship with particular phenotypic features was an integral feature of the sociobiological approach. However, many evolutionists emphasized that an individual gene is seldom directly exposed to selection. Such exposure is usually in the context of its entire genotype, and since a gene may have different selective values in different genomic contexts, it is highly unsuitable as the target of selection. Thus, the individual organism, the phenotype, would be the main target of selection, not the gene.[36] This issue has become only more salient as molecular biologists increase their understanding of the complex relationships between genes and phenotypes, and as the very concept of the gene has become ever more abstract (see chapter 17).
2. Sociobiologists tended to treat every evolved trait as if it were an adaptive trait, that is, as the product of selection. This panselectionism was the main target of Gould and Lewontin's influential paper of 1978, "The Spandrels of San Marco and the Panglossian Paradigm: A Critique of the Adaptationist Program."[37] To caution evolutionists from assuming that every phenotypic trait is an adaptive trait designed by natural selection, they drew an analogy with the architecture of the basilica of San Marco in Venice. The tapered spaces, that is, the spandrels (or pendentives) between the archways supporting the doomed roof, were beautifully decorated in a way that made splendid use of the triangular space, almost as if these spaces were made for that very artistic purpose. However, they were not designed for that purpose at all; they were just an architectural by-product of employing arches to support a domed room.

   One could not simply assume that a structure arose for its present purpose, nor that it had an adaptive purpose then or now and had therefore evolved by natural selection. Appreciating this point was also important for understanding the early evolution of complex organs such as the wings of a bird or insect: before they were used for flying, proto-wings had other purposes.[38] Evolutionists sensitive to the fallacy that all structures arose for their present purpose have invoked Voltaire's Dr. Pangloss: "Things cannot be other than they are, for since everything was made for a purpose, it follows that everything is made for the best purpose. Our noses were made to carry spectacles, so we have spectacles. Legs were clearly intended for breeches, and we wear them."[39] That every trait was adaptive and arose by natural selection had long been ridiculed by biologists advocating saltationist views of evolution, such as William Bateson, who pointed to the Panglossian paradigm of his adversaries in 1914. "Naturalists," he wrote, may still be found expounding teleological systems which would have delighted Dr. Pangloss himself, but at present few are misled."[40] Gould and Lewontin's

"Spandrels" paper led evolutionists to question the excesses of a purely adaptationist approach and to reconsider the importance of developmental constraints on evolution as well as the importance of external factors in mediating the relationship between genotype and phenotype.[41]

3. Sociobiologists' views, critics argued, were based on trying to find features common to all human societies, invoking genes to explain them, and then making up a story about how they evolved. Insisting that the gene was the primary target of selection, sociobiologists had often invoked a gene for this and a gene for that. Dawkins, for example referred to the "kin-altruism gene."[42] Gould launched one of the strongest critiques of sociobiologists' accounts when he argued that, in their art of storytelling, sociobiologists often fall prey to the temptation to tell "just-so stories" with no more validity than Rudyard Kipling's creative fairy tales about how the leopard lost its spots, how the camel got its hump, and how the elephant got its trunk.[43]

Sociobiology and its recent offspring, evolutionary psychology, have spawned societies, journals, and an ever-expanding program of research, and sociobiologists have insisted that they have outgrown the charges of "naive" genetic determinism.[44] Indeed, they and other panselectionists have continually addressed their critics.[45] In 1987, Dawkins ridiculed "naive critics" who believed that neo-Darwinists wasted their time looking for "genes for this and that trait." The genotype, he asserted was not a "blueprint" but a prescription, "a recipe" (as for baking a cake), "a set of instructions" for how to assemble the body (see chapter 17).[46]

Sociobiologists insisted that their views were not politically motivated, and that they were not fantasies but scientifically testable hypotheses, driven by data.[47] Despite such rebuttals, it was often difficult to see some sociobiologists' accounts as anything other than political, as exemplified in the following description of the origin of sex by Sarah Blaffer Hrdy, who, in her book the *Woman That Never Evolved* (1983), wrote about "the initial inequality":

> Sex, it is now thought, began as a simple act of hijacking when, some several billion years ago, a small cell waylaid and merged with a bigger one, richer in substance and nutrient. . . . Competition among small cells for access to the larger ones favored smaller, faster, and more manoeuvrable cells, analogous to sperm. The hostages we might as well call ova.[48]

The problem with this account is not simply that cells like ours were not around "several billion years ago," but with the language of "waylaid," "hijacking," and "hostage," and with the implication that male domination has a fundamental biological basis. Hrdy's account is an egregious illustration of how assumptions about the universal nature of human social relationships are used to explain nature, and then, coming full circle, this version of nature is used to explain human social inequalities.

Actually, other evolutionists argue that the origins and maintenance of sex may have evolved by selection operating at three different levels: the

> group, the individual, and the gene.[49] Sex is generally understood to have been of adaptive value at the population level because of the enormous genetic diversity it can generate by genetic recombination. It may have also allowed individuals to produce more offspring because the differences among offspring may increase chances that some will survive under changing circumstances. However, sex may not have originated for those reasons. The genesis of sex in eukaryotes involving haploid and diploid phases is still uncertain, but the textbook answer since the 1980s is that sex arose as a way to minimize DNA damage due to copying errors.[50] The early stages of meiosis, in which the two copies of each chromosome line up and pair with each other, may have originated as a mechanism for repairing double-strand damage to DNA by using the undamaged version of the chromosome as a template or guide to fix the damaged one.

Despite detailed critiques of genetic determinism over the past thirty years, the idea that our behavior and fate can be read in our genes (collectively or individually) remains as attractive to some today as astrology was for some Renaissance scholars. So too is the idea that, because of human nature, society is pretty much the way it has to be. Sociobiology sells. In a fin-de-siècle best-seller, *The Spirit of the Gene*, endorsed by leading biologists (including E. O. Wilson), it is argued that the notion of cultural influence on human behavior is an illusion, and talk of "cultural history is misleading."[51] Culture, it is asserted, is merely a "genetic feedback mechanism" designed to fool us into believing that genes are not in control. "Morality is a shrewdly fashioned genetic propaganda device designed to heighten our mystical gullibility and conceal from us the real source of our behavior, of both heroes and villains—our genes."[52] Capitalism is a "genetic imperative." Our species is thus destined to destroy itself through overpopulation, technological growth, and consumerism, and there is nothing we can do about it because it is not a question of values, politics, economics, or social history; it is in our genes, our evolution. The author intimates that we actually live in a virtual reality constructed by a computerlike program running in our DNA. These pernicious assertions, of course, amount to irrefutable metaphysics, since all arguments to the contrary would be neither mine nor yours, but merely the author's predicted response of our genes.

Symbiotic Ties

Sociobiology was mainly concerned with relationships between individuals of a species and with expanding biological explanation to human societies and human social history. But what about relationships between species? How could biologists account for the evolution of stable symbiotic relations between microbe and host? Ever since the word "symbiosis" was used for contiguous associations between different kinds of organisms in 1876, two competing definitions have persisted.[53] Some biologists have used it, with Anton de Bary, to include parasitic relationships, and others have used it to mean solely mutualistic relations (see chapter 19). By the early twentieth century, the word was most often used in opposition to parasitism to mean mutualism. Some experimental biologists were reluctant to use the term because it had an unscientific aura of teleology insofar as it seemed to imply "cooperation for a common good."[54] The Gaia hypothesis, proposed by James Lovelock and by Lynn Margulis in the 1970s, had the same aura when, in one of its early forms, it postulated that the biosphere itself was a superorganism based on reciprocal relationships between species cooperating for a common good.[55]

Certainly, no post-Darwinian evolutionist would suggest that natural selection could favor the development in one species of a behavior pattern that is beneficial to another, if that behavior were either detrimental or of no selective value to the species itself. Biologists have generally insisted that symbiosis is driven by conflict and competition, which sometimes may result in a balanced relationship. Nonetheless, there is no unified Darwinian view about the nature of symbiosis. From the nineteenth century to the present there have been two opposing conceptions of the relationships between symbiotic microbes and their host plants and animals: some saw the microbial partners as parasites, others saw them as slaves. The Harvard ecologist Roscoe Pound spoke for many when he wrote in 1893:

> Mutualism of the kind we meet with in the vegetable kingdom involves sacrifice on the part of the host. The parasite is not there gratuitously. It is there to steal from its host the living it is hereditarily and constitutionally indisposed to make for itself. If the host gains any advantage from the relation, it can only do so by sacrificing—by giving the parasite the benefit of its labor that it may subsist.[56]

That the ("lower") microbial partner was a thief, stealing the host's rightful inheritance and living at the expense of the ("higher") host's labor, was a common assumption. Even when relations seemed more stable and mutualistic, biologists emphasized that they could easily become parasitic, depending on environmental conditions. For example, the fungi attached to the roots of the Scotch heather brought nutrients it absorbed from the soil in exchange for sugars from the plant. But the relationship works this smoothly only when the soil is poor in nitrogen; increase the nitrogen and the fungus grows too vigorously, becomes a parasite, and may kill the plant.[57]

The alternative perspective—that the microbial partner was actually a slave (a view that contrasted with the germ theory of disease)—arose from the lack of evidence for any apparent benefit to the microbial partners in many cases of so-called mutualism.[58] Certainly there were lots of stories about microbes exchanging lodging for food, but these were just-so stories. Experimental evidence was lacking. As late as 1987, David Smith and Angela Douglas could remark that "there has been no complete and rigorous experimental proof that any symbiosis is mutualistic."[59]

Over the past century, a host of anthropomorphic metaphors—slavery, consortium, and the relations between men and women, among nations, and between humans and domesticated plants—have been used by biologists to describe and understand symbiosis.[60] But as the Sorbonne zoologist Maurice Caullery cautioned in 1952, "The double danger of research into this type of phenomenon lies, on the one hand, in bringing to them preconceived ideas of too subjective a nature, bordering on an illusory anthropomorphism, and on the other hand, trying to reduce complex facts to simple elementary reactions."[61] Nonetheless, descriptions of symbiosis in terms of human social relationships continued. When commenting about the symbiotic origin of mitochondria and chloroplasts in his well-known text *An Introduction to Molecular Genetics* (1971), Gunther Stent expressed his view of the politics of the cell, in terms of international cooperation, not as mutualism but as its exact social Darwinist antithesis: "Thus a eukaryotic cell may be thought of as an empire directed by a republic of sovereign chromosomes in the nucleus. The chromosomes preside over the outlying cytoplasm in which formerly independent but now subject and degenerate prokaryotes carry out a variety of specialized service functions."[62]

Such political analogies, as well as the "master-slave" interpretation of intracellular relations, were common among biologists. But in more popular writings, symbiosis was more often interpreted as mutualism. In his award-winning book *The Lives of a Cell* (1974), Lewis Thomas insisted that the perception of mitochondria and chloropasts as "enslaved creatures," captured to provide ATP or carbohydrate and oxygen for their hosts, was merely anthropocentricism.[63] Far from being a case of one-sided exploitation, he assured his readers,

> there is something intrinsically good-natured about all symbiotic relations, necessarily, but this one, which is probably the most ancient and most firmly established of all, seems especially equable. There is nothing resembling predation, and no pretense of an adversary stance on either side. If you were looking for something like natural law to take the place of the "social Darwinism" of a century ago, you would have a hard time drawing lessons from the sense of life alluded to by chloroplasts and mitochondria, but there it is.[64]

Others turned to computer modeling and game theory to understand microbial symbiosis. Concern over international relations and especially the risk of nuclear war led Robert Axelrod, a political scientist, to team up with William Hamilton in 1981 to model the evolution of cooperation.[65] They thought that the game-theoretic approach would also be useful for understanding how such a "mutually advantageous symbiosis" as the lichen could originate and be maintained in a world of egoists without central authority, and for understanding the "defection" of resident bacterial symbionts into pathogenic parasites, or latent viruses into cancer-causing agents, if a human host becomes old or seriously ill.[66]

Their conclusions were formed by pitting various strategies against one another in a round-robin computer tournament of the famous "prisoner's dilemma" game, based on two egoistic players responding to each other's previous moves of defecting and cooperating. For long-term, evolutionarily stable relationships, they advocated a program of "tit-for-tat" developed by Anatol Rapoport. It begins with a cooperative choice and from then on the second player does what the first player did on his or her previous move. Axelrod and Hamilton thought this strategy would account for such mutualisms as a hermit crab and its sea-anemone partner, a cicada and the colonies of microbes living in its body, and a tree and its mycorrhiza.[67] Axelrod also had advice for national leaders about developing a more cooperative world: "Don't be envious, don't be the first to defect, reciprocate both cooperation and defection, and don't be too clever."[68]

Based on mathematical and other formal analyses, leading ecologists and theoretical evolutionary biologists of the 1970s and 1980s, such as Robert May and George Williams, insisted that mutualistic interactions between two or more species were unstable, and consequently rare and evolutionarily unimportant.[69] It had often been assumed, particularly in medical texts, that the coevolution of host and parasite would naturally develop either toward avirulence or toward a state of peaceful coexistence.[70] But this view was also contradicted.[71] Several theoreticians argued that the way the microbes are transmitted, whether vertically (from parent to offspring) or horizontally (from host to unrelated individuals), was important in determining whether the coevolution of host and virus or bacterium would develop toward mutualism or parasitism.[72] A vertical mode of transmission through the reproduction of the host would result in selection pressure away from parasitism and toward mutualism because harming the host would reduce transmission of the microbe. Thus, one would expect bacteria transmitted vertically through the eggs of aphids, the gut bacteria of a newborn acquired from suckling its mother's breast, or the symbionts acquired by larval termites from licking their mother's anus to evolve toward peaceful coexistence because only those bacterial mutations that benefited the host would be selected.

In 1995, Maynard Smith and Szathmáry promoted the view that all endosymbionts inherited through the cytoplasm of their hosts are best understood as "encapsulated slaves."[73] To understand the evolution of mitochondria and chloroplasts they added another principle, "central control," pointing to the evidence that most of the genes once present in these symbiotic organelles have been transferred to the cell nucleus. They liken this to human societies in which cooperation is enforced by some form of central authority; the transfer of genes to the nucleus was comparable to "paying taxes."[74] Thus, concepts and analogies borrowed from human social relations remain embedded in discussions of the symbiotic relations between microbe and host, which experimentalists argue frequently defy static categorization in terms of mutualism and parasitism. Social relations of intracellular symbionts and their vestiges are added to the technological metaphors—the messages, codes, and circuitry of information society governed by a master molecule—as the analogies in science, often forgotten as such, come full circle.

Universal models are often as difficult to maintain for microbes as they are for humans. Microbial evolutionists themselves continue to discuss how and why exactly the bacteria that gave rise to mitochondria might have become encapsulated in the cell. The standard account was indeed that the captured bacterium that became the mitochondrion lost the bulk of its genes to the nucleus. The magnitude of the loss can be seen from comparing the number of genes between mitochondria and free-living bacteria. The number of genes located in mitochondria in different species range from 2 in *Plasmodium falciparum,* the human malarial parasite, to 67 in *Reclinomonas americana* (human mitochondria contains about 13 genes), whereas the smallest genome of any free-living bacterium (*Bartonella henselae*) is composed of about 1600 genes.[75] Gene loss was obvious, but their actual transfer to the nucleus was called into question during the 1990s when microbial evolutionists offered alternative scenarios.[76]

That all hereditary symbionts are encapsulated slaves is contradicted by the most widespread hereditary "infection" in animals: bacteria of the genus *Wolbachia* are inherited through the cytoplasm of the eggs of 20 to 70 percent of all insect species as well as many other invertebrates, from shrimp to worms (see chapter 19). They cause cytoplasmic incompatibility between strains and species. Far from being captives or being centrally controlled by a "parliament of genes" in the nucleus, *Wolbachia* manipulate the reproduction and development of their hosts so as to maximize their own reproduction and transmission through the egg: they cause parthenogenetic induction, they can convert genetic males into reproductive females, and in some species they systematically kill males they infect.[77] Biologists have only begun to understand the significance of microbial symbionts in development, heredity, and evolution. Life abhors inflexible rules.

Epilogue

MOST OF THE GREAT CONTROVERSIES and conceptual oppositions of the nineteenth century are still present at the beginning of the twenty-first century: religion and vitalism versus evolution and materialism, structuralism versus functionalism, reductionism versus holism, gradualism versus saltationism, selectionism versus nonadaptationism, the inheritance of acquired characteristics, and nurture versus nature. What has changed is not so much the nature of the ideas but the evidence supporting them and the intensity of the debates.

The great consensual advances in biology are equally obvious. Paramount among them are the transmutation of species and the cellular basis of life. The advancement of biology has taken various forms. It has proceeded by the removal of contradictions (e.g., blending inheritance and Darwinian theory) and has often been expressed in terms of the resolution of paradoxes (none more persistent than differentiation among cells possessing identical genomes). Biology developed by opening black boxes (the molecular biology of the gene) and by making better statistical predictions (exemplified in Mendelian laws and in population genetics). It also progressed through establishing causal connections between previously unconnected phenomena (in particular, the role of the cell and its constituent parts in development and heredity).

Biology's evolution took other shapes as well. Fast-paced, cutting-edge biology proceeded by reducing complex phenomena to simple explanations, and it maintained conceptual order through oversimplification. The validity of a hypothesis depended on the validity of the purpose, its usefulness, or its heuristic value. American pragmatists moved along these lines in their approach to the nature of the gene and gene function during the first half of the twentieth century. Whatever the merits of the concept of the organism as an integrated whole, one could question its fruitfulness as a research program. The limits of progress by reductionism and oversimplification have become ever more conspicuous in recent decades as concepts of evolution, individual organism, and gene function have become more complex. After all, by the early 1960s, most biologists might have agreed that Mendel's laws apply to all inheritance in eukaryotes, most geneticists would have agreed with the one-gene: one-enzyme theory, and molecular biologists espoused the central dogma, and the control of development by a genetic program. In evolutionary theory, common descent, natural selection, population genetics, and the biological-species concept were accepted as universal certainties.

Yet today, biologists would agree that Mendel's laws do not apply to all inheritance in eukaryotes, the one-gene: one-enzyme theory is incorrect, and so too is the central dogma. Proteins are acknowledged to play much more active roles in genetic regulation and heredity than the doctrines of classical molecular biology had ever allowed. Molecular biologists now speak of split genes, RNA editing of gene expression by spliceosomes, chaperon proteins necessary for correct folding, and prions that are self-perpetuating and infectious. Molecular biologists have come to recognize that genes are neither discrete entities nor simply regions of DNA. Instead, they return to the viewpoint common in the first decades of the century of genes as abstract, operational entities: a way of speaking. We also have to consider the role of cell structures as guidance mechanisms for the ordering and arrangement of new cell structures. And in multicellular organisms, one has to attend to effects of cellular interactions far removed from DNA.

In contemporary evolutionary theory, common descent, the evolutionary "tree," and the biological species concept do not hold for the bacteria (sensu lato), among which horizontal gene transfer is rampant and evolution is reticulated. Panselectionism is questioned in relation to characteristics at all levels, from molecules to body plans and adult traits. While paleontologists consider punctuated evolutionary change, others reconsider symbiosis and the inheritance of acquired microbes in the great transition from bacteria to eukaryotes and in the subsequent evolution of eukaryotes. The Weismannian doctrine of one-genome: one-organism is replaced by a multigenomic concept of the "individual." All of this extends far beyond the tenets of the neo-Darwinian synthesis.

Creationists often use debates in evolutionary theory to imply that a view of evolution operating by natural forces undirected by God is incorrect. Obviously, none of the issues discussed above implies a return to pre-Darwinian teleology. Evolutionary theory is evolving as it should, and so too is cell theory. Of course, not all biologists may agree, or even be aware that such changes and challenges to accepted doctrines have occurred; synthesis often lags years behind research within specialties. DNA is still taught as the orchestrator of development, despite evidence to the contrary, and evolutionary change is still taught as population genetics, despite evidence for other important processes.

Advances in one direction have been accompanied by losses in others. Embryologists have long protested against gene-centered views of development, arguing that there was a hierarchy of levels of determination, as did those who studied intracellular morphogenesis. The same holds true for symbiosis in heredity, development, and evolution. Symbiosis researchers had long complained that they were ignored or discounted by those upholding the doctrines of their specialties.

The ideas about microbial symbiosis that emerged at the end of the century had been around since the nineteenth century. (Even horizontal gene transfer in bacteria, recognized at the turn of the twenty-first century to be so important, had been known fifty years earlier.) At any particular time, there is a hierarchy of specialties with assertions about the importance of one approach over another. This struggle for authority in science is not simply a matter of rhetoric, but of efficacy of techniques and/or institutional power, which have combined in different ways to determine relationships between groups in various global and local settings; and they have shaped the modern life sciences ever since Lamarck.

Improvements in techniques and instrumentation have directed the development of many aspects of the life sciences, from paleontology to genomics. For example, the invention of the microscope in the late sixteenth century, the simple lens, the achromatic lens, the oil immersion lens, the electron microscope, refinements in staining and fixing, and the microtome have driven cytology. Molecular biology has evolved by technological advances from crystallography and chromatography to nucleotide sequencing, which has transformed so much of biology. Choice organisms have also been vital: *Ascaris* and sea urchins for studies of fertilization and embryology, *Drosophila* for classical genetics, *Neurospora* for biochemical genetics, and bacteria and its viruses for molecular biology. Experimental objects play a determining role in both the kinds of questions that can be posed and the kinds of answers received. The concept of self-assembly worked for viruses, but not for *Paramecium*. Concepts of gene regulation based on bacteria were complicated by studies in eukaryotes. What was true for an elephant was not true for a bacterium.

It would be a serious mistake, however, to believe that the history of biology is nothing more than a history of the effects of technical improvements. The relative importance assigned to questions that manage to be posed and the kinds of answers it is possible to obtain are determined by both private and state funding. Much of biology has reflected commercial ends, from genetics and the rise of biochemical genetics and molecular biology in the 1940s to the Human Genome Project of the 1990s. The successful funding of science for applied purposes is not a matter of directly determining theories, as in Nazi eugenics or reckless Lysenkoist legislation, but of establishing and reinforcing research programs and specialties that themselves determine theory. Hence, debates over theories have often involved tensions over funding research in "pure" science in opposition to that reflecting commercial interests.

Charges of foul play upon an unfair field of competing theories and approaches have become more and more the norm as biological knowledge has become knowledge for some applied purpose. They have been made by those investigating cellular morphogenesis against gene-centered approaches to cell biology, and by microbial evolutionists in competition with the medically justified Human Genome Project. They have also been made by symbiosis researchers against the heavily medically funded view of bacteria as disease-causing germs. While research on microbial phylogeny and symbiosis once struggled in relation to genetics and medical perspectives on microbes as germs, today the growth of molecular evolution and microbial phylogeny, as well as studies of symbiosis and horizontal gene transfer, relies in part on these specialties' association with the medical and agricultural industries. The perceived imbalance between pure and applied science may also reflect attitudes about the nature of science as a vocation and as a business.

Still a larger influence from society permeates biology. The social context of science has played a fundamental role in the development of concepts, sometimes implicitly, sometimes explicitly. This is obvious in the technological conceptions of the cell in terms of codes, circuitry, and cybernetics developed during the 1950s and 1960s in the age of communication. But biology is concerned with history, interaction, and organization, and as such, it is also closely related to the social sciences.

Generations of biologists and social theorists have shared central concepts. Adam Smith's concept of division of labor, used by Darwin for his concept of divergence and speciation and based on Malthusian principles, was abandoned by the architect of the evolutionary synthesis, who replaced it with geographic isolation and statistical changes of genes within populations. But the concept of the division of labor remains central to the contemporary concept of the organism, as articulated in the cell theory. Analogies drawn from sociopolitical organization and from theories about human socioeconomic progress and political structure are as evident deep within organisms as they are in theories about the relationships between them. An individual organism today is generally conceived as an integrated system characterized by relations of interdependence, with regulating checks and balances between entities, molecules, cells, and microbes. The existence of an organism depends on the cooperation of its parts.

The concept of the division of labor was thought to be as applicable to the organization of biology as it was to the organisms biologists studied. This liberal nineteenth-century view of progressive biology, with its close affinity with the doctrine of laissez-faire, was the product of a composed, self-confident outlook. Let everyone specialize and get on with his or her job, and the "hidden hand" would take care of the harmony of biology, as it would regulate the world at large. The fact of the animal organism itself was a demonstration of the supreme truth of a beneficent progress toward higher things.

Fierce specialization has accompanied the growth of the life sciences; a genealogical tree of specialties could be made. But the innocent nineteenth-century view of a harmonious whole was shattered in the twentieth century as one specialty overpowered another: genetics swept over experimental embryology and natural history, only to be succeeded by molecular biology, which in turn "intruded" into evolutionary biology. The relationships between specialties only resembled a functional division of labor, a mutual aid society working together for the benefit of the whole.

However, the growth of the life sciences has not simply been a matter of conflict and divergence; it has also involved convergence, integration, and cooperation. This interdependence is apparent at every level of scientific activity—from the collaborative work in the laboratory to the references in a scientific paper. This is why scientists and scholars often have so much difficulty assigning discoveries to individuals, and when they do, they risk creating fictitious chimeras. The distortion of history is not just a matter of looking at the past exclusively for contributions past scientists have made to the current (correct) interpretation. Victors slant the history of their opponents, and they slant the history of their heroes (Darwin, Mendel, Garrod, et al.), trimming off ideological taint, ambiguity, and false concepts. Scientists thus usually take a pragmatic approach to history. The value of their accounts depends on their merits in fortifying the veracity of current ideas and raising the spirits of the scientist. Of course, no such theory of history has been professed, but the practice in classroom and texts, where a simple tale rides roughshod over facts, is no less apparent.

Evolution by association is as apparent between specialties as it is between individuals. Specialties do not survive long in isolation. Information is not just passed on vertically from one generation of specialists to the next, it is also passed on horizontally between specialties. This has been a macromechanism of change in the evolution of biology. New disciplines and even entire fields have resulted from the integration of ideas and techniques from once-independent specialties. The emergence of evolutionary theory in the nineteenth century itself is a triumph of synthesis of comparative morphology, paleontology, geology, and natural history. The fusion of cytology, breeding, and statistical reasoning formed classical genetics. The merger of genetics, population biology, and natural history led to the Darwinian renaissance of the 1930s and 1940s. The intermingling of techniques from physics and chemistry with those of microbiology and genetics resulted in the field of molecular biology. Research on developmental and hereditary symbiosis has relied on molecular biological approaches, as have investigations of prokaryote diversity and evolution. And we can see further how studies at the molecular level have to be combined with (not replace) those at higher levels to advance biology. Diversity and mixing have been creative forces in the life sciences, just as they have been in our larger social history.

Notes

Note to Preface

1. Theodosius Dobzhansky, "Nothing in Biology Makes Sense Except in the Light of Evolution," *American Biology Teacher* 35 (1973): 125–129.

Notes to Chapter 1

1. See, e.g., Peter Bowler, *Evolution: The History of an Idea* (Berkeley: University of California Press, 1984), and Ernst Mayr, *The Growth of Biological Thought* (Cambridge, Mass.: Harvard University Press, 1982). For discussion of naturalism confronting theology in the British context surrounding Darwinism, see Robert M. Young, *Darwin's Metaphor: Nature's Place in Victorian Culture* (Cambridge: Cambridge University Press, 1985).

2. Arthur O. Lovejoy, *The Great Chain of Being: A Study of the History of an Idea* (Cambridge, Mass.: Harvard University Press, 1936).

3. See Richard Burkhardt, "Biology," in *Dictionary of Scientific Biography*, ed. W. F. Bynum et al. 18 vols. (London: Macmillan, 1981), 1:43. See also Gottfried Reinhold Treviranus, *Biologie, oder Philosphie der lebenden Natur*, 6 vols. (Göttingen: Johann Friedrich Röwer, 1802–22).

4. Richard Burkhardt, *The Spirit of the System: Lamarck and Evolutionary Biology* (Cambridge, Mass.: Harvard University Press, 1977); Madeleine Barthélemy- Madaule, *Lamarck, ou le mythe du précursor* (Paris: Seuil, 1979). See Leslie Burlingame, "Lamarck," in *Dictionary of Scientific Biography*, ed. W. F. Bynum et al. (London: Macmillan, 1981), 7:584–593. For the most thorough account to date, see Pietro Corsi, *The Age of Lamarck: Evolutionary Theories in France, 1790–1830*, trans. Jonathan Mandelbaum, rev. and updated (Berkeley: University of California Press, 1988).

5. Burkhardt, *The Spirit of the System*; Barthélemy-Madaule, *Lamarck*; and Corsi, *The Age of Lamarck.* See also Camille Limoges, "Lamarck in His Milieu," *Science* 199 (1978): 1427–1428, and "Lamarck in Context," *Science* 243 (1989): 243–244.

6. See Conway Zirkle, "The Early History of the Idea of the Inheritance of Acquired Characters and of Pangenesis," *Transactions of the American Philosophical Society* 35 (1946): 91–151.

7. Erasmus Darwin, *Zoonomia, or the Laws of Organic Life*, 2d ed. (1984; reprint, London: Johnson, 1796).

8. J. B. Lamarck, *Zoological Philosophy: An Exposition with Regard to the Natural History of Animals*, trans. Hugh Elliot, with a foreword by Richard Burkhardt (Chicago: University of Chicago Press, 1984), 109.

9. Ibid., 122.

10. J. B. Lamarck, *Système des animaux sans vertébres* (Paris: Detérville, 1801), 14. Translated by Burkhardt in foreword to Lamarck, *Zoological Philosophy,* xxx.

11. Quoted in Lamarck, *Zoological Philosophy,* xxx.

12. Ibid., 416.

13. Ibid., 265; my italics.

14. Ibid., 55. On page 236 he wrote that "life and organisation are products of nature, and at the same time results of the powers conferred upon nature by the Supreme Author of all things and of the laws by which she herself is constituted: this can no longer be called into question. Life and organisation are thus purely natural phenomena." Lamarck considered humans to be the highest and most perfect form of organization, but Richard Burkhardt has suggested that he did not consider them to be the endpoint of evolution. He recognized that humans possessed some of the worst qualities as well as some of the best. In 1817 he wrote: "One would say that [man] is destined to exterminate himself after having rendered the globe uninhabitable." Ibid., xxxiii.

15. Ibid., 113.

16. Ibid., 107.

17. This work was read by every educated European in French or in any one of the many translations. Nordensköild wrote of Buffon: "In the purely theoretical sphere he was the foremost biologists of the 18th century, the one who possessed the greatest wealth of ideas, of real benefit to subsequent ages and exerting an influence stretching far into the future." Eric Nordensköild, *The History of Biology* (New York: A. A. Knopf, 1928), 229. Similarly, Ernst Mayr has commented: "It is no exaggeration to claim that virtually all the well-known writers of the Enlightenment, and even later generations in France as well as in other countries were Buffonians, either directly or indirectly." Mayr, *The Growth of Biological Thought*, 330. Mayr also wrote that "it makes no difference which of the authors in the second half of the eighteenth century one reads, their discussions are, in the last analysis, merely commentaries on Buffon's work. Except for Aristotle and Darwin, there has been no other student of organisms who has had as far-reaching an influence." Ibid., 337.

18. See Georges Louis Leclerc, Comte de Buffon, "Dégénération des animaux," *Histoire naturelle*, in *Oeuvres complète de Buffon*, ed. Pierre Flourens. 12 vols. (Paris: Garnier, 1853–55). See also Robert J. Richards, *Darwin and the Emergence of Evolutionary Theories of Mind and Behavior* (Chicago: University of Chicago Press, 1987), 37–40; W. T. Stearn, *An Introduction to the Species Plantarum and Cognate Botanical Works of Carl Linnaeus* (London: Bartholomew Press, 1957), 156–158.

19. See Burkhardt, *The Spirit of the System*, 202–210.

20. Lamarck, *Zoological Philosophy*, 404.

21. Lamarck's career indeed was not just that of a biologist; it also involved extensive writings in an anti-Lavoisierian type of chemistry, in anti-Laplacian physics and meteo-

rology, in geology, and in psychology and philosophy. See Corsi, *The Age of Lamarck*, and Limoges, "Lamarck and his Milieu."

22. Burkhardt, *The Spirit of the System*.

23. Ibid., xxxv.

24. On the life of Cuvier, see William Coleman, *Georges Cuvier, Zoologist* (Cambridge, Mass.: Harvard University Press, 1964); Dorinda Outram, *Georges Cuvier: Vocation, Science, and Authority in Post-Revolutionary France* (Dover, N.H.: Manchester University Press, 1984); and Martin Rudwick, *Georges Cuvier* (Chicago: University of Chicago Press, 1997).

25. Ibid. See also Toby Appel, *The Cuvier-Geoffroy Debate: French Biology in the Decades before Darwin* (New York: Oxford University Press, 1987).

26. As Erik Nordensköid noted, he also "possessed something of the latter's genius for organization; his energy was inexhaustible, he could discharge many duties at the same time without neglecting a single detail, he was full of ideas touching problems of organization." Nordenskiöld, *The History of Biology*, 332.

27. Cuvier is notorious among historians for his great disservice to Lamarck's theory of evolution. He was one of the first to misread the role of *besoin* (need) in Lamarck's evolutionary theory to an extremely detrimental caricature of it. See Burkhardt, in Lamarck, *Zoological Philosophy*, xlvi.

28. In 1795 Cuvier published his *Memoir on the Classification of the Animals Named Worms*. The catch-all taxon recognized by Linneaus as *Vermes* and renamed "invertebrates" by Lamarck was divided by Cuvier into six new classes of equal rank: mollusks, crustaceans, insects, worms, echinoderms, and zoophytes (coral, jellyfish, and sea anemone).

29. The historian Frank Sulloway has argued that familial birth order plays an important role in one's willingness to rebel, and that later-borns such as Darwin and Lamarck are statistically more rebellious than firstborns. He sees Lamarck as a revolutionary. Cuvier, a first born, maintained the status quo. Sulloway, *Born to Rebel: Birth Order, Family Dynamics, and Creative Lives* (New York: Pantheon Books, 1996), 50–53.

30. For Ernst Mayr, Cuvier's revolution in comparative anatomy was really a technical revolution, using dissection in classification, not a conceptual one. Mayr, *The Growth of Biological Thought*, 371. Others have emphasized that Cuvier's work was strong on the conceptual side as well. See Henri Daudin, *Cuvier et Lamarck: les classes zoologiques et l'idée de série animales, 1790–1830* (Paris: Librairie Félix Alcan, 1926).

31. Michel Foucault, *The Order of Things: An Archaeology of the Human Sciences* (London: Tavistock Publications, 1974); François Jacob, *The Logic of Living Systems: A History of Heredity*, trans. Betty Spillmann (London: Allen Lane, 1974). See also Daudin, *Cuvier et Lamarck.*

32. Mayr, *The Growth of Biological Thought*, 185.

33. Appel, *The Cuvier-Geoffroy Debate.*

34. Quoted in Corsi, *The Age of Lamarck*, 255.

35. Ibid., 211.

36. See, e.g., Brian Hall, "Baupläne, Phylotypic Stages, and Constraint: Why There Are so Few Types of Animals?" *Evolutionary Biology* 29 (1996): 215–261; Ron Amundson, "Typology Reconsidered: Two Doctrines on the History of Evolutionary Biology," *Biology and Philosophy* 13 (1998): 153–177.

37. See Dov Ospovat, *The Development of Darwin's Theory: Natural History, Natural Theology, and Natural Selection, 1838–1859* (Cambridge: Cambridge University Press, 1981). See also Michael Ruse, *The Darwinian Revolution: Science Red in Tooth Claw*

(Chicago: University of Chicago Press, 1979); Philip Rehbock, *The Philosophical Naturalists: Themes in Early Nineteenth-Century British Biology* (Madison: University of Wisconsin Press, 1983).

38. See Appel, *The Cuvier-Geoffroy Debate*.

39. Ibid.

40. Appel argues that during the late 1820s, Cuvier had feared the return of revolution and rule by an uncontrolled mob. His willingness to descend into the public debate in 1830, she argues, was due in part to his anxiety in the months preceding the Paris revolution; ibid., 9. Others disagree with this interpretation. See Camille Limoges, "A Clash between Naturalists," *Science* 239 (1988): 21–22. Dorinda Outram has also argued that Cuvier was far from being the archconservative his enemies depicted. See Outram, *Georges Cuvier*.

41. See Adrian Desmond, *The Politics of Evolution: Morphology, Medicine, and Reform in Radical London* (Chicago: University of Chicago Press, 1989).

42. Ibid.

43. Ibid., 2.

44. This suggestion has been made by Desmond, ibid; and [Robert Grant], "Observations on the Nature and Importance of Geology," *Edinburgh New Philosophical Journal* (1826): 293–302, at 297 and 300. James Secord argues that article was more likely written by the editor Robert Jameson. See James Secord, "Edinburgh Lamarckians: Robert Jameson and Robert E. Grant," *Journal of the History of Biology* 24 (1991): 1–18.

45. See Desmond, *The Politics of Evolution*.

Notes to Chapter 2

1. Charles Darwin, *Structure and Distribution of Coral Reefs* (London: John Murray, 1842); Darwin, *Geological Observations on Volcanic Islands and Parts of South America* (London: John Murray, 1846).

2. Charles Darwin, *Monographs on the Cirripidae* (London: John Murray, 1851–54).

3. Charles Darwin, *The Variation of Animals and Plants under Domestication*, vols. 1–2 (London: 1868).

4. Charles Darwin, *On the Various Contrivances by Which British and Foreign Orchids Are Fertilized by Insects* (London: John Murray, 1862).

5. Charles Darwin, *The Descent of Man* (London: Murray, 1871).

6. Charles Darwin, *The Expression of the Emotions in Man and Animals* (Chicago: University of Chicago Press, 1965).

7. See Georges Canguilhem, *Les concepts de "lutte pour l'existence" et de "sélection naturelle" en 1858: Charles Darwin et Alfred Russel Wallace*. (Paris: Université de Paris, 1959); Wilma George, *Biologist Philosopher: A Study of the Life and Writings of Alfred Russel Wallace* (London: Abelard-Schuman, 1964); Harry Clements, *Alfred Russel Wallace: Biologist and Social Reformer* (London : Hutchinson, 1983).

8. See Adrian Desmond, *The Politics of Evolution: Morphology, Medicine, and Reform in Radical London* (Chicago: University of Chicago Press, 1989).

9. Erasmus Darwin, *Zoonomia, or the Laws of Organic Life*, 2d ed. (London: Johnson, 1796).

10. Robert Chambers, *Vestiges of Natural History of Creation* (London: Churchill, 1844). Reprinted with an introduction by Sir G. de Beer (Leicester: Leicester University Press, 1969).

11. See James Secord, "Behind the Veil: Robert Chambers and *Vestiges*," in *History, Humanity and Evolution: Essays in Honor of John C. Greene*, ed. James Moore (Cambridge: Cambridge University Press, 1989); M. J. S. Hodge, "The Universal Gestation of Nature: Chambers' Vestiges and Explanations," *Journal of the History of Biology* 5 (1972): 237–246.

12. Quoted in Ernst Mayr, *The Growth of Biological Thought* (Cambridge, Mass.: Harvard University Press, 1982), 385.

13. See F. Darwin, ed., *The Autobiography of Charles Darwin and Selected Letters* (New York: Dover, 1958).

14. Ibid., 13.

15. See Charles Darwin, *The Voyage of the "Beagle"* (Letchworth: Aldine Press, 1959); Alan Moorehead, *Darwin and the "Beagle"* (London: Hamish Hamilton, 1969); Janet Browne, *Charles Darwin, Voyaging: Volume 1 of a Biography* (New York: Knopf, 1995).

16. Charles Darwin, *Journal of Researches into the Natural History and Geology of the Countries Visited During the Voyage of the H. M. S. "Beagle" round the World*, new ed. (New York: D. Appleton, 1875).

17. Martin Rudwick, "The Strategy of Lyell's Principles of Geology," *Isis* 61 (1970): 5–33.

18. In Lyell's words, "To explain the observed phenomena, we may dispense with sudden, violent and general catastrophes, and regard the ancient and present fluctuations . . . as belonging to one continuous and uniform series of events." Charles Lyell, *Principles of Geology*, 6th ed. (London: John Murray, 1840), 1:preface, ix.

19. See Nicolaas Rupke, *Richard Owen: Victorian Naturalist* (New Haven: Yale University Press, 1994), 138–141.

20. Peter Bowler calls this "transcendental progressionism." See Peter Bowler, *Fossils and Progress: Paleontology and the Idea of progressive Evolution in the Nineteenth Century* (New York: Science History Publications, 1976), 47–53.

21. Some German naturalists, including the poet and philosopher Johann Wolfgang von Goethe (1749–1832), did consider the possibility of an actual temporal unfolding towards higher levels of organization from minerals to plants and animals and man. Ibid. On *Naturphilosophie* see, e.g., Robert Richards, *The Meaning of Evolution: The Morphological Construction and Ideological Reconstruction of Darwin's Theory* (Chicago: University of Chicago Press, 1992); Timothy Lenoir, *The Strategy of Life: Teleology and Mechanics in Nineteenth Century German Biology* (Dordrecht: D. Reidel, 1982).

22. This point is emphasized in Ernst Mayr, *The Growth of Biological Thought* (Cambridge, Mass.: Harvard University Press, 1982), 380. See also F. Darwin, *The Autobiogaphy of Charles Darwin and Selected Letters*, 177–178.

23. H. C. Watson to Charles Darwin, Dec. 21, 1859, in F. Darwin, *The Autobiogaphy of Charles Darwin and Selected Letters*, 178.

24. Charles Darwin, *On the Origin of Species*, introd. Ernst Mayr, facsimile ed. (1859; Cambridge, Mass.: Harvard University Press, 1964), 398–399.

25. Nora Barlow, ed., *The Autobiography of Charles Darwin* (London: Collins, 1958), 118–119.

26. Gavin de Beer, ed., "Darwin's Notebooks on the Transmutation of Species," *Bulletin of the British Museum* (Natural History), Historical series 2 (1960), nos. 2–5.

27. A great deal of attention has been given to Darwin's reluctance to publish his theory of evolution. It has been associated with Darwin's recurrent periods of illness from 1837, which some argue were psychological in origin due to the severe anxiety Darwin experi-

enced as he began to explore the ramifications of the theory of natural selection and how shocking it would seem to the society of his day. See Ralph Colp, *To Be an Invalid: The Illness of Charles Darwin* (Chicago: University of Chicago Press, 1977). Desmond intimates that Darwin did not publish during the revolutionary period of the 1830s and 1840s because evolutionary theory was indeed allied with revolutionary views of radicals. See Desmond, *The Politics of Evolution*, 2–3.

28. F. Darwin, *The Autobiography of Charles Darwin and Selected Letters*, 42.

29. Charles Lyell, *Principles of Geology*, 3d ed., vol. 2 (London: John Murray, 1834), 364, 63.

30. See Edward Blyth, "An Attempt to Classify the 'Varieties' of Animals," *Magazine of Natural History* 8 (1835): 40–53; Blyth, "On the Psychological Distinctions between Man and All Other Animals; and the Consequent Diversity of Human Influence over the Inferior Ranks of Creation, from Any Mutual and Reciprocal Influence Exercised among the Latter," *Magazine of Natural History* 10 (1937): 1 ff. See also Loren Eiseley, "Charles Darwin, Edward Blyth, and the Theory of Natural Selection," *Proceedings of the American Philosophical Society* 103 (1959): 94–111; Eiseley, *Darwin and the Mysterious Mr. X* (London: J. M. Dent and Sons, 1979).

31. See C. C. Gillispie, *Genesis and Geology: The Impact of Scientific Discoveries upon Religious Beliefs in the Decades before Darwin* (New York: Harper and Brothers, 1951); R. Hooykaas, *Natural Law and Divine Miracle* (Leiden: Brill, 1959); T. McPherson, *The Argument from Design* (London: Macmillan, 1972).

32. Quoted in Donald Worster, *Nature's Economy: A History of Ecological Ideas* (Cambridge: Cambridge University Press, 1977), 37.

33. The remittent works were Thomas Chalmers, *The Adaptation of External Nature to the Moral and Intellectual Constitution of Man* (1833); William Prout, *Chemistry, Meteorology, and Digestion* (1834); William Kirby, *History, Habits, and Instincts of Animals* (1835); Charles Bell, *The Hand, as Evincing Design* (1837); Dean Buckland, *Geology and Mineralogy* (1837); J. Kidd, *The Adaptation of External Nature to the Physical Condition of Man* (1837); William Whewell, *Astronomy and General Physics* (1839), and P. M. Roget, *Animal and Vegetable Physiology* (1840).

34. See Camille Limoges, "Darwinism et adaptation," *Revue des questions scientifiques* 141 (1970): 353–374; Limoges, *La sélection naturelle* (Paris: Presse Univérsitaire de France, 1970).

35. Loren Eisley and Camille Limoges have noted that Darwin had given considerable attention to Blyth's arguments when formulating his own theory. Eisley charged that Blyth was among the forerunners to whom Darwin paid insufficient acknowledgment, and indeed whose debt he may have tried to conceal. See Eisely, *Darwin and the Mysterious Mr. X*. Limoges, on the other hand, emphasized that Darwin's theory represented an epistemological break away from the whole tradition of natural history. For a refutation of the so-called precursors of Darwin, see Limoges, *La sélection naturelle*, 86–116. See also Antonello La Vergata, "Images of Darwin: A Historiographic Overview," in *The Darwinian Heritage*, ed. David Kohn (Princeton: Princeton University Press, 1985), 901–972.

36. Thus Limoges argued that Darwin's so-called precursors did "not constitute in any way an anticipation of the Darwinian theory of natural selection." Limoges, *La sélection naturelle*, 71.

37. Ibid. See also Eisley, *Darwin and the Mysterious Mr. X*, 57, 69.

38. See Limoges, *La sélection naturelle*; David Kohn, "Theories to Work By: Rejected Theories, Reproduction, and Darwin's Path to Natural Selection," *Studies in History of*

Biology 4 (1980): 67–170; M. J. S. Hodge and David Kohn, "The Immediate Origins of Natural Selection," in *The Darwinian Heritage*, ed. David Kohn (Princeton: Princeton University Press, 1985), 185–206; Dov Ospovat, *The Development of Darwin's Theory: Natural History, Natural Theology, and Natural Selection, 1838–1859* (Cambridge: Cambridge University Press, 1981); Philip Rehbock, *The Philosophical Naturalists: Themes in Early Nineteenth-Century British Biology* (Madison: University of Wisconsin Press, 1983); M. J. S. Hodge, "The Development of Darwin's General Biological Theorizing," in *Evolution from Molecules to Men*, ed. D. S. Bendall (Cambridge: Cambridge University Press, 1983), 43–62; Ernst Mayr, *The Growth of Biological Thought.* (Cambridge, Mass.: Harvard University Press, 1984).

39. See Thomas Malthus, *An Essay on the Principle of Population, as It Affects the Future Improvement of Society with Remarks on the Speculations of Mr. Godwin, M. Condorcet, and Other Writers* (London: J. Johnson, 1798).

40. See Limoges, *La sélection naturelle*, 77.

41. Charles Darwin to Emma Darwin, July 5, 1844, in F. Darwin, *The Autobiography of Charles Darwin and Selected Letters*, 180–182.

42. Alfred Russel Wallace, "On the Tendency of Varieties to Depart Indefinitely from the Original Type," reprinted in *Darwin*, ed. Philip Appleman (New York: W. W. Norton, 1970), 89–97, 90.

43. Alfred Russel Wallace to A. Newton, Dec. 3, 1887, in F. Darwin, *The Autobiography of Charles Darwin and Selected Letters*, 200.

44. Charles Darwin to Charles Lyell, June 18, 1858, in F. Darwin, *The Autobiography of Charles Darwin and Selected Letters*, 196–197.

45. Charles Darwin and Alfred Russel Wallace, "On the Tendency of Species to Form Varieties; and on the Perpetuation of Varieties and Species by Means of Selection," *Journal of the Linnaean Society*, Zoology III (1858): 45–62.

46. On Wallace see, e.g., John Langdon Brooks, *Just Before the Origin: Alfred Russel Wallace's Theory of Evolution* (New York: Columbia University Press, 1984); Harry Clements, *Alfred Russel Wallace: Biologist and Social Reformer* (London : Hutchinson, 1983); Martin Fichman, *Alfred Russell Wallace* (Boston, Mass. : Twayne, 1981).

47. Alfred Russel Wallace to A. Newton, Dec. 3, 1887, in F. Darwin, *The Autobiography of Charles Darwin and Selected Letters*, 200–201.

48. He also was awarded the Copley Medal (1908), the Order of Merit (1908), the Linnean Society of London's Gold Medal (1892), the Darwin-Wallace Medal (1908), and the Royal Geographical Society's Founder's Medal (1892), as well as honorary doctorates from the Universities of Dublin (1882) and Oxford (1889), and in 1893 he won election to the Royal Society.

49. Charles Darwin to Charles Lyell, June 25, 1958, in F. Darwin, *The Autobiography of Charles Darwin and Selected Letters*, 196.

50. Alfred Russel Wallace to A. Newton, Dec. 3, 1887, in F. Darwin, *The Autobiography of Charles Darwin and Selected Letters*, 200–201, at 201.

51. Darwin, *The Origin*, 420.

52. Ibid., 484.

53. See David Kohn, "Darwin's Principle of Divergence as Internal Dialogue," in *The Darwinian Heritage*, ed. David Kohn (Princeton: Princeton University Press, 1985), 245–263.

54. Darwin, *The Origin*, 471

55. Ibid., 130

56. Ibid., 95.
57. Ibid.
58. "Nature does not make leaps"; ibid., 471; see also 194.
59. Ibid., 45.
60. Ibid., 46.
61. Ernst Mayr, in *Evolution and Anthropology* (Washington, D.C.: Anthropology Society of Washington, 1959), 2. See also, Mayr, *The Growth of Biological Thought.*
62. Darwin, *The Origin*, 64.

Notes to Chapter 3

1. See Adrian Desmond, *Huxley: From Devil's Disciple to Evolution's High Priest* (Reading, Mass.: Perseus Books, 1994).
2. T. H. Huxley to C. Darwin, Nov. 23, 1859, in *The Autobiography of Charles Darwin and Selected Letters*, ed. F. Darwin (New York: Dover, 1958), 60–66.
3. Desmond, *Huxley*, 4.
4. See Ronald Clark, *The Huxleys* (New York: McGraw-Hill, 1968), 51.
5. T. H. Huxley to C. Darwin, Nov. 23, 1859, in F. Darwin, *The Autobiography of Charles Darwin and Selected Letters*, 226.
6. Thomas H. Huxley, *Man's Place in Nature* (Ann Arbor: University of Michigan Press, 1959).
7. Charles Darwin, *On The Origin of Species*, facsimile ed. with an introduction by Ernst Mayr (Cambridge, Mass.: Harvard University Press, 1964), 488.
8. Charles Darwin, *The Descent of Man* (London: John Murray, 1971).
9. Huxley, *Man's Place in Nature*, 71.
10. See Desmond, *Huxley*.
11. Charles Darwin autobiography of 1876, in F. Darwin, *The Autobiography of Charles Darwin and Selected Letters*, 66.
12. T. H. Huxley to Francis Darwin, June 27, 1861, in F. Darwin, *The Autobiography of Charles Darwin and Selected Letters*, 253–254.
13. Huxley to F. Dyster, Sept. 9, 1860, in Desmond, *Huxley*, 279.
14. See Nicolaas Rupke, *Richard Owen: Victorian Naturalist* (New Haven, Conn.: Yale University Press, 1994), 239.
15. See Adrian Desmond, *Archetypes and Ancestors: Paleontology in Victorian London, 1850–1875* (London: Blond and Briggs, 1982); Evelyn Richards, "A Question of Property Rights: Richard Owen's Evolutionism Reassessed", *British Journal for the History of Science* 20 (1987): 129–171; Michael Ruse, "Were Owen and Darwin *Naturphilosophen*? *Annals of Science* 50 (1993): 383–388; Lynn Nyhart, *Biology Takes Form* (Chicago: University of Chicago Press, 1995); Mario Di Gregorio, "A Wolf in Sheep's Clothing: Carl Gegenbaur, Ernst Haeckel, the Vertebral Theory of the Skull, and the Survival of Richard Owen," *Journal of the History of Biology* 28 (1995): 247–280; Ron Amundson, "Typology Reconsidered: Two Doctrines on the History of Evolutionary Biology," *Biology and Philosophy* 13 (1998): 153–177.
16. Richard Owen, *Report on the Archetype and Homologies of the Vertebrate Skeleton* (London: Voorst, 1848); Owen, *On the Nature of Limbs* (London: Voorst, 1849).
17. Rupke, *Richard Owen: Victorian Naturalist.*

18. Owen's pronouncements on evolution puzzled Darwin, who depicted Owen as someone who alternately denied its validity, professed ignorance on the matter, and claimed to have discovered the principle of natural selection himself before Darwin had. Darwin wrote, in his historical sketch of the fifth edition of *The Origin*, "As far as the mere enunciation of the principle of Natural Selection is concerned, it is quite immaterial whether or not Professor Owen preceded me, for both of us . . . were long ago preceded by Dr. Wells and Mr. Matthew." See F. Darwin, *The Autobiography of Charles Darwin and Selected Letters,* 290. It is not surprising that, as is often the case when a new discovery is successfully made, Darwin was charged with not sufficiently crediting others who may have contributed to his theories. On forerunners who had some idea of natural selection before Darwin, see Loren Eiseley, *Darwin and the Mysterious Mr. X* (New York: E. P. Dutton, 1979).

19. See Desmond, *Archetypes and Ancestors*, 72–83.

20. Erik Nordenskiöld, *The History of Biology* (New York: Tudor, 1929), 505, 506: "There are not many personalities who have so powerfully influenced the development of human culture—and that, too, in many spheres—as Haeckel."

21. See Darwin, *On the Origin of Species*, 338, 440–442.

22. See Brian K. Hall, "Balfour, Garstang and de Beer: The First Century of Evolutionary Embryology," *American Zoologist* 40 (2000): 718–728.

23. See Stephen Jay Gould, *Ontogeny and Phylogeny* (Cambridge, Mass.: Harvard University Press, 1977).

24. See Peter Bowler, *Evolution: The History of an Idea* (Berkeley: University of California Press, 1984), 247.

25. See Robert Richards, *The Meaning of Evolution: The Morphological Construction and Ideological Reconstruction of Darwin's Theory* (Chicago: University of Chicago Press, 1992), 40. See also Timothy Lenoir, *The Strategy of Life: Teleology and Mechanics in Nineteenth-Century German Biology* (Dordrecht: D. Reidel, 1978).

26. Richards, *The Meaning of Evolution.*

27. Ernst Haeckel, *The History of Creation,* vol. 1 (New York: Appleton, 1880), 10–11.

28. W. W. Ballard, "Problems of Gastulation: Real and Verbal," *Bioscience* 26 (1976): 36–39, 38.

29. See, e.g., Brian Hall, "Baupläne, Phylotypic Stages, and Constraint: Why There Are So Few Types of Animals," *Evolutionary Biology* 29 (1996): 215–261; Gould, *Ontogeny and Phylogeny.*

30. See Gould, *Ontogeny and Phylogeny*; Ernst Mayr, *The Growth of Biological Thought* (Cambridge, Mass.: Harvard University Press, 1982).

31. Richards, *The Meaning of Evolution.*

32. Ibid., 15, 73.

33. Darwin, *The Origin*, 338.

34. Ibid.

35. See Mark Ridley, "Triumph of the Embryo?" *Nature* 357 (1992): 203–204.

36. See Nicolass Rasmussen, "The Decline of Recapitulationism in Early Twentieth Century Biology," *Journal of the History of Biology* 24 (1991): 51–89.

37. See Hall, "Balfour, Garstang and de Beer"; Hall, *Evolutionary Developmental Biology*, 2d ed. (Dordrecht: Kluwer Academic Publishers, 1998); Gould, *Ontogeny and Phylogeny*. See also Gavin de Beer, *Embryology and Evolution* (Oxford: Oxford University Press, 1930); de Beer, *Embryology and Ancestors* (Oxford: Oxford University Press, 1940).

38. W. Garstang, "The Theory of Recapitulation: A Critical Restatement of the Biogenetic Law," *Proceedings of the Linnaean Society of London* 35 (1922): 81–101, at 81.

39. Haeckel, *The History of Creation*, 11.

40. Ibid., 27.

41. Ibid., 23.

42. Ibid., 36.

43. Ibid.

44. Ibid., 36–37.

45. Richard Dawkins, *The Blind Watch Maker: Why the Evidence of Evolution Reveals a Universe without Design* (New York: Norton, 1996), 6.

46. See Ronald L. Numbers, *The Creationists: The Evolution of Scientific Creationism* (New York: Knopf, 1992); Michael Ruse, *Can a Darwinian Be a Christian? The Relationship between Science and Religion* (New York: Cambridge University Press, 2000); Stephen Jay Gould, *Rocks of Ages: Science and Religion in the Fullness of Life* (New York: Ballantine, 1999).

Notes to Chapter 4

1. There is a great body of literature on social Darwinism. See, e.g., John Greene, "Biology and Social Theory in the Nineteenth Century," in *Critical Problems in the History of Science*, ed. Marshall Clagett (Madison: University of Wisconsin Press, 1962), 416–446; Richard Hofstadter, *Social Darwinism in American Thought*, rev. ed. (Boston: Beacon Press, 1955); Greta Jones, *Social Darwinism in English Thought* (London: Harvester, 1980); Robert Young, *Darwin's Metaphor: Nature's Place in Victorian Culture* (Cambridge: Cambridge University Press, 1985); Young, "Darwinism Is Social," in *The Darwinian Heritage*, ed. David Kohn (Princeton: Princeton University Press, 1985), 609–638. For a critique of Young's approach, see Ingemar Bohlin, "Robert M. Young and Darwin Historiography," *Social Studies of Science* 21 (1991): 597–648; Jim Moore, "Socializing Darwinism: Historiography and the Fortunes of a Phrase," in *Science as Politics*, ed. Les Levidow (London: Free Association Books, 1986), 38–75.

2. Some of Spencer's critics exploited the apparent contradiction between his individualism and his organismic view of society. For example, socialist reformers such as Beatrice Webb and the reform-minded American sociologist Lester Frank Ward argued that Spencer's metaphor of the "social organism" would imply the subordination of the individual to the needs of the whole and the need for the state as the "organ of integration" to coordinate individuals into the social organism. See Hofstadtler, *Social Darwinism*, 80.

3. See H. I. Sharlin, "Herbert Spencer and Scientism," *Annals of Science* 33 (1976): 457–480.

4. H. Spencer, "A Theory of Population, Deduced from the General Law of Animal Fertility," *Westminister Review* 57 (1852): 468–501.

5. See, e.g., Hofstadter, *Social Darwinism*.

6. See Robert Bannister, *Social Darwinism: Science and Myth in Anglo-American Social Thought* (Philadelphia: Temple University Press, 1979); Howard L. Kaye, *The Social Meaning of Modern Biology: From Social Darwinism to Sociobiology* (New Haven: Yale University Press, 1986), 26–31.

7. Herbert Spencer, *The Principles of Ethics* (New York: D. Appleton, 1892), 189, 201.

8. Herbert Spencer, *The Principles of Biology,* rev. and enlarged ed., vol. 2 (New York: D. Appelton, 1899), 533.

9. Ibid.

10. See Ernst Mayr, *The Growth of Biological Thought* (Cambridge, Mass.: Harvard University Press, 1983), 386.

11. On Spencer's organismic conceptions and symbiosis, see Jan Sapp, *Evolution by Association: A History of Symbiosis* (New York: Oxford University Press, 1994).

12. See Adrian Desmond, *Huxley: From Devil's Disciple to Evolution's High Priest*, (Reading, Mass.: Perseus Books, 1994), 184. It should be noted, however, that Desmond wrongly refers to Spencer as an "anarchist."

13. S. Persons, ed., *Social Darwinism: Selected Essays of William Sumner* (Englewood Cliffs, N.J.: Prentice-Hall, 1963), 157.

14. Ibid., 76–77.

15. E. E. Kirkland, ed., *The Gospel of Wealth and Other Timely Essays by Andrew Carnegie* (Cambridge, Mass.: Belknap, 1962), 16–17.

16. Daniel Gasman, *The Scientific Origins of National Socialism: Social Darwinism in Ernst Haeckel and the German Monist League* (London: MacDonald, 1971).

17. H. von Treitschke, *Politics*, trans. Blanche Dugdale and Torben de Bille, introd. A. J. Balfour, 2 vols. (London: Constable, 1916), 1:21.

18. Adolf Hitler, *Mein Kampf*, trans. Ralph Manheim, introd. D. C. Watt (London: Hutchison, 1969), 258–260.

19. K. Marx and F. Engels, *Selected Correspondence*, 2d ed. (Moscow: Foreign Language Publishing House, 1965), 115.

20. At one time, it was believed that Marx offered to dedicate the second volume of Das *Capital* to Darwin. See Ralph Colp Jr., "The Contacts between Charles Darwin and Karl Marx," *Journal of the History of Ideas* 35 (1974): 329–338; Colp, "The Contacts of Charles Darwin and Edward Aveling and Karl Marx," *Annals of Science* 33 (1976): 387–394; Colp, "The Myth of the Darwin-Marx Letter," *History of Political Economy* 14 (1982): 461–482; M. A. Fay, "Did Marx Offer to Dedicate *Capital* to Darwin? A Reassessment of the Evidence," *Journal of the History of Ideas* 39 (1978): 133–146. Fay argues that Darwin's literary contact was Edward Aveling (the de facto husband of Marx's daughter Eleanor), not Marx himself.

21. K. Marx and F. Engels, *Selected Works in One Volume* (Moscow: Lawrence and Wishart, 1968), 429.

22. See E. H. Carr, *What Is History?* (New York: Vintage Books, 1961), 133.

23. Leon Trotsky, *My Life* (London: Butterworth, 1930), 422.

24. See Antonello La Vergata, "Images of Darwin: A Historiographic Overview," in *The Darwinian Heritage*, ed. D. Kohn (Princeton: Princeton University Press, 1985). For a concise review and commentary, see Stephen Shapin and Barry Barnes, "Darwin and Social Darwinism: Purity in History," in *Natural Order*, ed. B. Barnes and S. Shapin (London: Sage, 1979), 125–139. Shapin and Barnes offer an anthropological explanation about scientific disciplines and myths of origins for why Darwin's motives have become so important:

> The scientific discipline of evolutionary biology had its font and origin in the person of Charles Darwin and in the text of 1859. Darwin is a sacred totem by virtue of his "foundership" of modern biology: science is sacred, so must Dar-

> win and his Book be sacred; both must be protected from the contamination by the profane. As the author of the Origin he must himself be pure; his thought must be unmingled with worldly pollutions and incapable of satisfactorily blending or combining with the suspect formulations of social Darwinism. Thus "influences" from the "profane" Malthus can only be the spiritual emanations of mathematics and genuine science, or nonessential stimuli or manners of speech. And implications for social Darwinism can only be misunderstandings. (134)

See also Moore, "Socializing Darwinism."

25. E. Manier, *The Young Darwin and His Cultural Circle* (Dordrecht: Reidel, 1978), 138–146.

26. See, e.g., Ernst Mayr, *The Growth of Biological Thought* (Cambridge, Mass.: Harvard University Press, 1982), 492.

27. For a discussion of Darwin's social Darwinist proclivities, see J. C. Greene, "Darwin as a Social Evolutionist," *Journal of the History of Biology* 10 (1977): 1–27; Greene, *Science, Ideology, and World View: Essays in the History of Evolutionary Ideas* (Berkeley: University of California Press, 1981). See also Moore, "Socializing Darwinism."

28. F. Darwin, *The Autobiography of Charles Darwin and Selected Letters*, 65.

29. Charles Darwin, letter to W. Graham, July 3, 1881, in F. Darwin, *The Autobiography of Charles Darwin and Selected Letters*, 69.

30. S. Herbert, "The Place of Man in the Development of Darwin's Theory of Transmutation," *Journal of the History of Biology* 10 (1977): 155–227, at 194–195; Moore, "Socializing Darwinism."

31. Moore, "Socializing Darwinism," 62.

32. Robert Young, *Darwin's Metaphor* (Cambridge: Cambridge University Press, 1985), 3. See also R. C. Lewontin, "The Concept of Evolution," in *International Encyclopedia of Social Science*, ed. D. L. Sills, vol. 5 (New York: Macmillan, Fress Press, 1968), 2002–2010; Lewontin, *The Genetic Basis of Evolutionary Change* (New York: Columbia University Press), 1974.

33. David Kohn, "Darwin's Principle of Divergence as Internal Dialogue," in *The Darwinian Heritage*, ed. David Kohn (Princeton: Princeton University Press, 1985), 245–263, at 245.

34. As Camille Limoges has shown, the concept went through a series of transformations from the social sciences to the biology back to the social sciences. Limoges, "Milne-Edwards, Darwin, Durkheim and Division of Labor: A Case Study in Reciprocal Conceptual Exchanges Between the Social and Natural Sciences," in *The Relations between the Natural Sciences and the Social Sciences*, ed. I. B. Cohen (Princeton: Princeton University Press, 1994), 317–343. See also Camille Limoges and Claude Ménard, "Organization and Division of Labor: Biological Metaphors at Work in Alfred Marshall's Principles of Economics," in *Natural Images in Economics*, ed. Philip Mirovski (Cambridge: Cambridge University Press), 336–359.

35. Charles Darwin, *On the Origin of Species*, with an introduction by Ernst Mayr, facsimile ed. of 1859 (Cambridge, Mass.: Harvard University Press, 1964), 115.

36. Quoted in Limoges, "Milne-Edwards, Darwin, Durkheim, and Division of Labour," 321.

37. The concept of the division of labor was subsequently developed as the famous doctoral dissertation of the social theorist Emile Durkheim (1858–1917), who referred to Darwin and argued that the labor will become increasing divided in societies as the struggle

for existence becomes more strenuous; ibid. See also Emile Durkheim, *The Division of Labor in Society*, trans. W. D. Halls, introd. Lewis Coser (New York: Free Press, 1984), 208–209.

38. His attack on the poor laws and criticism of public charity became more prominent in the more widely read second edition of his *Essay*.

39. See Thomas Malthus, *An Essay on the Principle of Population, as It Affects the Future Improvement of Society with Remarks on the Speculations of Mr. Godwin, M. Condorcet, and Other Writers* (London: J. Johnson, 1798), chap. 5.

40. Ibid.

41. See Thomas Malthus, "A Summary View of the Principle of Population," reprinted in *Three Essays on Population: Thomas Malthus, Julian Huxley and Frederick Osborn* (New York: Mentor Books, 1960), 13–59, at 59.

42. Frank W. Notestein, introduction to *Three Essays on Population: Thomas Malthus, Julian Huxley and Frederick Osborn* (New York: Mentor Books, 1960), x.

43. Much of the interest in the issue of the Malthusian thought on Darwin was stimulated by Robert Young, "Malthus and the Evolutionists: The Common Context of Biological and Social Theory," *Past and Present* 43 (1969): 114; reprinted in Young, *Darwin's Metaphor: Nature's Place in Victorian Culture* (Cambridge: Cambridge University Press, 1985). See also Young, "The Historiographic and Ideological Contexts of the Nineteenth Century Debate on Man's Place in Nature," in *Darwin's Metaphor*, 164–247; Young, "Darwinism Is Social," in *The Darwinian Heritage*, ed. David Kohn (Princeton: Princeton University Press, 1985), 609–638.

44. See Camille Limoges, *La sélection naturelle* (Paris: Presse Univérsitaire de France, 1970); Sandra Herbert, "Darwin, Malthus, and Selection," *Journal of the History of Biology* 4 (1971): 209–217; David Kohn, "Theories to Work By: Rejected Theories, Reproduction, and Darwin's Path to Natural Selection," *Studies in History of Biology* 4 (1980): 67–170; M. J. S. Hodge and David Kohn, "The Immediate Origins of Natural Selection," in *The Darwinian Heritage*, ed. David Kohn (Princeton: Princeton University Press, 1985), 185–206; M. J. S. Hodge, "The Development of Darwin's General Biological Theorizing," in *Evolution from Molecules to Men*, ed. D. S. Bendall (Cambridge: Cambridge University Press, 1983), 43–62; Ernst Mayr, *The Growth of Biological Thought* (Cambridge, Mass.: Harvard University Press, 1984).

45. F. Darwin, *The Autobiography of Charles Darwin and Selected Letters*, 42–43.

46. Alfred Russel Wallace, *My Life: A Record of Events and Opinions*, 2 vols. (1905 reprint, Farnborough: Gregg International, 1969), 1: 232.

47. Alfred Russel Wallace to A. Newton, Dec. 3, 1887, in F. Darwin, *The Autobiography of Charles Darwin and Selected Letters*, 200.

48. Darwin, *The Origin*, 63–64.

49. Ibid., 75.

50. Darwin, *The Origin*, 490.

•

Notes to Chapter 5

1. Ernst Haeckel, *The History of Creation* (New York: Appelton, 1880), 19–20.

2. See Roger Baldwin, ed., *Kropotkin's Revolutionary Pamphlets* (New York: Dover, 1970); Martin A. Miller, *Kropotkin* (Chicago: University of Chicago Press, 1976).

3. Baldwin, *Kropotkin's Revolutionary Pamphlets*, 20–25.

4. Peter Kropotkin, *Mutual Aid: A Factor of Evolution*, popular ed. (London: William Heinemann, 1915), 13.

5. See T. H. Huxley, *Evolution and Ethics and Other Essays* (New York: Columbia University Press, 1911).

6. Kropotkin, *Mutual Aid,* 2.

7. Kropotkin was first alerted to the view of mutual aid as a factor in evolution when attending a lecture, "On the Law of Mutual Aid," delivered at a Russian Congress of Naturalists in January 1880 by the zoologist K. F. Kessler, then the dean of St. Petersburg University; ibid., 3.

8. See Daniel P. Todes, *Darwin without Malthus: The Struggle for Existence in Russian Evolutionary Thought* (New York: Oxford University Press, 1989), 168.

9. See Jan Sapp, *Evolution by Association: A History of Symbiosis* (New York: Oxford University Press, 1994).

10. Charles Darwin, *On the Origin of Species by Means of Natural Selection* (London: John Murray, 1859), reprinted with an introduction by Ernst Mayr (Cambridge, Mass.: Harvard University Press, 1966), 62.

11. Ibid., 63, 92–93.

12. Kropotkin, *Mutual Aid,* 13.

13. Ibid., 14.

14. Ibid., 20.

15. Ibid., 17.

16. A. Kemna, *P. J. van Beneden: la vie et l'oeuvre d'un zoologiste* (Anvers: J.-E. Buschmann, 1897), 20.

17. Pierre-Joseph van Beneden, "Un mot sur la vie sociale des animaux inférieurs," *Bulletin de l'Académie Royale de Belgique*, série 2, 36 (1873): 779–796; van Beneden, *Animal Parasites and Messmates* (London: Henry S. King, 1876).

18. van Beneden, *Animal Parasites*, 85.

19. Ibid., 1.

20. Ibid., 53.

21. Ibid., 55.

22. Ibid., 56.

23. Ibid., 68.

24. Ibid., 81, 107.

25. See D. H. Boucher, "The Idea of Mutualism, Past and Future," in *The Biology of Mutualism: Ecology and Evolution*, ed. D. H. Boucher (London: Croom Helm, 1983), 1–28, at 12–13.

26. See E. Schulkind, ed., *The Paris Commune of 1871: The View from the Left* (London: Jonathan Cape, 1972), 164; quoted in Boucher, "The Idea of Mutualism," 12.

27. Boucher, "The Idea of Mutualism," 12.

28. van Beneden, *Animal Parasites*, xvii.

29. Boucher, "The Idea of Mutualism," 14.

30. See Frank Egerton, "Changing Concepts of the Balance of Nature," *Quarterly Review of Biology* 48 (1973): 322–350, at 326.

31. Ibid., 329.

32. Ibid., 186.

33. Ibid.

34. Ibid. See also Carolyn Merchant, *The Death of Nature: Women, Ecology, and the Scientific Revolution* (New York: Harper and Row, 1983).

35. See Donald Worster, *Nature's Economy*, 45.
36. van Beneden, *Animal Parasites*, xviv.
37. Ibid., 29.
38. van Beneden, *Animal Parasites*, xxvii.
39. Kemna, *P. J. van Beneden*, 132.
40. See Sapp, *Evolution by Association.*

Notes to Chapter 6

1. See Julian Huxley, *Evolution: The Modern Synthesis* (London: George Allen and Unwin, 1942); Peter J. Bowler, *The Eclipse of Darwinism: Anti-Darwinian Evolution Theories in the Decades around 1900* (Baltimore: Johns Hopkins University Press, 1983); David Hull, *Darwin and His Critics: The Reception of Darwin's Theory of Evolution by the Scientific Community* (Cambridge, Mass.: Harvard University Press, 1983).

2. Thomas F. Glick, ed., *The Comparative Reception of Darwinism* (Chicago: University of Chicago Press, 1988). See Peter Bowler, "Scientific Attitudes to Darwinism in Britain and America," in *The Darwinian Heritage*, ed. David Kohn (Princeton: Princeton University Press, 1985), 654–682; Pietro Corsi and Paul Weindling, "Darwinism in Germany, France and Italy," in *The Darwinian Heritage*, ed. David Kohn (Princeton: Princeton University Press, 1985), 683–730; Francesco M. Scudo and Michele Acanfora, "Darwin and Russian Evolutionary Biology," in *The Darwinian Heritage*, ed. David Kohn (Princeton: Princeton University Press, 1985), 731–754;Yvette Conry, *L'introduction du darwinisme en France* (Paris: Librarie Philosophique J. Vrin, 1974); Cédric Grimoult, *Evolutionisme et fixisme en France* (Paris: CNRS Editions, 1998); Ronald Numbers, *Darwinism Comes to America* (Cambridge, Mass.: Harvard University Press, 1999); Daniel Todes, *Darwin without Malthus: The Struggle for Existence in Russian Evolutionary Thought* (New York: Oxford University Press, 1989).

3. J. D. Burchfield, *Lord Kelvin and the Age of the Earth* (New York: Science History Publications, 1975).

4. See Fleeming Jenkin, "The Origin of Species," *North British Review* 46 (1867): 277–318; David Hull, *Darwin and His Critics* (Cambridge, Mass.: Harvard University Press, 1973); Peter Bowler, "Darwin's Concept of Variation," *Journal of the History of Medicine* 29 (1974): 196–212; Ernst Mayr, *The Growth of Biological Thought* (Cambridge, Mass.: Harvard University Press, 1982), 513–514; Garland Allen, "Hugo de Vries and the Reception of the 'Mutation Theory,'" *Journal of the History of Biology* 1 (1968): 55–87, at 71.

5. Georges Buffon, *Histoire naturelle*, vol 2. (Paris: *Imprimerie Royale*, 1749), 10; quoted in Arthur O. Lovejoy, "Buffon and the Concept of Species," in *Forerunners of Darwin*, ed. B. Glass (Baltimore: Johns Hopkins University Press), 84–113, at 93.

6. See, e.g., Ernst Mayr, *The Growth of Biological Thought* (Cambridge, Mass.: Harvard University Press), 1982.

7. Louis Agassiz, *Contributions to the Natural History of the United States of America*, vol. 3 (Boston: Little Brown, 1860), 89–90.

8. Charles Darwin to Asa Gray, Aug. 11, 1860, in *The Life and Letters of Charles Darwin*, ed. Francis Darwin, 3 vols. (London: Murray, 1887), 2:124.

9. See John Beatty, "What Is in a Word? Coming to Terms in the Darwinian Revolution," *Journal of the History of Biology* 15 (1982): 215–239.

10. Ibid., 231–232. See also D. Hull, "Are Species Really Individuals?" *Systematic Zoology* 25 (1976): 174–191.

11. See Frank J. Sulloway, "Geographic Isolation in Darwin's Thinking: The Vicissitudes of a Crucial Idea," *Studies in History of Biology* 3 (1979): 23–65.The controversy was opened again in the 1880s by Georges John Romanes and John Thomas Gulick, who argued that reproductive isolation was necessary because Darwin's theory of divergence could not overcome the effects of blending inheritance. See John E. Lesch, "The Role of Isolation in Evolution: George John Romanes and John T. Glulick," *Isis* 66 (1975): 483–503.

12. Darwin pointed to polymorphic species with an inordinate amount of variation, which he believed was of "no service to the species and not seized on by natural selection." Such characteristics would persist independently of the conditions of life. Charles Darwin, *On the Origin of Species*, introd. Ernst Mayr, facsimile ed. of 1859 (Cambridge, Mass.: Harvard University Press, 1964), 46.

13. See Malcom J. Kottler, "Darwin, Wallace, and the Origin of Sexual Dimorphism," *Proceedings of the American Philosophical Society* 124 (1980): 203–226; Peter Bowler, *Evolution: The History of an Idea* (Berkeley: University of California Press, 1984), 196–197.

14. R. Smith, "Alfred Russel Wallace: Philosophy of Nature and Man," *British Journal of the History of Science* 6 (1972): 177–199; Malcom Jay Kottler, "Alfred Russel Wallace, the Origin of Man, and Spiritualism," *Isis* 65 (1974): 145–192; Frank Miler Turner, *Between Science and Religion: The Reaction to Scientific Naturalism in Late Victorian England* (New Haven: Yale University Press, 1974).

15. St. George Jackson Mivart, *The Genesis of Species* (New York: Macmillan, 1871). See Jacob Gruber, *A Conscience in Conflict: The Life of St. George Jackson Mivart* (New York: Columbia University Press, 1960).

16. See, e.g., E. D. Cope, "The Energy of Evolution," *American Naturalist* 28 (1894): 205.

17. R. C. Punnett, *Mimicry in Butterflies* (Cambridge: Cambridge University Press, 1915).

18. Darwin, *On the Origin of Species*, 342.

19. See Martin Rudwick, *The Meaning of Fossils: Episodes in the History of Paleontology*, 2d ed. (New York: Science History Publications, 1976); Peter Bowler, *Fossils and Progress: Paleontology and the Idea of Progressive Evolution in the Nineteenth Century* (New York: Science History Publications, 1976); Adrian Desmond, *The Hot-Blooded Dinosaurs: A Revolution in Paleontology* (New York: Dial Press, 1976); Desmond, *Archetypes and Ancestors: Paleontology in Victorian London, 1850–1875* (London: Blond and Briggs, 1982).

20. Othniel C. Marsh, *Odontornithes: A Monograph on the Extinct Toothed Birds of North America*, Report of the Geological Exploration of the Fortieth Parallel 8 (Washington, 1880).

21. Richard Aulie," The Origin of the Idea of the Mammal-Like Reptile," *American Biology Teacher* 36 (1974): 476–484, 545–553; Desmond, *Archetypes and Ancestors*.

22. Some of these fossils were later reinterpreted as extinct side chains rather than as representing a linear series. On the continued mistake of citing such "lineages" as "textbook cases" of evolution, see Stephen Jay Gould, *Wonderful Life* (London: Hutchinson Radius, 1989), 35–37.

23. Raymond Dart, "*Australopithecus africanus*: The Man-Ape of South Africa," *Nature* 115 (1925): 195–199. For a lucid account of the common narratives about human evolution, see Misia Landau, *Narratives of Human Evolution* (New Haven: Yale University Press, 1991).

24. Peter Bowler, "Edward Drinker Cope and the Changing Structure of Evolutionary Theory," *Isis* 68 (1977): 249–265; Bowler, *The Eclipse of Darwinism.*

25. See Stuart Persell, *Neo-Lamarckism and the Evolution Controversy in France, 1870–1920* (Lewston: Edwin Mellen Press, 1999); Ernest Boesiger, "Evolutionary Biology in France at the Time of the Evolutionary Synthesis," in *The Evolutionary Synthesis: Perspectives on the Unification of Biology*, ed. William B. Provine and Ernst Mayr (Cambridge, Mass.: Harvard University Press, 1980), 309–321; Camille Limoges, "A Second Glance at Evolutionary Biology in France," in *The Evolutionary Synthesis: Perspectives on the Unification of Biology*, ed. William B. Provine and Ernst Mayr (Cambridge, Mass.: Harvard University Press, 1980), 322–329; Jan Sapp, *Beyond the Gene* (New York; Oxford University Press, 1987).

26. George M. Cook, "Neo-Lamarckian Experimentalism in America: Origins and Consequences," *Quarterly Review of Biology* 74 (1999): 417–437.

27. See Robert Richards, *Darwin and the Emergence of Evolutionary Theories of Mind and Behavior* (Chicago: University of Chicago Press, 1987).

28. See Bowler, *The Eclipse of Darwinism*; Boesiger, "Evolutionary Biology in France"; Limoges, "A Second Glance at Evolutionary Biology in France."

29. See Richards, *Darwin and the Emergence of Evolutionary Theories of Mind and Behavior.*

30. Herbert Spencer, *The Factors of Organic Evolution* (New York: D. Appelton, 1887), vi. See also H. F. Osborn,"The Present Problem of Heredity," *Atlantic Monthly* 67 (1891), 353–364, at 354.

31. Charles Darwin, *The Descent of Man and Selection in Relation to Sex*, 2d ed. (New York: Hurst, 1874), 145.

32. See Theodor Eimer, *Organic Evolution as the Result of the Inheritance of Acquired Characters according to the Laws of Organic Growth*, trans. J. T. Cunningham (London: Macmillan, 1890), 18–19.

33. Today, evolutionists interpret the huge ninety-pound antlers as having had an adaptive function, not in combat, but rather for the purpose of avoiding it, by ritualistic display among males. The extinction of the Irish elk is now held to be the result of a rather commonplace mechanism: it succumbed to the changing conditions of climate. See Stephen J. Gould, "The Origin and Function of 'Bizarre' Structures: Antler Size and Skull Size in the 'Irish Elk,' *Megaloceros giganteus*," *Evolution* 28 (1974): 191–200.

34. See Ron Rainger, *An Agenda for Antiquity: Henry Fairfield Osborn and Vertebrate Paleontology at the American Museum of Natural History, 1890–1935* (Tuscaloosa: University of Alabama Press, 1991). See also Bowler, *Evolution*, 255–256.

35. Of course, other interpretations of the evidence of orthogenesis itself could be offered. Perhaps the phenomenon itself was an illusion—that is, the straight lines were simply an artifact resulting from an incomplete fossil record. In the mid- and late twentieth century, most of the fossil evidence used to support orthogenesis was reinterpreted in a manner compatible with natural selection. George Gaylord Simpson, *Tempo and Mode in Evolution* (New York: Columbia University Press, 1944); Simpson, *The Major Features of Evolution* (New York: Columbia University Press, 1953).

36. Carl von Nägeli, "Mechanisch-physiologishe Abstammungslehre" (Munich, 1884).

37. Quoted in Eimer, *Organic Evolution*, 14.

38. Ibid., 15.

39. Ibid., 18, 53.

40. Ibid., 21.

41. Ibid., 33.
42. Ibid., 34.
43. Ibid., 45.
44. Ibid., 49.
45. St. George Jackson Mivart, *Genesis of Species* (London: John Murray 1871).
46. Nicolaas A. Rupke, *Richard Owen: Victorian Naturalist* (New Haven: Yale University Press, 1994), 239.
47. Darwin, *On the Origin of Species*, 471.
48. See William Bateson, *Materials for the Study of Variation: Treated with Special Regard to Discontinuity in the Origin of Species* (London: Macmillan, 1894); Hugo de Vries, *The Mutation Theory: Experiments and Observations on the Origin of Species in the Vegetable Kingdom*, trans. J. B. Farmer and A. D. Darbyshire, 2 vols. (London, 1910); Garland Allen, "Hugo de Vries and the Reception of the 'Mutation Theory,'" *Journal of the History of Biology* 1 (1968): 55–87; Allen, "Naturalists and Experimentalists: The Genotype and the Phenotype," *Studies in History of Biology* 3 (1979): 179–209.

Notes to Chapter 7

1. E. B. Wilson, *The Cell in Development and Heredity*, 3d ed. (New York: Macmillan, 1925), 1.
2. See, e.g., David J. Depew and Bruce Weber, *Darwinism Evolving* (Cambridge, Mass.: MIT Press, 1997), 176. On the history of cell theory, see François Duchesneau, *Genèse de la théorie cellulaire* (Montréal: Bellarmin, 1987); Marcel Florkin, *Naissance et déviation de la théorie cellulaire dans l'oeuvre de Théodore Schwann* (Paris: Hermann, 1960); Marc Klein, *Histoire de la théorie cellulaire* (Paris: Hermann, 1936); Arthur Hughes, *A History of Cytology* (London: Abelard-Schuman, 1959), 42–44; Henry Harris, *The Birth of the Cell* (New Haven: Yale University Press, 1999).
3. See, e.g., E. G. Conklin, "Cell and Protoplasm Concepts: Historical Account," in *The Cell and Protoplasm*, ed. Forest Ray Moulton (Washington: Science Press, 1940).
4. See J. H. Gerould, "The Dawn of the Cell Theory," *Scientific Monthly* 14 (1922): 267–276; E. G. Conklin, "Predecessors of Schleiden and Schwann," *American Naturalist* 73 (1939): 538–546.
5. Conklin, "Predecessors of Schleiden and Schwann."
6. Lamarck wrote:

> It has indeed been long recognized that the membranes which form the investments of the brain, nerves, vessels of all kinds, glands, viscera, muscles and their fibres, and that even the skin itself are all the produce of cellular tissue. . . . Yet, in this multitude of harmonizing facts, nothing more appears to have been seen than the mere facts themselves; and no one that I know of has yet perceived that cellular tissue is the universal matrix of all organization, and that without this tissue no living body could continue to exist. Since the year 1796 I have been accustomed to set forth these principles in the first lessons of my course. (J. B. Lamarck, *Zoological Philosophy* [1809], trans. Hugh Elliot [Chicago: University of Chicago Press, 1984], 230)

Lamarck emphasized that he had been teaching these principles in his course since 1796.

7. Quoted in Conklin, "The Cell and Protoplasm Concepts," 11.

8. R. J. H. Dutrochet, "Recherches anatomiques et physiologiques sur la structure intime des animaux et des végétaux et sur leur motilité," trans. M. L. Gabriel, in *Great Experiments in Biology*, ed. Mordecai L. Gabriel and Seymour Fogel (Englewood Cliffs, N.J.: Prentice-Hall, 1955), 6–9; quoted in L. W. Sharp, *An Introduction to Cytology*, 2d ed. (New York: McGraw-Hill, 1926), 9.

9. Conklin, "Cell and Protoplasm Concepts," 6–19.

10. Robert Brown, "The Organs and Mode of Fecundation in *Orchideae* and *Asclepiadeae*," *Transactions of the Linnean Society* 16 (1833): 710–713. Brown is also remembered for his description of the natural continuous motion of minute particles in solution, which came to be called Brownian movement.

11. See Harris, *The Birth of the Cell*, 98.

12. Theodor Schwann, *Microscopic Researches into the Accordance in the Structure and Growth of Animals and Plants*, trans. Henry Smith (London: Sydenham Society, 1847), 231–263.

13. See Harris, *The Birth of the Cell.*

14. Ibid., 65.

15. Hughes, *A History of Cytology*, 42–44.

16. Harris, *The Birth of the Cell*, 110–113.

17. Ibid, 114–115.

18. See H.-P. Schmiedebach, *Robert Remak (1815–1865)* (Stuttgart: Gustav Fischer, 1995).

19. On anti-Semitism and nineteenth-century science, see also Harris, *The Birth of the Cell*, 129–137; Paul Weindling, "Theories of the Cell State in Imperial Germany," in *Biology, Medicine, and Society, 1840–1940,* ed. Charles Webster (Cambridge: Cambridge University Press, 1981), 99–154, at 114–115.

20. Ibid.

21. R. Virchow, *Cellular Pathology* (New York: Dover, 1971). For an account of Virchow's life, see A. Boyd, *Rudolf Virchow: The Scientist as Citizen* (New York: Garland, 1991).

22. See Weindling, "Theories of the Cell State," 118.

23. Ibid., 24.

24. Eric Nordenskiöld, *The History of Biology*, trans. Leonard B. Eyre (New York: Alfred A. Knopf, 1927), 392.

25. Julius von Sachs, *History of Botany (1530–1860)*, trans. Henry E. F. Garnsey (Oxford: Oxford University Press, 1890), 188.

26. Hughes, *A History of Cytology*, 38.

27. Conklin, "Cell and Protoplasm Concepts."

28. See Hughes, *A History of Cytology*. Some have pointed to the influence of the German romantic tradition or *Naturphilosophie*, which emphasized synthesis, bringing together a large number of phenomena within a unified framework; see L. S. Jacyna, "Romantic Thought and the Origins of Cell Theory," in *Romanticism and the Sciences*, ed. A. Cunningham and N. Jardine (Cambridge: Cambridge University Press, 1990), 161–168.

29. See Natasha Jacobs, "From Unity to Unity: Protozoology, Cell Theory, and the New Concept of Life," *Journal of the History of Biology* 22 (1989): 215–242.

30. See John Farley, *The Spontaneous Generation Controversy from Descartes to Oparin* (Baltimore: Johns Hopkins University Press, 1977).

31. See James Strick, *Sparks of Life* (Cambridge, Mass.: Harvard University Press, 2000).

32. Ibid., 108.

33. For a discussion of Schleiden's methodology, see Gerd Buchdahl, "Leading Principles and Induction: The Methodology of Matthias Schleiden," in *Foundations of Scientific Method: The Nineteenth Century*, ed. R. N. Giere and Richard Westfall (Bloomington: Indiana University Press, 1973), 23–52.

34. T. Schwann, "Microscopische Untersuchungen uber die Uebereinstimmung in der Struktur und dem Wachsthum der Thiere und Pflanzen," trans. Henry Smith, in *Great Experiments in Biology*, ed. Mordecai L. Gabriel and Seymour Fogel (Englewood Cliffs, N.J.: Prentice-Hall, 1955), 12–17, at 14.

Notes to Chapter 8

1. E. W. Brücke, "Die Elementarorganismen," *Sitzungsberichte der Kaiserlichen Akademie Wien* 44, no. 2 (1861): 381–406.

2. R. Virchow, *Die Cellularpathologie in ihrer Begrundung auf physiologische und pathologische Gewebelehre,* 1858, English Trans. *Cellular Pathology*, (New York: Dover, 1971), 12.

3. See Donald Worster, *Nature's Economy: A History of Ecological Ideas* (Cambridge: Cambridge University Press, 1977). On complexity and stability in ecological debates, see Jan Sapp, *What Is Natural? Coral Reef Crisis* (New York: Oxford University Press, 1998).

4. Herbert Spencer, "Professor Weismann's Theories," *The Contemporary Review* 63 (1893): 743–760.

5. See Hebert Spencer, *First Principles* (London: Williams and Norgate, 1904), 263.

6. Paul Weindling, "Theories of the Cell State in Imperial Germany," in *Biology, Medicine, and Society, 1840–1940*, ed. Charles Webster (Cambridge: Cambridge University Press, 1981), 99–154, at 100.

7. Ibid. See also Gregg Mitman, "Defining the Organism in the Welfare State: The Politics of Individuality in American Culture, 1890–1950," *Sociology of the Sciences Yearbook* 18 (1994): 249–280.

8. Theodor Schwann, "Microscopical Researches," in *Great Experiments in Biology*, ed. Mordecai L. Gabriel and Seymour Fogel (Englewood Cliffs, N.J.: Prentice-Hall, 1955), 12–17, at 15.

9. See Weindling, "Theories of the Cell State," 119.

10. Immanuel Kant, *Foundations of the Metaphysics of Morals*, trans. Lewis White Beck and Robert Paul Wolff (New York: Bobbs-Merrill, 1969), 4–5. "All crafts, handiworks, and arts have gained by the division of labor, for when one person does not do everything but each limits himself to a particular job which is distinguished from all the others by the treatment it requires, he can do it with greater perfection and with more facility. Where work is not thus differentiated and divided, where everyone is a jack-of-all-trades, the crafts remain at a barbaric level."

11. See Jane Maienschein, "Whitman at Chicago: Establishing a Chicago Style of Biology?" in *The American Development of Biology*, ed. Ronald Rainger, Keith. R. Benson, and Jane Maienschein (Philadelphia: University of Pennsylvania Press, 1988), 151–184; Maienschein, *One Hundred Years Exploring Life, 1888–1898: The Marine Biological Laboratory at Woods Hole* (Boston: Jones and Bartlett, 1989).

12. Whitman, "Socialization and Organization," 21.

13. E. B. Wilson, *The Cell in Development and Inheritance* (New York: MacMillan, 1896), 41. In the third edition (retitled *The Cell in Development and Heredity* [1925]), Wilson continued to celebrate the conception of the cell state stating that, as "elaborated by Milne-Edwards, Virchow and Haeckel, this conclusion offered a simple and natural point of attack for the problems of cytology, embryology, and physiology, and revolutionized the problems of organic individuality." E. B. Wilson, *The Cell in Development and Heredity* (New York: Macmillan, 1925), 102.

14. Wilson, *The Cell in Development and Heredity*, 3d ed., 5.

15. See Natasha Jacobs, "From Unity to Unity: Protozoology, Cell Theory, and the New Concept of Life," *Journal of the History of Biology* 22 (1989): 215–242.

16. See Marsha L. Richmond, "Protozoa as Precursors of Metazoa: German Cell Theory and Its Critics at the Turn of the Century," *Journal of the History of Biology* 22 (1989): 243–276.

17. See Lynn J. Rothschild, "Protozoa, Protista, Protoctista: What's in a Name?" *Journal of the History of Biology* 22 (1989): 277–305.

18. Ibid., 277.

19. Richard Hertwig, "Die Protozoen und die Zelltheorie," *Archiv für Protistenkunde* 1 (1902): 1–40.

20. E. A. Minchin, "The Evolution of the Cell," *Report of the Eighty-Fifth Meeting of the British Association for the Advancement of Science, Sept. 7–11* (1915): 437–444, at 439.

21. See Alfred I. Tauber and Leon Chernyak, *Metchnikoff and the Origins of Immunology* (New York: Oxford University Press, 1991).

22. Robert Hooke, "Of the Schematisme or Texture of Cork, and of the Cells and Pores of Some other such Frothy Bodies," from *Microphagia or Some Physiological Descriptions of Minute Bodies Made by Magnifying Glasses* (London, 1665); reprinted in *Great Experiments in Biology*, ed. Mordecai L. Gabriel and Seymour Fogel (Englewood Cliffs, N.J.: Prentice-Hall, 1955), 3–6.

23. Arthur Hughes, *A History of Cytology* (London: Abelard-Schuman, 1959), 40.

24. See T. H. Huxley, "On the Physical Basis of Life," *Fortnightly Review* (1869): 129–145; Huxley, "The Cell Theory," *British and Foreign Medico-Chirurgical Review* (1853): 221–243. See also Gerald Geison, "The Protoplasmic Theory of Life and the Vitalist-Mechanist Debate," *Isis* 60 (1969): 273–292.

25. See Wilson, *The Cell in Development and Inheritance*, 1st ed., 14; E. G. Conklin, "Cell and Protoplasm Concepts: An Historical Account," in *The Cell and Protoplasm*, ed. F. R. Moulton (Washington: Science Press, 1940), 6–19, at 15.

26. Wilson, *The Cell in Development and Inheritance*, 1st ed., 13–14.

27. Conklin, "Cell and Protoplasm Concepts," 15.

28. Wilson, *The Cell in Development and Inheritance*, 1st ed., 14.

29. Wilson, *The Cell in Development and Heredity*, 3d ed., 642.

30. See William Coleman, "Cell, Nucleus, and Inheritance: An Historical Study," *Proceedings of the American Philosophical Society* 109 (1965): 124–158.

31. Wilson, *The Cell in Development and Inheritance*, 1st ed., 185.

32. See A. H. Sturtevant, *A History of Genetics* (New York: Harper and Row, 1965), 33.

33. Wilhem Roux, "Ueber die Bedeutung der Kerntheilungsfiguren," in *Gesammelte Abhandlungen über Entwickelungsmechanik der Organismen* (Leipzig: Engelmann, 1883), 125.

34. See Henry Harris, *The Birth of the Cell* (New Haven: Yale University Press, 1999), 170–173.

35. See Jan Sapp, *Evolution by Association: A History of Symbiosis* (New York: Oxford University Press, 1994); Sapp, "Free-Wheeling Centrioles," *History and Philosophy of the Life Sciences* 20 (1999): 255–290. In 1898, Camillo Golgi (1843–1926) reported a weblike apparatus, and in 1910, Aldo Perroncito (1882–1929) reported that structures called Golgi bodies also reproduced by division. See also Paolo Mazzerello and Marina Bentivoglio, "The Centenarian Golgi Apparatus," *Nature* 392 (1998): 543–544. One of the main problems of cytology during this period was sorting out what was real from artefactual, and determining how many of the images were actually due to fixation and stating procedures. The reality of many cytoplasmic structures including centrioles, spindles and the Golgi apparatus were debated for decades, until the deployment of the electron microscope in the mid twentieth century. On the history of electron microscopy see, Nicolas Rasmussen, *Picture Control : The Electron Microscope and the Transformation of Biology in America, 1940–1960* (Stanford: Stanford University Press, 1997).

36. Richard Altmann, *Die Elementarorganismen* (Leipzig: Veit and Company, 1890).

37. See Sapp, *Evolution by Association.*

38. A. F. W. Schimper, "Untersuchungen über die Chlorophyllkörner und die ihnen homologen Gebilde," *Jahrbücher für wissenschaftliche Botanik* 16 (1885): 1–247, at 202.

39. Sapp, "Free-Wheeling Centrioles," The problem of distinguishing the centriole from the centrosome was not resolved until the deployment of the electron microscope.

40. See Wilson, *The Cell in Development and Inheritance*, 1st ed., 53–56.

41. L. F. Henneguy, "Sur les rapports des cils vibratiles avec les centrosomes," *Archives d'anatomie microscopique et de morphologie expériemntale* 1 (1898): 481–496, at 481. M. von Lenhossék, "Ueber Flimmerzellen", *Verhandlungen der Anatomischen Gesellschaft* 12 (1898), 106–128. See Sapp, "Free-Wheeling Centrioles."

42. E. B. Wilson, *The Cell in Development and Inheritance*, 2d ed. (New York: Macmillan, 1900), 434.

43. Wilson, *The Cell in Development and Inheritance*, 1st ed., 211. As Wilson commented in 1896, "Both [nucleus and cytoplasm] are necessary to *development;* the nucleus alone suffices for the *inheritance* of specific possibilities of development" (327).

44. F. B. Churchill, "August Weismann and a Break from Tradition," *Journal of the History of Biology* 1 (1968): 91–112.

45. Weismann began his career as a medical doctor before becoming interested in evolution. He carried out cytological studies on water fleas, but suffered from eye trouble and finally had to give up microscopic work. Gregor Mendel and Carl Correns also suffered from eye problems: eye trouble brought on by excessive work with strong light was an occupational hazard of nineteenth-century microscopy. See A. H. Sturtevant, *A History of Genetics*, (New York: Harper and Row, 1965), 18.

46. Charles Darwin, *Variations of Plants and Animals under Domestication* (London: John Murray, 1868).

47. August Weismann, *The Germ Plasm: A Theory of Heredity*, trans. W. N. Parker and H. Rofeldt (London: Walter Scott, 1983), xv.

48. Ibid., 11.

49. See Wilson, *The Cell in Development and Inheritance*, 1st ed., 300–301.

50. Weismann, *The Germ Plasm*, 38.

51. Ibid., xi.

52. Wilhem Roux, "Ueber die Bedeutung der Kerntheilungsfiguren," in *Gesammelte Abhandlungen über Entwickelungsmechanik der Organismen* (Leipzig: Engelmann, 1883),

125. See also Henry Harris, *The Birth of the Cell* (New Haven: Yale University Press, 1999), 167–68.

53. Hugo de Vries, *Intracellular Pangenesis*, trans. C. Stuart Gager (Chicago: Open Court, 1910).

54. Weismann, *The Germ Plasm*, 22.

55. Ibid., 27.

56. Ibid., 29.

57. Ibid., 26.

58. Ibid.

59. Ibid., 63.

60. August Weisman, "Continuity of the Germ-Plasm as the Foundation of a Theory of Heredity." 1885. Republished in *Essays Upon Heredity and Kindred Biological Problems*, trans. E. B. Poulton, 2 vols., 2nd ed., Oxford: Clarendon Press, 1891–1892, vol. 1, pp. 163–255; 195.

Notes to Chapter 9

1. This point has been emphasized by many historians of biology. See, e.g., Bernardino Fantini, "The Sea Urchin and the Fruit Fly: Cell Biology and Heredity, 1900–1910," *Biological Bulletin* 164 (1985): 99–105. See also Adele Clarke and Joan Fujimura, eds., *The Right Tools for the Job* (Princeton: Princeton University Press, 1992).

2. Theodor Boveri, cited in F. Baltzer, *Theodor Boveri* (Berkeley: University of California Press, 1967), 121.

3. See Ronald Rainger, Keith R. Benson, and Jane Maienschein, eds., *The American Development of Biology* (New Brunswick: Rutgers University Press, 1988); Jane Maienschein, *Transforming Traditions in American Biology, 1880–1915* (Baltimore: Johns Hopkins University Press, 1991).

4. Ethel Brown Harvey, *The American Arbacia and other Sea Urchins* (Princeton: Princeton University Press, 1956), vii.

5. On the career of Jacques Loeb, see Philip Pauly, *Controlling Life: Jacques Loeb and the Engineering Ideal in Biology* (New York: Oxford University Press, 1987).

6. T. H. Huxley, "The Cell Theory," *British and Foreign Medico-Chirurgical Review* 12 (1853): 221–243, at 243.

7. L. W. Sharp, *Introduction to Cytology*, 2d ed. (New York: McGraw-Hill, 1926), 17–18.

8. Adam Sedgwick, "On the Inadequacy of the Cellular Theory of Development, and on the Early Development of Nerves, Particularly of the Third Nerve and of the Sympathetic in Elasmobranchii," *Quarterly Journal of Microscopic Science* 37 (1895): 87–101.

9. See, e.g., Gary Calkins, *The Biology of Protozoa* (Philadelphia: Lea and Febiger, 1933), 19; Martha Richmond, "Protozoa as Precursors of Metazoa: German Cell Theory and Its Critics at the Turn of the Century," *Journal of the History of Biology* 22 (1989): 243–276; John Corliss, "The Protozoon and the Cell: A Brief Twentieth Century Overview," *Journal of the History of Biology* 22 (1989): 307–323.

10. C. Clifford Dobell, "The Principles of Protistology," *Archiv für Protistenkunde* 23 (1911): 269–310, at 270.

11. Ibid., 284–285.

12. Ibid., 302–303.

13. The notion that protozoa were noncellular organisms was supported by some prominent protozoologists, including the Nobel Prize–winning protozoologist André Lwoff; Lwoff, *Problems of Morphogenesis in Ciliates* (New York: Wiley and Sons, 1950), 3. See also John Corliss, "The Protozoon and the Cell," *Journal of the History of Biology* 22 (1989): 307–323, at 316. Peter Medawar and Jean Medawar were more incisive when they commented:

> Protozoa are whole organisms whose structure corresponds to that of a single cell in a Metazoan tissue. Whether they are described as single-celled or 'non-cellular' is a matter of indifference to all except those nature-philosophers who have ensured their livelihood by inventing pseudo-problems and discussing them earnestly and at length, sometimes in specialist publications exclusively devoted to exercises of this kind. (Peter Medawar and Jean Medawar, *Aristotle to Zoos: A Philosophical Dictionary of Biology* [Cambridge, Mass.: Harvard University Press, 1983], 287.

14. See E. A. Minchin, "The Evolution of the Cell," *Report of the Eighty-Fifth Meeting of the British Association for the Advancement of Science, Sept. 7–11* (1915): 437–464, at 437. This paper was published the next year in *American Naturalist* 50 (1916): 3 ff. See also E. B. Wilson, *The Cell in Development and Heredity,* 3d ed. (New York: Macmillan, 1925), 103.

15. For example, simple unicellular forms of the green alga *Chlamydomonas* have a stage during which each cell encysts and divides asexually. The clone is enclosed in a gelatinous case and is regarded as a colonial form. Other protists such as *Pleodorina illinoiensis* and *Pleodorina californica* (close relatives to *Chlamydomonas*) show further stages in the development of a colony. A group of cells within the colony becomes differentiated from the others and unable to take part in reproduction, in which respect they resemble somatic cells. Many dinoflagellates (green algae with whirling flagella) also have colonial stages. *Ceratium* may form long chains of individuals joined together in a temporary attachment. Other dinoflagellates were thought to be animals, such as such as *Haplozoön* when it was discovered in 1906. But in 1920, Eduoard Chatton showed that it was a parasitic dinoflagellate and that its organization into a multicellular body is a process of colony formation. See Libbie Henrietta Hyman, *The Invertebrates: Protozoa through Ctenophora* (New York: McGraw-Hill, 1940), 243–244. See also G. A. Kerkut, *Implications of Evolution* (New York: Pergamon Press, 1960).

16. Kerkut, *Implications of Evolution*, 36–49.

17. For discussion of competing theories, see L. V. Salvini-Plawen, "On the Origin and Evolution of the Lower Metazoa," *Zeitschrift für zoologische Systematik und Evolutionsforschung* 16 (1978): 40–88; P. Wilmer, *Invertebrate Relationships* (Cambridge: Cambridge University Press, 1990); L. Wolpert, "Gastrulation and the Evolution of Development," *Development*, supplement (1992): 7–13; Wolpert, "The Evolutionary Origin of Development: Cycles, Patterning, Privilege, and Continuity," *Development*, supplement (1994): 79–84; L. Buss, *The Evolution of Individuality* (Princeton: Princeton University Press, 1987).

18. C. O. Whitman, "The Inadequacy of the Cell-Theory of Development," *Journal of Morphology* 8 (1893): 639–658, 649.

19. Ibid., 653.

20. Criticisms of cell theory were the highlight of its centenary. See essays in American Association for the Advancement of Science, *The Cell and Protoplasm*, ed. Forest

Ray Moulton (Washington: Publications for the American Association for the Advancement of Science Press, 1940), 6–19, at 11.

21. F. R. Lillie, "Observations and Experiments Concerning the Elementary Phenomena of Embryonic Development in *Chaetopterus*," *Journal of Experimental Zoology* 3 (1906): 153–267, at 252.

22. Wilson, *The Cell in Development and Heredity*, 3d ed., 103 (my italics).

23. H. Driesch, "Entwickslungsmechanische Studien. I. Der Werth der beiden ersten Furchungszellen in der Echinodermentwickslung. Experimentelle Erzeugung von Theil- und Doppel-bildungen," *Zeitschrift für wissenschaftliche Zoologie* 53 (1891): 160–178. See translation: Hans Driesch, "The Potency of the First Two Cleavage Cells in Echinoderm Development," in *Foundations of Experimental Embryology*, ed. Benjamin Willier and Jane M. Oppenheimer (Englewood Cliffs, N.J.: Prentice Hall, 1964), 38–59.

24. Wilhelm Roux, "Contributions to the Developmental Mechanics of the Embryo," in *Foundations of Experimental Embryology*, ed. Benjamin Willier and Jane M. Oppenheimer (Englewood Cliffs, N.J.: Prentice Hall, 1964), 2–37.

25. August Weismann, *The Germ Plasm: A Theory of Heredity* (London: Walter Scott, 1893), xv, 137–138.

26. Quoted in Wilson, *The Cell in Development and Heredity*, 3d ed., 1056.

27. Spemann had become fascinated with questions of embryology after reading Weismann's *The Germ Plasm: A Theory of Heredity* in the winter of 1896, as he sat isolated in a sanitarium recovering from tuberculosis. See Victor Hamburger, "Hans Spemann on Vitalism in Biology: Translation of a Portion of Spemann's Autobiography," *Journal of the History of Biology* 32 (1999): 231–243.

28. H. Spemann and H. Mangold, Ueber Induktion von Embryonalanlagen durch Implantation artfremder Organisatoren," *Archiv für Mikroscopische Anatomie* 100 (1924): 599–638. See also Victor Hamburger, *The Heritage of Experimental Embryology: Hans Spemann and the Organizer* (New York: Oxford University Press, 1988).

29. See Hans Spemann, *Embryonic Development and Induction* (New Haven: Yale University Press, 1938). Later, in the 1950s, the mechanism of induction was linked to protein molecules called growth factors, now known to play an important role in development.

30. See Jane Maienschein, "Competing Epistemologies and Developmental Biology," in *Biology and Epistemology*, ed. Richard Creath and Jane Maienschein (Cambridge: Cambridge University Press, 2000), 122–137. See also S. F. Gilbert, "The Embryological Origins of the Gene Theory," *Journal of the History of Biology* 11 (1978): 307–351.

Notes to Chapter 10

1. Jan Sapp, *Beyond The Gene: Cytoplasmic Inheritance and the Struggle for Existence in Genetics* (New York: Oxford University Press, 1987).

2. This was true for both mosaic and regulative embryos. On cell lineage studies, see Jane Maienschein, "Cell Lineage, Ancestral Reminiscence, and the Biogenetic Law," *Journal of the History of Biology* 11 (1978): 129–58; Maienschein, *Transforming Traditions in American Biology, 1880–1915* (Baltimore: Johns Hopkins University Press, 1991).

3. Embryologists were able to trace inverse symmetry in a certain species of snails back to an inverted symmetry in the unsegmented egg. See E. G. Conklin, "The Cause of Inverse Symmetry," *Anatomischer Anzeiger Centralblatt* 23 (1903): 577–588.

4. See E. G. Conklin, *Heredity and Environment in the Development of Men*, 3d ed. (Princeton: Princeton University Press, 1920), 190–197.

5. See E. B. Wilson, *The Cell in Development and Heredity*, 3d ed. (New York: Macmillan, 1925), 1102–1108.

6. By the 1920s results of hybridization experiments had become confusing. While in some hybrids early development showed only maternal characteristics, other showed some paternal characteristics as well. The results seemed to depend on the kind of organisms used and the specific techniques for making the hybrids. Nonetheless, in all experiments there were characteristics that were determined exclusively by the cytoplasm of the fertilized egg. See L. W. Sharp, *An Introduction to Cytology*, 2d ed. (New York: McGraw-Hill, 1926), 413.

7. Conklin, *Heredity and Environment,* 3d ed., 112.

8. M. F. Guyer, "Nucleus and Cytoplasm in Heredity," *American Naturalist* 45 (1911): 284–305.

9. J. W. Jenkinson, *Vertebrate Embryology* (Oxford: Clarendon Press, 1913), 92–93.

10. Albert Brachet, *L'oeuf et les facteurs de l'ontogenèse* (Paris: Gaston Doin, 1917), 176–179.

11. Conklin, *Heredity and Environment*, 2d ed., 176.

12. J. Loeb, *The Organism as a Whole* (New York: Putnam and Sons, 1916), 8.

13. See Sharp, *Introduction to Cytology*, 17–18; see also Wilson, *The Cell in Development and Heredity*, 3d ed., 1085.

14. K. Toyama, "Maternal Inheritance and Mendelism," *Journal of Genetics* 2 (1913): 351–405; A. E. Boycott and C. Diver, "On the Inheritance of Sinistrality in *Limnaea peregra*," *Proceedings of the Royal Society of London*, series B, 95 (1923): 207–213; A. H. Sturtevant, "Inheritance of Direction of Coiling in *Limnaea*," *Science* 58 (1923): 269–270.

15. L. C. Dunn, "Nucleus and Cytoplasm as Vehicles of Heredity," *American Naturalist* 51 (1917): 286–300, at 296.

16. E. G. Conklin, "The Share of Egg and Sperm in Heredity," *Proceedings of the National Academy of Sciences* 3 (1917): 101–105.

17. Quoted in Fritz Baltzer, *Theodor Boveri*, trans. Dorthea Rudnick (Berkeley: University of Chicago Press, 1967), 83–84.

18. V. B. Wigglesworth, "Growth and Form in an Insect," in *Essays on Growth and Form* (Oxford: Oxford University Press, 1945), 24–40.

19. Hans Spemann, "Die Entwicklung seitlicher und dorso-ventraler Keimhäftlen bei verzöogeter Kernversorgung," *Zeitschrift für wissenschaftliche Zoologie* 132 (1928): 105–134.

20. Hans Spemann, *Embryonic Development and Induction* (New Haven: Yale University Press, 1938). When such cloning experiments were carried out in the late 1990s, Spemann was celebrated as the founder of the specialty.

21. T. H. Morgan, *The Physical Basis of Heredity* (Philadelphia: J. B. Lippincott, 1919), 241.

22. Conklin, *Heredity and Environment*, 3d ed., 403.

23. Spemann himself, however, did not rule out other kinds of nuclear differentiation besides that proposed by Weismann; as he put it in 1938, "Genes may be lost or become ineffective in other ways besides elimination out of the cell." See Spemann, *Embryonic Development and Induction*, 210.

24. F. B. Churchill, "From Machine Theory to Entelechy: Two Studies in Developmental Teleology," *Journal of the History of Biology* 2 (1969): 165–185.

25. A. P. Mathews, *Physiological Chemistry* (New York: William Wood, 1915), 11–12.

26. Wilson, *The Cell in Development and Heredity*, 3d ed., 670.

27. Loeb, *The Organism as a Whole*, 39.

28. Ibid., 8.

29. Embryologists had come to recognize that the pigment substances in the eggs themselves were not always crucial for development. Some species of amphibians, for example, did not have any pigment in their eggs. See Wilson, *The Cell in Development and Heredity*, 3d ed., 1093.

30. The actual chemical structure of this cortical layer was unknown. But it was known that the cortex was different physically from the rest of the cytoplasm. Most of the egg cytoplasm was in a physical state of suspension containing large and small components ranging from inorganic ions to protein molecules, to larger inclusions such as mitochondria and centrioles. The cortex, on the other hand, was understood to be in a gelated, or at least a much more viscous state so that it is components are not readily displaced by moderately strong centrifugation. See Sharp, *An Introduction to Cytology*, 416.

31. That the polarity resided in the cell cortex was supported by other experiments with sea squirts. Rearrangement of the egg substances by centrifugation resulted in extremely abnormal development. It could lead to a severe dislocation of organs in which the animal could be turned inside out so that its endoderm was on the outside and its ectoderm on the inside. However, if the eggs were centrifuged well in advance of the onset of egg-cell division, or if the first divisions of the egg were artificially delayed, one could observe the redistribution of the egg contents and the return to normal conditions. The most plausible interpretation of this phenomenon was that the displaced cytoplasmic particles returned to the proximity of certain regions of the egg cortex that had remained in their respective positions all the time the egg was being centrifuged. See E. G. Conklin, "The Development of Centrifuged Eggs of Ascidians," *Journal of Experimental Zoology* 60 (1931): 1–119.

32. E. E. Just was well known for his advocacy of the importance of the egg cortex. See Just, *The Biology of the Cell Surface* (Philadelphia: P. Blakison's Son, 1939); Just, "On the Origin of Mutations," *American Naturalist* 66 (1932): 61–74. See also T. H. Morgan, *Experimental Embryology* (New York: Columbia University Press, 1927). The notion that the cell cortex is responsible for the pattern of the early embryo was maintained throughout the twentieth century. See B. I. Balinsky, *An Introduction to Embryology* (Philadelphia: W. B. Saunders, 1970), 172.

33. Paul Weiss, *Principles of Development* (New York: Henry Holt, 1939). See also Donna Haraway, *Crystals, Fabrics, and Fields: Metaphors of Organicism in Twentieth Century Biology*. (New Haven: Yale University Press, 1976).

34. Spemann, *Embryonic Development*, 322.

35. See Jan Sapp, "Cytoplasmic Heretics," *Perspectives in Biology and Medicine* 41 (1998): 225–290; Sapp, "Concepts of Organization: The Leverage of Protozoa," in *Developmental Biology: A Comprehensive Synthesis*, ed. Scott Gilbert (New York: Plenum Press, 1991), 229–258.

36. In 1941, Conklin compared the idea of localizing the properties of life in cell chromosomes or other specific cell organs to ancient views about the localization of certain emotions in the heart, kidneys, or bowels, or René Descartes's proposal that the pineal gland was the seat of the soul. See Conklin, "Cell and Protoplasm," 17.

37. See Viktor Hamburger, "Embryology and the Modern Synthesis in Evolutionary Theory," in *The Evolutionary Synthesis: Perspectives on the Unification of Biology*, ed.

William B. Provine and Ernst Mayr (Cambridge, Mass.: Harvard University Press, 1980), 97–111.

38. Viktor Hamburger, "Evolutionary Theory in Germany: A Comment," in *The Evolutionary Synthesis: Perspectives on the Unification of Biology*, ed. William B. Provine and Ernst Mayr (Cambridge, Mass.: Harvard University Press, 1980), 303–308, at 306.

39. C. O. Whitman, *Orthogenetic Evolution in Pigeons: Posthumous Works of Charles Otis Whitman*, ed. Oscar Riddle, Carnegie Institute of Washington Publication 257, vol. 1 (1919), 12.

40. E. G. Conklin, "The Mechanism of Heredity," *Science* 27 (1908): 89–99, at 98.

41. E. G. Conklin, "The Mutation Theory from the Standpoint of Cytology," *Science* 21 (1905): 525–529.

42. Wilhelm Johannsen, "Some Remarks about Units in Heredity," *Hereditas* 4 (1923): 133–141, at 140.

43. Ibid., 137.

44. Ibid.

45. Ibid.

46. William Bateson, "Address of the President of the British Association for the Advancement of Science," *Science* 40 (1926): 201–235, at 235.

47. Ernst Mayr, "Prologue: Some Thoughts on the History of the Evolutionary Synthesis," in *The Evolutionary Synthesis: Perspectives on the Unification of Biology*, ed. William B. Provine and Ernst Mayr (Cambridge, Mass.: Harvard University Press, 1980), 16; E. Boesiger, "Evolutionary Biology in France at the Time of the Evolutionary Synthesis," in *The Evolutionary Synthesis: Perspectives on the Unification of Biology*, ed. William B. Provine and Ernst Mayr (Cambridge, Mass.: Harvard University Press, 1980), 309–320; Camille Limoges, "A Second Glance at Evolutionary Biology in France," in *The Evolutionary Synthesis: Perspectives on the Unification of Biology*, ed. William B. Provine and Ernst Mayr (Cambridge, Mass.: Harvard University Press, 1980), 322–328; Sapp, *Beyond the Gene: Cytoplasmic Inheritance and the Struggle for Authority in Genetics* (New York: Oxford University Press, 1987).

48. Maurice Caullery, *Les conceptions modernes de l'hérédité* (Paris: Flammarion, 1935), 263; my translation.

49. T. H. Morgan, *The Physical Basis of Heredity* (Philadelphia: Lippincott, 1919), 226.

50. T. H. Morgan, "Genetics and the Physiology of Development," *American Naturalist* 60 (1926): 489–515, at 491.

Notes to Chapter 11

1. Gregor Mendel, "Experiments on Plant Hybrids," trans. Eva Sherwood, in *The Origin of Genetics: A Mendel Source Book*, ed. Curt Stern and Eva Sherwood (San Francisco: F. H. Freeman, 1966), 47; see also William Bateson, *Mendel's Principles of Heredity* (Cambridge: Cambridge University Press, 1909), 379.

2. For a critical account of the context of that rediscovery, see Augustine Brannigan, *The Social Basis of Scientific Discovery* (Cambridge: Cambridge University Press, 1981).

3. Loren Eisley, *Darwin's Century: Evolution and the Men Who Discovered It* (New York: Anchor Books, 1961), 211.

4. Jan Sapp, "The Nine Lives of Gregor Mendel," in *Experimental Inquiries*, ed. H. E. Le Grand (Dordrecht: Kluwer, 1990), 137–166. See also Douglas Allchin, "Mending Mendelism," *American Biology Teacher* 62 (2000): 633–639.

5. W. Bateson, *Mendel's Principles of Heredity*.

6. R. A. Fisher, "Has Mendel's Work Been Rediscovered?" *Annals of Science* 1 (1936): 115–137.

7. E. B. Gasking, "Why Was Mendel's Work Ignored?" *Journal of the History of Ideas* 20 (1959): 60–84.

8. L. A. Callender, "Gregor Mendel: An Opponent of Descent with Modification," *History of Science* 26 (1988): 41–75.

9. C. Zirkle, "Gregor Mendel and his Precursors," *Isis* 42 (1951): 97–104; Ernst Mayr, *The Growth of Biological Thought* (Cambridge, Mass.: Harvard University Press, 1982).

10. Augustine Brannigan, "The Reification of Gregor Mendel," *Social Studies of Science* 9 (1979): 423–454; Brannigan, *The Social Basis of Scientific Discoveries*; R. Olby, "Mendel No Mendelian?" *History of Science* 17 (1979): 53–72.

11. Fisher, "Has Mendel's Work Been Rediscovered?"

12. G. W. Beadle, "Mendelism, 1965," in *Heritage from Mendel*, ed. R. A. Brink and E. D. Styles (Madison: University of Wisconsin Press, 1967), 335–350; Robert Olby, *The Origins of Mendelism* (New York: Schocken Books, 1966); Sewall Wright, "Mendel's Ratios," in *Origin of Genetics*, ed. C. Stern and E. Sherwood (San Francisco: W. H. Freeman, 1966), 173–175; L. M. van Valen, "Mendel Was No Fraud," *Nature* 325 (1985): 395; A. W. F. Edwards, "Defending Mendel Merely Perpetuating a Myth," *Nature* 326 (1987): 449; I. Pilgrim, "The Too Good to Be True Paradox and Gregor Mendel," *Journal of Heredity* 75 (1984): 501–502; J. M. Thoday, "Mendel's Work as an Introduction to Genetics," *Advancement in Science* 23 (1966): 120–134.

13. W. Bateson, *Mendel's Principles of Heredity*.

14. L. C. Dunn, *A Short History of Genetics* (New York: McGraw-Hill, 1965), 19.

15. Ibid.; see also W. Bateson, *Mendel's Principles of Heredity,* 2.

16. See, e.g., Gasking, "Why Was Mendel's Work Ignored?"; Dunn, *A Short History of Genetics*.

17. As T. H. Morgan put it in 1917: "The genial abbot's work was not entirely heaven-born, but had a background of one hundred years of substantial progress that made it possible for his genius to develop to its full measure." T. H. Morgan, "The Rise of Genetics," *Science* 77 (1932): 261–267, at 262–263. Conway Zirkle showed that Mendel's statistical approach was not unorthodox compared to existing traditions. See Zirkle, "Mendel and His Precursors."

18. See Olby, "Mendel no Mendelian?"; Callender, "Gregor Mendel"; Brannigan, "The Reification of Gregor Mendel"; Brannigan, *The Social Basis of Scientific Discoveries.*

19. W. T. Stearn, *An Introduction to the Species Plantarum and Cognate Botanical Works of Carl Linnaeus*, 2 vols. (London: Bartholomew Press, 1957), vol. 1, pp. 156–158.

20. See Robert Olby, *The Origins of Mendelism*, 2d ed. (Chicago: University of Chicago Press, 1985), 266–267.

21. Brannigan, "The Reification of Gregor Mendel," 440.

22. Alexander Weinstein, "How Unknown Was Mendel's Paper?" *Journal of the History of Biology* 10 (1977): 341–364, at 361.

23. A. H. Sturtevant, *A Short History of Genetics* (New York: Harper and Row), 27.

24. Weinstein, "How Unknown Was Mendel's Paper?" On whether de Vries knew of Mendel's laws before he published his rediscovery of them, see Erik Zevenhuizen, "Keeping and Scrapping: The Story of a Mendelian Lecture Plate of Hugo de Vries," *Annals of Science* 57 (2000): 329–352.

25. Brannigan, *The Social Basis of Scientific Discovery*, 94.

26. Scholarly debates over Mendel's objectives have only increased over the past decade. See, e.g., B. E. Bishop, "Mendel's Opposition to Evolution and to Darwin," *Journal of Heredity* 87, no. 3 (1996): 205–213; Daniel L. Hartl and Vitezslav Orel, "What Did Gregor Mendel Think He Discovered?" *Genetics* 131 (1992): 245–253; Floyd Monaghan and Alain Corcos, "The Real Objective of Mendel's Paper," *Biology and Philosophy* 5 (1990): 267–292; R. Falk and S. Sarkar, "The Real Objective of Mendel's Paper: A Response to Monaghan and Corcos," *Biology and Philosophy* 6 (1991): 447–451. See also Douglas Allchin, "Mending Mendelism," *American Biology Teacher* 62 (2000): 633–639; Robin Marantz Henig, *The Monk in the Garden* (New York: Houghton Mifflin, 2002).

27. Olby, *The Origins of Mendelism.*

28. See Curt Stern and Eva Sherwood, *The Origin of Genetics: A Mendel Source Book*, (San Francisco: F. H. Freeman, 1966), 47; see also W. Bateson, *Mendel's Principles of Heredity*, 379.

29. Fisher, "Has Mendel's Work Been Rediscovered?" 118.

30. Gavin de Beer, "Mendel, Darwin and Fisher (1865–1965)," *Notes and Records of the Royal Society* 19 (1964): 192–226, at 208.

31. See Callender, "Gregor Mendel," 54. Callender points out that a concept of mutation is completely lacking in Mendel's paper (72). He insists, however, that in fact, Mendel had adopted a sophisticated form of the doctrine of special creation as proposed by Linnaeus, who, as mentioned earlier, proposed that the great majority of species we see today arose by hybridization.

32. Paul Forman has exposed various myths in scientists' accounts of the discovery of X-ray crystallography. He interpreted these myths as attempts to strengthen the tradition of X-ray crystallography by tracing it to a higher, better, more supernatural reality of initial events." Forman, "The Discovery of the Diffraction of X-Rays by Crystals: A Critique of Myths," *Archive for the History of Exact Sciences* 5 (1969): 38–71, at 68. Olby suggested the same explanation for understanding the aggrandizement of Mendel in scientists' accounts of the origin of genetics; Olby, "Mendel no Mendelian?" See also Sapp, "The Nine Lives of Gregor Mendel."

33. See B. Barber, "Resistance by Scientists to Scientific Discovery." In *The Sociology of Science*, ed. B. Barber and W. Hirsch (New York: Free Press. 1962), 539–556, at 540; Brannigan, "The Reification of Gregor Mendel," 453–454.

34. See D. Kevles, "Genetics in the United States and Great Britain, 1890–1930: A Review with Speculations," *Isis* 71 (1980): 441–455; W. B. Provine, *The Origins of Theoretical Population Genetics* (Chicago: University of Chicago Press, 1971); D. MacKenzie, *Statistics in Britain, 1865–1930* (Edinburgh: Edinburgh University Press, 1981).

35. B. Bateson, *William Bateson, F.R.S., Naturalist: His Essays and Addresses Together with a Short Account of His Life* (Cambridge: Cambridge University Press, 1928), 93.

36. Ibid.

37. W. Bateson, *Mendel's Principles of Heredity*, 289.

38. Quoted in Kevles, "Genetics in the United States and Great Britain," 442.

39. Ibid., 445.

40. Francis Galton, *Natural Inheritance* (London: Macmillan, 1889).

41. In the first decade of the century, some statisticians did argue that ancestral inheritance could be understood in terms of Mendelian principles operating in a random-breeding population. See Provine, *The Origins of Theoretical Population Genetics.*

42. See William Bateson and E. R. Saunders, "Experimental Studies in the Physiology of Heredity," *Reports to the Evolution Committee of the Royal Society* 1 (1902): 1–16, at 6.

43. W. Bateson, *Mendel's Principles of Heredity*, 2. The view that Darwin's influence was partly responsible for the neglect of Mendel's work was promoted by other geneticists who had pioneered the development of Mendelian analysis at a time when many biologists believed it was a minor curiosity with little bearing on the grand problems of evolution. See, e.g., L. C. Dunn, *A Short History of Genetics* (New York: McGraw-Hill, 1965), 19.

44. W. Bateson, *Mendel's Principles of Heredity*, 311.

45. Ibid., 316.

46. Fisher, "Has Mendel's Work Been Rediscovered?" 116.

47. Ibid.

48. Ibid., 117.

49. Ibid., 123:

> In 1930, as a result of a study of the development of Darwin's ideas, I pointed out that the modern genetic system, apart from such special features as dominance and linkage, could have been inferred by any abstract thinker in the middle of the nineteenth century if he were led to postulate that inheritance was particulate, that the germinal material was structural, and that the contributions of the two parents were equivalent. I had no idea that Mendel had arrived at his discovery in this way. From an examination of Mendel's work it now appears not improbable that he did so and that his ready assumption of the equivalence of the gametes was a potent factor in leading him to his theory. In this way his experimental programme becomes intelligible as a carefully planned demonstration of his conclusions.

50. The question of whether Mendel deliberately faked some of his data was given great attention at centennial celebrations of Mendel's paper and centennial symposiums of the genetics clan. See Dunn, *A Short History of Genetics*, 12; Olby, *The Origins of Mendelism*; Thoday, "Mendel's Work"; Wright, "Mendel's Ratios," 173–175; Beadle, "Mendelism 1965," 337–338; H. Iltis, *The Life of Mendel* (New York: W. W. Norton, 1966).

51. W. Bateson, *Mendel's Principles of Heredity*, 305.

52. Fisher, "Has Mendel's Work Been Rediscovered?" 119.

53. See, e.g., Richard Yeo and John Schuster, eds., *The Politics and Rhetoric of Scientific Method* (Dordrecht: Reidel, 1989).

54. Peter Medawar, "Is the Scientific Paper a Fraud?" *Listener* (Sept. 12, 1963), 377–378, at 377.

55. Ibid.

56. Stephen Shapin has examined conventions for writing scientific papers. He shows that the celebrated seventeenth-century experimentalist Robert Boyle set out rules to distinguish authenticated scientific knowledge from mere belief. This was done, in part, by what Shapin calls "the literary technology of virtual witnessing." This "*literary technology* by means of which the phenomena produced . . . were made known to those who were not direct witnesses" involved providing protocols for experiments, recounting unsuccess-

ful experiments, displaying humility so as not to look self-interested and untrustworthy, citing other writers not as judges but as witnesses to attest matters of fact, and so on; S. Shapin, "Pump and Circumstance: Robert Boyle's Literary Technology," *Social Studies of Science* 14 (1984): 481–520.

57. See Michael Polanyi, *The Tacit Dimension* (New York: Anchor Books, 1967). For studies of problems of replication based on methods described in scientific papers, see, e.g., H. M. Collins, *Changing Order: Replication and Induction in Scientific Practice* (Beverly Hills: Sage, 1985); Jan Sapp, *Where the Truth Lies: Franz Moewus and the Origins of Molecular Biology* (Cambridge: Cambridge University Press, 1990).

58. B. Nemec, "Before Mendel," in *Fundamenta Genetica*, ed. J. Krizenecky (Oosterhout: Anthropological Publications, 1965), 7–13, at 13.

Notes to Chapter 12

1. E. B. Wilson, "The Bearing of Cytological Research on Heredity," *Science* 88 (1914): 333–352.

2. E. G. Conklin, "The Mechanism of Heredity," *Science* 27 (1908): 89–99, at 90.

3. "Heredity: Creative Energy," *American Breeder's Magazine* 1 (1910): 79.

4. Earlier, in his influential book *Hereditary Genius* (1869), Galton had proposed that "it would be quite practical to produce a highly gifted race of men" over several generations by arranged marriages between men of distinction and women of wealth. Francis Galton, *Hereditary Genius*, ed. C. D. Darlington, rev. ed. (Cleveland: Meridian Books, 1962).

5. The chair was occupied by Karl Pearson, founder of biometry, who maintained that heredity was largely responsible for mental abilities and emotional qualities, and that the high birthrate of the poor was a threat to humanity.

6. Mark B. Adams, ed., *The Wellborn Science: Eugenics in Germany, France, Brazil, and Russia* (New York: Oxford University Press, 1990); Daniel Kevles, *In the Name of Eugenics: Genetics and the Uses of Human Heredity* (New York: Knopf, 1985); R. Falk, D. Paul, G. Allen, and S. Gissis, eds., "Eugenic Thought and Practice: A Reappraisal," *Science in Context* 11 (1998): 329–637.

7. See Benno Muller-Hill, *Murderous Science, 1933–1945* (Cold Spring Harbor, N.Y.: Cold Spring Harbor Laboratory Press, 1998).

8. See Adams, *The Wellborn Science,* 219–221. The idea that society is determined by our genes continues to arouse popular and scientific imagination (see chapter 20). Moreover, the modern practice of genetic counseling is a eugenic activity; the genetic basis of alcoholism and violent crime has been and continues to be seriously discussed, and genetic screening for genes associated with such maladies is visible on the horizon.

9. Richard Lewontin analyzed the diversity of blood types within populations and showed that 85 percent of the variation was within human populations, while only 15 percent was between populations. See Richard C. Lewontin, "The Apportionment of Human Diversity," *Evolutionary Biology* 6 (1972): 381–398; Lewontin, *Human Diversity* (San Francisco: Scientific American Books, 1982). See also Joshua Lederberg, "The Genetics of Human Nature," *Social Research* 40 (1973): 375–406; Luigi Luca Cavalli-Storza, *Genes, Peoples, and Languages* (New York: North Point Press, 2000).

10. This notion of the struggle for authority between specialties is adapted from Pierre Bourdieu, "The Specificity of the Scientific Field and the Social Conditions of the Progress of Reason," *Social Science Information* 1 (1975): 19–47. See also Jan Sapp, *Beyond the*

Gene: Cytoplasmic Inheritance and the Struggle for Authority in Genetics (New York: Oxford University Press, 1987).

11. William Provine, *The Origins of Theoretical Population Genetics* (Chicago: University of Chicago Press, 1971), 31.

12. Karl Pearson, " Mathematical Contributions to the Theory of Evolution: On the Law of Ancestral Heredity," *Proceedings of the Royal Society of London*, series B, 62 (1898): 368–412, at 397.

13. Conklin, "The Mechanism of Heredity," 89–90.

14. William Bateson, "Address of the President of the British Society for the Advancement of Science," *Science* 40 (1914): 287–302, at 288.

15. Ibid., 289.

16. Ibid., 293.

17. Wilhem Johannsen, "The Genotype Conception of Heredity," *American Naturalist* 45 (1911): 129–159, at 129. See also F. B. Churchill, "William Johannsen [*sic*] and the Genotype Concept," *Journal of the History of Biology* 7 (1974): 5–30; G. E. Allen, "Naturalists and Experimentalists: The Genotype and the Phenotype," in *Studies in the History of Biology* 3 (1979): 179–209; Jan Sapp, "The Struggle for Authority in the Field of Heredity, 1900–1932: New Perspectives on the Rise of Genetics," *Journal of the History of Biology* 16 (1983): 311–342.

18. "The laws governing inheritance," he argued, "are quite unknown." Charles Darwin, *On the Origin of Species*, facsimile ed. of 1859, with an introduction by Ernst Mayr (Cambridge, Mass.: Harvard University Press, 1964), 13.

19. Conklin, "The Mechanism of Heredity," 90.

20. T. H. Morgan, "Chromosomes and Heredity," *American Naturalist* 65 (1910): 449–496, at 449.

21. Johannsen, "The Genotype Conception of Heredity," 139.

22. Ibid.

23. Ibid., 130.

24. Ibid., 141.

25. Ibid.

26. Ibid., 142–143.

27. Ibid., 132.

28. Wilhelm Johannsen, "Some Remarks about Units in Heredity," *Hereditas* 4 (1923): 133–141, at 139.

29. T. H. Morgan, *The Theory of the Gene* (New Haven: Yale University Press, 1926), 26.

30. Frank Lillie to Julian Huxley, Mar. 19, 1928, Lillie Papers, Marine Biological Laboratory, Woods Hole, Mass.

31. William Bateson, "Address to the Agricultural Sub-Section," *British Association for the Advancement of Science* (1911): 1–10.

32. T. H. Morgan, A. H. Sturtevant, H. J. Muller, and C. B. Bridges, *The Mechanism of Mendelian Heredity* (London: Constable, 1915).

33. See G. E. Allen, *Thomas Hunt Morgan: The Man and His Science* (Princeton: Princeton University Press, 1978).

34. See Stephen G. Brush, "Nettie M. Stevens and the Discovery of Sex Determination by Chromosomes," *Isis* 69 (1978): 163–172; Marilyn B. Ogilvie and Clifford J. Choquette, "Nettie Maria Stevens (1861–1912): Her Life and Contributions to Cytogenetics," *Proceedings of the American Philosophical Society* 125 (1981): 292–311. Historians have emphasized how credit for the discovery had often been given to Wilson, not

because of the sexism often rampant in science, as elsewhere, but because of Wilson's more substantial contributions to cytology generally. Thus they emphasize what sociologist the Robert Merton called the "Matthew Effect": "Unto everyone that hath shall be given, and he shall have abundance; but from him that hath not shall away even that which he hath"(Matthew 25:29). In short, the rich get richer in the economy of scientific recognition—as elsewhere. See Robert Merton, "The Mathew Effect in Science," *Science* 1968 (159): 56–63.

35. T. H. Morgan, *The Physical Basis of Heredity* (Philadelphia: J. B. Lippincott,1919), 247.

36. See Daniel Kevles, "Genetics in the United States and Great Britain, 1890–1930: A Review with Speculations," *Isis* 71 (1980): 441–455, at 451.

37. Ibid.

38. See Charles Rosenberg, *No Other Gods* (Baltimore: Johns Hopkins University Press), 1976; Barbara Kimmelman, "The American Breeder's Association: Genetics and Eugenics in an Agricultural Context, 1903–1913," *Social Studies of Science* 13 (1983): 163–204.

39. J. G. Needham, "Methods for Securing Better Cooperation between Government and Laboratory Zoologists in the Solution of Problems of General or National Importance," *Science* 49 (1919): 455–458.

40. See remarks by Frederick O. Bowler, *Evolution in Light of Modern Knowledge* (London: Blacklie and Son, 1925), 208.

41. W. K. Gregory, "Genetics versus Paleontology," *American Naturalist* 51 (1917): 622–635, at 623.

42. W. M. Wheeler, "The Dry-Rot of Our Academic Biology," *Science* 57 (1923): 61–71, at 62.

43. Ibid.

44. L. C. Cole to members of the executive committee, American Society of Naturalists, Oct. 25, 1929; Herman J. Muller file, Manuscripts Department, Lilly Library, Indiana University, Bloomington, Ind.

45. F. R. Lillie, "The Gene and the Ontogenetic Process," *Science* 64 (1927): 361–368, at 362.

46. L. C. Dunn, "Nucleus and Cytoplasm as Vehicles of Heredity," *American Naturalist* 51 (1917): 286–300, at 299.

47. E. G. Conklin, "The Mechanism of Evolution in Light of Heredity and Development," *Scientific Monthly* 7 (1919): 481–506, at 487.

48. F. R. Lillie, "The Gene and the Ontogenetic Process," 362.

49. Albert Brachet, *Traité d'embryologie des vertébrés*, 2d ed. (Brussels: Masson, 1935), 3.

50. T. H. Morgan, *The Physical Basis of Heredity* (Philadelphia: J. B. Lippincott, 1919), 241.

51. T. H. Morgan, "Genetics and the Physiology of Development," *American Naturalist* 60 (1926): 489–515, at 491–496.

52. Lillie, "The Gene and the Ontogenetic Process," 367–368.

53. Ross Harrison, "Embryology and Its Relations," *Science* 85 (1937): 369–374, at 372.

54. T. H. Morgan, "The Rise of Genetics II," *Science* 76 (1932): 261–267, at 264.

55. T. H. Morgan, "The Relation of Genetics to Physiology and Medicine," *Nobel Lectures in Molecular Biology* (New York: Elsevier, 1977), 3–18, at 5.

56. See Elof Carlson, *Genes, Radiation, and Society: The Life and Work of H. J. Muller* (Ithaca: Cornell University Press, 1981).

57. H. J. Muller, "Variations Due to Change in the Individual Gene," *American Naturalist* 56 (1922): 32–50, at 48.

58. H. J. Muller, "The Gene as the Basis of Life," in *Proceedings of the International Congress of Plant Sciences*, vol. 1 (Menasha, Wisc.: George Banta, 1929), 897–921, at 897.

59. Ibid., 921.

60. On the history of the gene concept, see Raphael Falk, "What Is a Gene?" *Studies in History and Philosophy of Science* 17 (1986): 133–173; Peter Beurton, Raphael Falk, and Hans-Jörg Rheinberger, eds., *The Concept of the Gene in Development and Evolution: Historical and Epistemological Perspectives* (Cambridge: Cambridge University Press, 2000); Evelyn Fox Keller, *The Century of the Gene* (Cambridge, Mass.: Harvard University Press, 2000).

61. H. J. Muller, "Artificial Transmutation of the Gene," *Science* 46 (1927): 84–87.

62. L. J. Stadler, "Mutations in Barley Induced by X-Rays and Radium," *Science* 68 (1928): 186–187.

63. H. J. Muller, "Radiation and Genetics," *American Naturalist* 64 (1930): 220–250, at 224.

Notes to Chapter 13

1. Julian Huxley, *Evolution: The Modern Synthesis* (London: George Allen and Unwin, 1942), 24–25.

2. In England, Huxley's *Evolution: The Modern Synthesis* and Michael White's *Animal Cytology and Evolution* (1945); in the United States, Ernst Mayr's *Systematics and the Origin of Species* (1942), George Gaylord Simpson's *Tempo and Mode in Evolution* (1944). See Jan Sapp, "Geographical Isolation and *The Origin of Species*: The Migrations of Michael White," in *International Science and National Scientific Identity* (Dordrecht, the Netherlands: Kluwer, 1991), 233–254.

3. See Ernst Mayr and William B. Provine, eds., *The Evolutionary Synthesis: Perspectives on the Unification of Biology* (Cambridge, Mass.: Harvard University Press, 1980); Vassiliki B. Smocovitis, *Unifying Biology: The Evolutionary Synthesis and Evolutionary Biology* (Princeton: Princeton University Press, 1996).

4. See Theodosius Dobzhansky, *Evolution* (San Francisco: W. H. Freeman, 1977), 14.

5. Huxley, *Evolution*, 20.

6. W. Johannsen, *Über Erblichkeit in Populationen und in reinen Linien* (Jena: Fischer, 1903).

7. As Dobzhansky commented (*Evolution*, 16):

> Being missionaries for a new theory, [de Vries, Johannsen, and Bateson] naturally belittled any observations or phenomena that might appear to detract from its importance. To [them], Darwin's explanation that slight variations between individuals are due chiefly or entirely to phenotypic modifications of a constant genotype appeared to be most compatible with the concept of Mendelism as a universal theory of particulate heredity. Even when experiments of Nilsson-Ehle demonstrated that multiple factor inheritance could explain quantitative inheritance on a particulate basis, other geneticists failed to explore this possibility, and many of them completely ignored it.

8. See A. H. Sturtevant, *A History of Genetics* (New York: Harper and Row, 1965), 62–65.

9. See, e.g., W. E. Castle and J. C. Phillips, "Piebald Rats and Selection," *Carnegie Institution of Washington Publication* 195 (1914): 1–82; R. Payne, 'Selection for Increased and Decreased Bristle Number," *Anatomical Record* 17 (1920): 335–336; A. H. Sturtevant, "An Analysis of the Effects of Selection," *Carnegie Institution of Washington Publication* 264 (1918): 1–68.

10. T. H. Morgan, *The Scientific Basis of Evolution* (New York: W. W. Norton, 1932), 110. Peter Bowler suggested that Morgan's reluctance to go fully Darwinian was due at least in part to his stand on human evolution: that individual conflict and competition (the basis of social Darwinism) were not the means for advancing the human race. See Peter Bowler, "Hugo de Vries and Thomas Hunt Morgan: The Mutation Theory and the Spirit of Darwinism," *Annals of Science* 35 (1978): 55–73; Bowler, *The Mendelian Revolution* (Baltimore: Johns Hopkins University Press, 1989).

11. See G. Ledyard Stebbins, *Processes of Organic Evolution*, 2d ed. (Englewood Cliffs, N.J.: Prentice-Hall, 1971).

12. H. Nilsson-Ehle, "Kreuzungsuntersuchungen an Hafer und Weizen," *Lunds Universitets Årsskrift*, n.s., ser. 2, vol, 5, no. 2 (1909): 1–122; E. M. East, "The Role of Reproduction in Evolution," *American Naturalist* 52 (1918): 273–289.

13. In 1908 the British mathematician G. H. Hardy and the German physicist Wilhelm Weinberg independently offered a formula for a stable nonevolving population against which conditions for evolutionary change could be measured. Assuming Mendel's laws, the stable frequency of genotypes is p^2 (AA) + $2pq$ (Aa) + q^2 (aa), where p is the frequency of the dominant A genes in a population, q the initial frequency of the recessive a, and (since there are only two alleles at the locus) $p + q = 1$. This formula was valid only if several conditions were met: the population must be large enough so that sampling errors can be disregarded; there must be no mutation, since the change of A to a or of a to A will alter the values of p and q; there must be no selective mating; and there must be no selection, that is, A and a must have no selective advantage on the reproductive capacity of individuals bearing them. These were extraordinarily stringent requirements and may never have been fully met in nature. The aim of evolutionary population genetics was to determine the forces or processes that disturb the Hardy-Weinberg frequencies of genotypes in a population. See E. Sober, *The Nature of Selection: Evolutionary Theory in Philosophical Focus* (Cambridge, Mass.: MIT Press, 1984).

14. See William Provine, *The Origins of Theoretical Population Genetics* (Chicago: Chicago University Press, 1971); Provine, *Sewall Wright and Evolutionary Biology* (Chicago: University of Chicago Press, 1986).

15. R. A. Fisher, *The Genetical Theory of Natural Selection* (Oxford: Clarendon Press, 1930).

16. Ibid., 15–16.

17. J. B. S. Haldane, "Natural Selection," *Nature* 124 (1929): 444; Haldane, *The Cause of Evolution* (London: Longmans, Green, 1932).

18. See a superb account in Judith Hooper, *Of Moths and Men: An Evolutionary Tale* (New York: Norton, 2002).

19. See Ernst Mayr, *The Growth of Biological Thought: Diversity, Evolution, and Inheritance* (Cambridge, Mass.: Harvard University Press, 1982), 554.

20. See Provine, *The Origins of Theoretical Population Genetics*; Provine, *Sewall Wright and Evolutionary Biology*.

21. Sewall Wright, "Evolution in Mendelian Populations," *Genetics* 16 (1931): 97–159, at 99. As he wrote:

> The rediscovery of Mendelian heredity in 1900 came as a direct consequence of de Vries' investigation. Major Mendelian differences were naturally the first to attract attention. It is not therefore surprising that the phenomena of Mendelian heredity were looked upon as confirming de Vries' theory. . . . Johannsen's study of pure lines was interpreted as meaning that Darwin's selection of small random variations was not a true evolutionary factor.

22. See Provine, *Sewall Wright and Evolutionary Biology.*

23. Mark B. Adams, "Sergei Chetverikov, the Kol'tov Institute, and the Evolutionary Synthesis," in *The Evolutionary Synthesis: Perspectives on the Unification of Biology*, ed. William B. Provine and Ernst Mayr (Cambridge, Mass.: Harvard University Press, 1980), 242–278.

24. Chetverikov was awarded the Darwin medal by the German Academy Leopoldina just before his death. See Theodosius Dobzhansky, "The Birth of the Genetic Theory of Evolution in the Soviet Union in the 1920s," in *The Evolutionary Synthesis: Perspectives on the Unification of Biology*, ed. William B. Provine and Ernst Mayr (Cambridge, Mass.: Harvard University Press, 1980), 229–241, at 234–235.

25. Theodosius Dobzhansky, *Genetics and The Origin of Species*, 3d ed. (New York: Columbia University Press, 1951), 16.

26. Dobzhansky, *Genetics and The Origin of Species*, 1st ed., (New York: Columbia University Press:1937), 11.

27. See John Beatty, "Dobzhansky and Drift: Facts, Values, and Chance in Evolutionary Biology," in *The Probabilistic Revolution*, ed. L. Krüger, G. Gigerenzer, and M. Morgan (Cambridge, Mass.: MIT Press, 1987), 271–311; John Beatty, "Weighing the Risks: Stalemate in the Classical? Balance Controversy," *Journal of the History of Biology*, 20 (1987): 289–319.

28. See Provine, *Sewall Wright and Evolutionary Biology*; Beatty, "Dobzhansky and Drift"; Beatty, "Weighing the Risks."

29. See, e.g., Niles Eldridge, *Unfinished Synthesis. Biological Hierarchies and Modern Evolutionary Thought* (New York: Oxford University Press, 1985); Eldridge, *Macroevolutionary Dynamics* (New York: McGraw-Hill, 1989).

30. See Ernst Mayr, "Prologue: Some Thoughts on the History of the Evolutionary Synthesis," in *The Evolutionary Synthesis: Perspectives on the Unification of Biology*, ed. William B. Provine and Ernst Mayr (Cambridge, Mass.: Harvard University Press, 1980), 1–48.

31. G. Ledyard Stebbins summarized the modern synthetic theory as consisting of five basic types of processes: gene mutation (1) and genetic recombination (2) were the sources of variability, in a cross-fertilizing population in nature. But they did not produce direction. Chromosome organization (3) and its variation, which affect linkage, produce orderly arrangements of variation in the gene pool. Then natural selection (4) and reproductive isolation (5) put limits on the direction evolution can take and guide populations into adaptive channels. Stebbins, *Processes of Organic Evolution*, 2–17.

32. C. D. Darlington, *Recent Advances in Cytology* (London: Churchill, 1932); Darlington, *The Evolution of Genetic Systems* (Cambridge: Cambridge University Press, 1939).

33. M. J. D. White, *Animal Cytology and Evolution* (Cambridge: Cambridge University Press, 1945).

34. Ernst Mayr, *Systematics and the Origin of Species* (New York: Columbia University Press, 1942); G. Ledyard Stebbins, *Variation and the Evolution of Plants* (New York: Columbia University Press, 1950). See also Vassiliki B. Smocovitis, *Unifying Biology: The Evolutionary Synthesis and Evolutionary Biology* (Princeton: Princeton University Press, 1996).

35. Dobzhansky, *Genetics and the Origin of Species*, 1st ed., 1937, 312.

36. See discussion in Mayr, "Prologue: Some Thoughts on the History of the Evolutionary Synthesis," 34.

37. George Gaylord Simpson, "The Species Concept," *Evolution* 5 (1951): 285–298, at 289.

38. Frank J. Sulloway, "Geographical Isolation in Darwin's Thinking: The Vicissitudes of a Crucial Idea," *Studies in the History of Biology* 3 (1979): 23–65.

39. Mayr, *Systematics and the Origin of Species*, 155.

40. Mayr, "Prologue: Some Thoughts on the History of the Evolutionary Synthesis," 35. The relative importance of sympatric speciation is still debated among neo-Darwinians. See L. G. Bush, "Modes of Animal Speciation," *Annual Review of Ecology and Systematics* 6 (1975): 339–364; M. J. D. White, *Modes of Speciation* (San Francisco: W. H. Freeman, 1978).

41. Mayr, "Prologue: Some Thoughts on the History of the Evolutionary Synthesis," 35.

42. See Vassiliki B. Smocovitis, "Unifying Biology: The Evolutionary Synthesis and Evolutionary Biology," *Journal of the History of Biology* 25 (1992): 1–65; Smocovitis, "Organizing Evolution: Founding the Society for the Study of Evolution (1939–1950)," *Journal of the History of Biology* 27 (1994): 241–309; Smocovitis, *Unifying Biology;* Joseph Cain, "Ernst Mayr as Community Architect: Launching the Society for the Study of Evolution and the Journal *Evolution*," *Biology and Philosophy* 9 (1994): 387–427.

43. Richard Lewontin commented in 1980 that there was no inherently logical need to assume that mechanisms for the origin of species had to be the same as those operating in the production of higher taxa: "As an evolutionary geneticist, I do not see how the origin of higher taxa are the necessary consequence of neo-Darwinism. They are sufficiently explained, but they are not the necessary consequences." Richard C. Lewontin, "Theoretical Population Genetics in the Evolutionary Synthesis," in *The Evolutionary Synthesis: Perspectives on the Unification of Biology*, ed. William B. Provine and Ernst Mayr (Cambridge, Mass.: Harvard University Press, 1980), 58–68, at 60.

44. See Michael Dietrich, "Richard Goldschmidt's 'Heresies' and the Evolutionary Synthesis," *Journal of the History of Biology* 28 (1995): 431–461. See also Dietrich, "From Gene to Genetic Hierarchy: Richard Goldschmidt and the Problem of the Gene," in *The Concept of the Gene in Development and Evolution. Historical and Epistemological Perspectives*, ed. Peter Beurton, Raphael Falk, and Hans-Jörg Rheinberger (Cambridge: Cambridge University Press, 2000), 91–114.

45. Richard Goldschmidt, *The Material Basis of Evolution* (New Haven: Yale University Press, 1940), 6–7.

46. "The neo-Darwinian conception, which works perfectly well within the limits of the species, encounters difficulties and is not sustained by the actual facts when the step from species to species has to be explained." Ibid., 139.

47. Ibid., 183.

48. See Eldridge, *Macroevolutionary Dynamics*.

49. Dobzhansky, *Genetics and the Origins of Species*, 1st ed., 1937, 12.

50. Percy Raymond, "Invertebrate Paleontology," in *Geology, 1888–1938*, Geological Society of America fiftieth anniversary vol. (Washington, D.C.: Geological Society of America, 1938), 73–103, at 98–99.

51. See Stephen Jay Gould, "G. G. Simpson, Paleontology, and the Modern Synthesis," in *The Evolutionary Synthesis: Perspectives on the Unification of Biology*, ed. William B. Provine and Ernst Mayr (Cambridge, Mass.: Harvard University Press, 1980), 153–172. Huxley 's *Evolution: The Modern Syntheses* helped to narrow the breach between Mendelian genetics and macroevolutionary events. See also Bernhard Rensch, *Neuere Probleme de Abstammungslehre* (Stuttgart: Enke, 1947).

52. G. G. Simpson, *Tempo and Mode of Evolution* (New York: Columbia University Press, 1944).

53. Ibid., xv–xvi.

54. See Gould, "G. G. Simpson, Paleontology, and the Modern Synthesis," 97.

55. See Stephen Jay Gould, "The Origin and Function of 'Bizarre' Structures: Antler Size and Skull Size in the 'Irish Elk,' *Megaloceros giganteus*," *Evolution* 28 (1974): 191–220.

56. Simpson, *Tempo and Mode of Evolution*, 180.

57. G. G. Simpson, *The Major Features of Evolution* (New York: Columbia University Press, 1953). See also Gould, "G. G. Simpson, Paleontology, and the Modern Synthesis," 168.

58. Thomas Kuhn, *The Structure of Scientific Revolutions* (Chicago: Chicago University Press, 1962). See also Mayr, "Prologue: Some Thoughts on the History of the Evolutionary Synthesis," 43.

59. See Larry Laudan, *Progress and Its Problems: Toward a Theory of Scientific Growth* (Berkeley: University of California Press, 1977); Gary Gutting, ed., *Paradigms and Revolutions: Appraisals and Applications of Thomas Kuhn's Philosophy of Science* (Notre Dame: University of Notre Dame Press, 1980); Barry Barnes, *T. S. Kuhn and Social Science* (New York: Columbia University Press, 1982); Howard Margolis, *Paradigms and Barriers: How Habits of Mind Govern Scientific Belief* (Chicago: University of Chicago Press, 1993); Trevor Pinch, "Kuhn: The Conservative and Radical Interpretations," *Social Studies of Science* 27 (1997): 465–482; S. Fuller, *Thomas Kuhn: A Philosophical History of Our Times* (Chicago: University of Chicago Press, 2000).

60. Mayr, "Prologue: Some Thoughts on the History of the Evolutionary Synthesis," 40.

61. See, e.g., Brian K. Hall, *Evolutionary Developmental Biology*, 2d ed. (Dordrecht: Kluwer, 1999).

62. Jan Sapp, *Beyond the Gene: Cytoplasmic Inheritance and the Struggle for Authority in Genetics* (New York: Oxford University Press, 1987).

Notes to Chapter 14

1. George Beadle and Edward L. Tatum, "Genetic Control of Biochemical Reactions in *Neurospora*," *Proceedings of the National Academy of Sciences* 27 (1941): 499–506.

2. George Beadle, "Genes and Chemical Reactions in Neurospora," *Les Prix Nobel en 1958* (Stockholm, 1959), 147–159, at 156.

3. See, e.g., Matt Ridley, *Genome: The Autobiography of a Species in Twenty-Three Chapters* (New York: Harper Collins, 1999), 39–40.

4. Bentley Glass, "A Century of Biochemical Genetics," *Proceedings of the American Philosophical Society* 109 (1965): 227–236, at 231.

5. George Beadle, "Biochemical Genetics: Some Recollections," in *Phage and the Origins of Molecular Biology*, ed. J. Cairns, G. S. Stent, and J. D. Watson (Cold Spring Harbor, N.Y.: Cold Spring Harbor Press, 1966), 23–32, at 32.

6. See Jan Sapp, *Where the Truth Lies: Franz Moewus and the Origins of Molecular Biology* (New York: Cambridge University Press, 1990).

7. Ibid.; Alexander Bairn, *Archibald Garrod and the Individuality of Man* (New York: Oxford University Press, 1993).

8. Archibald Garrod, *Inborn Errors of Metabolism* (Oxford: Oxford University Press, 1909).

9. A. E. Garrod, "Inborn Errors of Metabolism," *Lancet* 2 (July 4, 1908): 1–7, at 5.

10. See Bairn, *Archibald Garrod.*

11. See Robert Kohler, *From Medical Chemistry to Biochemistry: The Making of a Biomedical Discipline* (Cambridge: Cambridge University Press, 1982).

12. Garrod, "Inborn Errors," 1–2.

13. H. J. Muller, "The Gene as the Basis of Life," *Proceedings of the International Congress of Plant Science* 1 (1929): 897–921, at 914.

14. Today it is understood that dominance and recessive traits are not about the transmission of genetic material at all, but about phenotypic expression. See Douglas Allchin, "Mending Mendelism," *American Biology Teacher* 62 (2000): 633–639; Allchin, "Dissolving Dominance," in *Medical Genetics: Conceptual Foundations and Classical Questions*, ed. Lisa Parker and Rachel Ankey (Pittsburgh: University of Pittsburgh Press, 2001).

15. R. G. Swinburne, "The Presence-and-Absence Theory," *Annals of Science* 18 (1962): 131–146.

16. See J. B. S. Haldane, "The Biochemistry of the Individual," in *Perspectives in Biochemistry*, ed. J. Needham and D. E. Green (Cambridge: Cambridge University Press, 1937), 1–10.

17. T. H. Morgan, *The Theory of the Gene* (New Haven: Yale University Press, 1926), 257.

18. See Sewall Wright, "Color Inheritance in Mammals," *Journal of Heredity* 8 (1917): 224–235; William B. Provine, *Sewall Wright and Evolutionary Biology* (Chicago: University of Chicago Press, 1986).

19. L. Troland, "Biological Enigmas and the Theory of Enzyme Action," *American Naturalist* 51 (1917): 321–350.

20. E. B. Wilson, "The Physical Basis of Life," *Science* 57 (1923): 277–286, at 285.

21. Beadle and Tatum, "Genetic Control of Biochemical Reactions in *Neurospora*," 499.

22. George Beadle, "Chemical Genetics," in *Genetics in the Twentieth Century*, ed. L.C. Dunn (New York: Macmillan, 1951), 221–240, at 228.

23. George Beadle, "Mendelism 1965," in *Heritage from Mendel*, ed. R. A. Brink and E. D. Styles (Madison: University of Wisconsin Press, 1967), 335–350, at 341.

24. Sewall Wright, "The Physiology of the Gene," *Physiological Reviews* 21 (1941): 487–527; J. B. S. Haldane, *New Paths in Genetics* (New York: Harper and Brothers, 1942).

25. Beadle, "Chemical Genetics," 222.

26. Bentley Glass, "A Century of Biochemical Genetics," *Proceedings of the American Philosophical Society* 109 (1965): 227–236, at 233.

27. Beadle, "Biochemical Genetics: Some Recollections," 31.

28. Max Delbrück, discussion following David Bonner, "Biochemical Mutations in *Neurospora*," *Cold Spring Harbor Symposia on Quantitative Biology* 11 (1946): 14–24, at 23.

29. Bonner, "Biochemical Mutations," 23.

30. Beadle, "Chemical Genetics," 234.

31. Evelyn Fox Keller, *A Feeling for the Organism* (New York: Freeman, 1983); Nathaniel Comfort, *The Tangled Field* (Cambridge, Mass.: Harvard University Press, 2000).

32. Boris Ephrussi, "Aspects of the Physiology of Gene Action," *American Naturalist* 72 (1938): 5–23, at 6.

33. George Beadle and Boris Ephrussi, "The Differentiation of Eye Pigments in *Drosophila* as Studied by Transplantation," *Genetics* 21 (1936): 225–247.

34. Beadle, "Chemical Genetics," 224.

35. Beadle, "Genes and Chemical Reaction in Neurospora"; Beadle, "Biochemical Genetics: Some Recollections," 66; Joshua Lederberg, "Memoir on Edward L. Tatum," *Biographical Memoirs of the National Academy of Sciences* (1986).

36. Carl Lindegren, "The Genetics of *Neurospora* I: The Inheritance of Response to Heat-Treatment," *Bulletin of the Torrey Botanical Club* 59 (1932): 85–102; Lindegren, "The Genetics of *Neurospora* II: Segregation of the Sex Factors in Asci of *N. crassa*, *N. sitophila*, and *N. tetrasperma*," *Bulletin of the Torrey Botanical Club* 59 (1932): 119–138; Lindegren, "The Genetics of *Neurospora* IV: The Inheritance of Tan versus Normal," *American Journal of Botany* 21 (1934): 55–58.

37. See Sapp, *Where the Truth Lies*.

38. Ibid.

39. H. S. Jennings, "Inheritance in Protozoa," in *Protozoa in Biological Research*, ed. G. N. Calkins and F. M. Summers (1941; reprint, New York: Macmillan, 1964), 710–771, at 750.

40. Beadle, "Biochemical Genetics," 43.

41. Quoted in Sapp, *Where the Truth Lies*, 283.

42. See Ruth Sager, *Cytoplasmic Genes and Organelles* (New York: Academic Press, 1972).

43. T. M. Sonneborn, "Sex, Sex Inheritance, and Sex Determination in *Paramecium aurelia*," *Proceedings of the National Academy of Sciences* 23 (1937): 378–395.

44. See Jan Sapp, *Beyond the Gene. Cytoplasmic Inheritance and the Struggle for Authority in Genetics* (New York: Oxford University Press, 1987); Judy Johns Schloegel, "From Anomaly to Unification: Tracy Sonneborn and the Species Problem in Protozoa, 1954–57," *Journal of the History of Biology* 32 (1999): 93–132.

45. Otto Winge, "On Haplophase and Diplophase in Some Saccaromycetes," *Comptes-rendus des travaux du Laboratoire Carlsberg, série physiologie* 21 (1935): 77–111; Otto Winge and O. Lausten, "Artificial Species Hybridization in Yeast," *Comptes-rendus des travaux du Laboratoire Carlsberg, série physiologie* 22 (1938): 235–244.

46. See Thomas D. Brock, *The Emergence of Bacterial Genetics* (Cold Spring Harbor, N.Y.: Cold Spring Harbor Laboratory Press, 2000).

47. C. D. Darlington, *The Evolution of Genetics Systems* (Cambridge: Cambridge University Press, 1939), 70.

48. Julian Huxley, *Evolution: The Modern Synthesis* (London: Allen and Unwin, 1942), 131–132. As Lederberg wrote in one of his first reviews on microbial genetics, "The lack of outward differentiation of bacteria and viruses does give the appearance of holo-cellular

propagation and of identity between direct transmission and inheritance. Geneticists and bacteriologists alike have . . . shown justifiable hesitation in accepting unanalyzed genetic variations as gene mutations." Joshua Lederberg, "Problems in Microbial Genetics," *Heredity* 11 (1948): 145–198, at 153.

49. Joshua Lederberg, "Gene Recombination in *Escherichia coli*," *Nature* 158 (1946): 558; Joshua Lederberg and Edward Tatum, "Novel Genotypes in Mixed Cultures of Biochemical Mutants of Bacteria," *Cold Spring Harbor Symposia on Quantitative Biology* 11 (1946): 113–114.

50. H. J. Muller, "Variation Due to Change in the Individual Gene," *American Naturalist* 56 (1922): 32–50, at 48.

51. Tobey Appel, *Shaping Biology: The National Science Foundation and American Biological Research, 1952–1975* (Baltimore: Johns Hopkins University Press, 2000).

52. See R. E. Kohler, "The Management of Science: The Experience of Warren Weaver and the Rockefeller Foundation Program in Molecular Biology," *Minerva* 14 (1976): 279–306; Kohler, "A Policy for the Advancement of Science: The Rockefeller Foundation, 1924–29," *Minerva* 16 (1978): 480–515, at 481; Pnina Abir-Am, "The Discourse of Physical Power and Biological Knowledge in the 1930s: A Reappraisal of the Rockefeller Foundation's 'Policy' in Molecular Biology," *Social Studies of Science* 12 (1982): 341–382; J. A. Fuerst, "The Definition of Molecular Biology and the Definition of Policy: The Role of the Rockefeller Foundations's Policy for Molecular Biology," *Social Studies of Science* 14 (1984): 225–237; D. Bartels, "The Rockefeller Foundation's Funding Policy for Molecular Biology: Success or Failure?" *Social Studies of Science* 14 (1984): 238–247; R. Olby, "The Sheriff and the Cowboys, or Weaver's Support of Astbury and Pauling," *Social Studies of Science* 14 (1984): 244–247; E. J. Yoxen, "Scepticism about Centrality of Technology Transfer in the Rockefeller Foundation Programme in Molecular Biology," *Social Studies of Science* 14 (1984): 248–252; Pnina Abir-Am, "Beyond Deterministic Sociology and Apologetic History: Reassessing the Impact of Research Policy upon New Scientific Disciplines" [reply to Fuerst, Bartels, Olby, and Yoxen], *Social Studies of Science* 14 (1984): 252–263. See also the excellent study of Lily Kay, *The Molecular Vision of Life: Caltech, the Rockefeller Foundation, and the Rise of the New Biology* (New York: Oxford University Press, 1993).

53. In 1937, Weaver noted the subsequent work of George Beadle and Ephrussi on eye-color mutations: "This new technique opens up a very wide field of experimentation in genetic and apparently is considered by the group as the third significant advance made in genetics during the last few years, the other two being the work on salivary chromosomes, and the effects of radiation on germ cells." Warren Weaver, Diary, Mar. 21, 1937, Rockefeller Foundation Archives, Record Group 1.1, Series 205D. On the development of genetics in France, see Sapp, *Beyond the Gene*; Richard Burian, Jean Gayon, and Doris Zallen, "The Singular Fate of Genetics in France, 1900–1940," *Journal of the History of Biology* 21 (1988): 357–402.

54. Warren Weaver, Diary, Dec. 14, 1941, Rockefeller Foundation Archives, Record Group 1.1, Series 205D.

55. Kohler, 'The Management of Science," 296.

56. See George W. Gray, "Pauling and Beadle" (1949), reprinted in *Linus Pauling: Scientist and Peacemaker*, ed. Clifford Mead and Thomas Hager (Corvalis: Oregon State University Press, 2001), 119–126, at 125.

57. F. B. Hansen, Diary, Dec. 18, 1941, Rockefeller Foundation Archives, Tarrytown, New York, Record Group 1.1, Series 205D.

58. Beadle, "Biochemical Genetics: Some Recollections," 29–30.

59. See N. C. Mullins, "The Development of a Scientific Specialty: The Phage Group and the Origins of Molecular Biology," *Minerva* 10 (1971): 51–82; Lily Kay, "Conceptual Models and Analytic Tools: The Biology of Physicist Max Delbrück," *Journal of the History of Biology* 18 (1985): 207–246.

60. Quoted in M. F. Perutz, "Erwin Schrödinger's *What Is Life?* and Molecular Biology," in *Schrödinger: Centenary Celebration of a Polymath*, ed. C. W. Kilmister (Cambridge: Cambridge University Press, 1987), 234–251, at 238. See also Robert Olby, "Schrödinger's Problem: What Is Life?" *Journal of the History of Biology* 4 (1971): 119–148.

61. Quoted in Perutz, "Erwin Schrödinger's *What Is Life?*" 238.

62. Perutz, "Erwin Schrödinger's *What Is Life?*" 243.

63. See Donald Fleming, "Emigré Physicists and the Biological Revolution," in *The Intellectual Migration: Europe and America, 1930–1960*, ed. D. Fleming and Barnard Bailyn (Cambridge, Mass.: Harvard University Press, 1969), 152–189.

64. Gunther Stent, "Introduction: Waiting for the Paradox," in *Phage and the Origins of Molecular Biology*, ed. J. Cairns, G. S. Stent, and J. D. Watson (Cold Spring Harbor, N.Y.: Cold Spring Harbor Press, 1966), 3.

65. Biographies of Pauling include Anthony Serafini, *Linus Pauling: A Man and His Science* (New York: Paragon House, 1989); Thomas Hager, *Force of Nature: The Life of Linus Pauling* (New York: Simon and Schuster, 1995); Hager, *Linus Pauling and the Chemistry of Life* (New York: Oxford University Press, 1998); Mead and Hager, *Linus Pauling: Scientist and Peacemaker*.

66. G. W. Beadle and Linus Pauling, "A Proposed Progam of Research on the Fundamental Problems of Biology and Medicine," 1946, Rockefeller Foundation Archives, Tarrytown, Record Group 1.2, Series 205.

Notes to Chapter 15

1. See David Joravsky, *The Lysenko Affair* (Cambridge, Mass.: Harvard University Press, 1970); Z. A. Medvedev, *The Rise and Fall of T. D. Lysenko*, trans. I. Michael Lerner (New York: Anchor Books, 1971); Dominique Lecourt, *Proletarian Science? The Case of Lysenko* (London: New Left Books, 1977); Richard Lewontin and Richard Levins, "The Problem of Lysenkoism," in *The Radicalization of Science*, ed. H. Rose and S. Rose (London: Macmillan, 1976), 32–65; Robert Young, "Getting Started on Lysenkoism," *Radical Science Journal* 6–7 (1978): 81–106.

2. On the history of vernalization and Lysenkoism, see Nils Rolls-Hansen, "A New Perspective on Lysenko?" *Annals of Science* 42 (1985): 261–278.

3. Ivan Vladimorovich Michurin, *Selected Works* (Moscow: Foreign Languages Publishing House, 1949), xvii. The Soviet government conferred on him the Order of Lenin and the Order of Red Banner of Labour.

4. I. I. Schmalhausen, *Factors of Evolution: The Theory of Stabilizing Selection,* trans. Isadore Dordick, ed. Theodosius Dobzhansky (Chicago: University of Chicago Press, 1986), v–xii, at v.

5. Ibid., xvi.

6. Some Mendelian geneticists and molecular biologists went underground, working unofficially in physical science institutions. See M. B. Adams, "Biology after Stalin: A Case Study," *Survey: A Journal of East/West Studies* 23 (1977–78): 53–80.

7. See Joravsky, *The Lysenko Affair*; Medvedev, *The Rise and Fall of T. D. Lysenko.*

8. C. H. Waddington, "Canalization of Development and Genetic Assimilation of Acquired Characters," *Nature* 183 (1959): 1654–1655.

9. C. H. Waddington, "The Epigenotype," *Endeavour* 1 (1942): 18–20, at 18. He proposed the term "epigenotype" for the complex of developmental processes between genotype and phenotype (19).

10. See C. H. Waddington, *The Evolution of an Evolutionist* (Edinburgh: Edinburgh University Press, 1975).

11. See Jonathan Harwood, *Styles of Scientific Thought: The German Genetics Community, 1900–1933* (Chicago: University of Chicago Press, 1993); Jan Sapp, *Beyond the Gene: Cytoplasmic Inheritance and the Struggle for Authority in Genetics* (New York: Oxford University Press, 1987); Margaret Samosi Saha, "Carl Correns and an Alternative Approach to Genetics: The Study of Heredity in Germany between 1880 and 1930" (Ph.D. diss., Michigan State University, 1984).

12. Fritz von Wettsetin, "Über plasmatische Vererbung, sowie Plasma- und Genwirkung," *Nachrichten von Geselleshaft der Wissenschaft zu Göttingen, Math-Phys. Kl.* (1926): 250–281. See Hans Winkler, *Verbreitung und Ursache der Parthenogenesis im Pflanzen- und Tierreiche* (Jena: Verlag Fischer, 1920). On page 165, he wrote: "I propose the expression Genom for the haploid chromosome set, which, together with the pertinent protoplasm, specifies the material foundations of the species." See also Hans Winkler, "Abbr. die Rolle von kern und Protoplasma bei Vererbung," *Zeitschrift fur induktive Abstammungs und Vererbungslehre* 33 (1924): 238–253.

13. Fritz von Wettstein, "Die genetische und entwicklungsphysiologie Bedeutung des Cytoplasmas," *Zeitschrift für induktive Abstammungs und Vererbungslehre* 73 (1937): 345–366, at 345–346.

14. A. H. Sturtevant and G. W. Beadle, *An Introduction to Genetics* (New York: Dover, 1939), 331–332.

15. E. M. East, "The Nucleus-Plasma Problem," *American Naturalist* 68 (1934): 289 ff., at 431.

16. Sapp, *Beyond the Gene.*

17. See Jan Sapp, *Evolution by Association: A History of Symbiosis* (New York: Oxford University Press, 1994).

18. Ibid.

19. S. Spiegelman and M. D. Kamen, "Genes and Nucleoproteins in the Synthesis of Enzymes," *Science* 104 (1946): 581–584.

20. See T. M. Sonneborn, "The Cytoplasm in Heredity," *Heredity* 4 (1950): 11–36.

21. See Carl Lindegren, *The Cold War in Biology* (Ann Arbor, Mich.: Planarian Press, 1966). See also Sapp, *Beyond the Gene.*

22. Phillippe L'Héritier, "Sensitivity to CO_2 in Drosophila: A Review," *Heredity* 2 (1948): 325–348.

23. T. M. Sonneborn, "Heredity, Environment, and Politics," *Science* 111 (1950): 529–539, at 535.

24. Ibid., 536.

25. See J. B. S. Haldane, *The Biochemistry of Genetics* (London: Allen and Unwin, 1954), 87. See also C. D. Darlington, *The Evolution of Genetic Systems*, 2d ed. (New York: Basic Books 1958), 211.

26. Haldane, *The Biochemistry of Genetics*, 87.

27. See Jan Sapp, "Jean Brachet, *L'hérédité générale*, and the Origins of Molecular Embryology," *History and Philosophy of the Life Sciences* 19 (1997): 69–87; "Interview of Jean Brachet by Jan Sapp Arc Felice, Italy, December 10, 1980," transcribed by Denis Thieffry and Richard Burian, *History and Philosophy of the Life Sciences* 19 (1997): 113–140.

28. Speech by I. I. Prezent, in *The Situation in Biological Science: Proceedings of the Lenin Academy of Agricultural Science of the U.S.S.R., July 31–August 7, Moscow* (Moscow: Foreign Language Publishing House, 1949), 596–597.

29. Chip Boutell, "Is There Any Scientific Basis for the Lysenko Theory?" *Daily Compass*, Apr. 1950, 8.

30. T. M. Sonneborn, unpublished autobiography, Sonneborn Papers, Lillie Library, Indiana University, 1978, 19.

31. See Arthur Koestler, *The Case of the Midwife Toad* (London: Hutchinson, 1971); Lester Aronson, "The Case of the Midwife Toad," *Behavioural Genetics* 5 (1975): 115–125.

32. Sonneborn, "Heredity, Environment, and Politics."

33. See H. J. Muller, "The Development of the Gene Theory," in L.C. Dunn, *Genetics in the Twentieth Century, ed. L. C. Dunn* (New York: Macmillan, 1951), 77–100; G. W. Beadle, "Genes and Biological Enigmas," in *Science in Progress*, 6th ser., ed. G. A. Baitsell (New Haven: Yale University Press, 1948), 184–248.

34. See Sapp, *Beyond the Gene.*

35. It was a response to the Lysenkoist text *The Situation in Biological Science*, which had appeared two years earlier and contained papers by fifty-nine authors. *The Situation in Biological Science*: *Proceedings of the Lenin Academy of Agricultural Sciences of the USSR* (Moscow: Foreign Languages Publishing House, 1949).

36. By 1939, 23 percent of corn grown in the United States was hybrid corn; that figure had increased to 99 percent by the 1960s.

37. See Sapp, *Beyond the Gene.*

38. Boris Ephrussi, *Nucleo-Cytoplasmic Relations in Microorganisms* (Oxford: Clarendon Press, 1953), v.

39. Ibid., 119.

40. See Ernest Boesiger, "Evolutionary Biology in France at the Time of the Evolutionary Synthesis," in *The Evolutionary Synthesis: Perspectives on the Unification of Biology*, ed. William B. Provine and Ernst Mayr (Cambridge, Mass.: Harvard University Press, 1980), 309–320; Camille Limoges, "A Second Glance at Evolutionary Biology in France," in *The Evolutionary Synthesis: Perspectives on the Unification of Biology*, ed. William B. Provine and Ernst Mayr (Cambridge, Mass.: Harvard University Press, 1980), 322–328; see also Sapp, *Beyond the Gene.*

41. See Henri Bergson, *Creative Evolution* (New York: Dover Books, 1998).

42. Maurice Caullery, *Les conceptions modernes de l'hérédité* (Paris: Flammarion, 1935), 263.

43. Joseph Ben David, *The Scientist's Role in Society: A Comparative Study* (Englewood Cliffs, N.J.: Prentice-Hall, 1971).

44. Junior faculty often received their income solely from students' fees, and they tended to be left in charge of smaller, more specialized classes. Their only recourse to a better income and security to carry out innovations of any kind was to acquire their own chairs and institutes and train their own following of students. This took a long time—an average of sixteen years by 1909—and only about half of those who completed their doctor-

ates ever advanced so far. See M. G. Ash, "Academic Politics in the History of Science: Experimental Psychology in Germany, 1879–41," *Central European History* 13 (1980): 255–286.

45. Jonathan Harwood, *Styles of Scientific Thought: The German Genetics Community, 1900–1933* (Chicago: University of Chicago Press, 1993).

46. Terrance Clark, *Prophets and Patrons: The French University System and the Emergence of the Social Sciences* (Cambridge, Mass.: Harvard University Press, 1973), 18.

47. See Sapp, *Beyond the Gene*, 190.

48. On the emergence of genetics in France, see Sapp, *Beyond the Gene*, 123–162, 181–191; Richard Burian, Jean Gayon, and Doris Zallen, "The Singular Fate of Genetics in France, 1900–1940," *Journal of the History of Biology* 21 (1988): 357–402; Richard Burian and Jean Gayon, "The French School of Genetics: From Physiological and Population Genetics to Regulatory Molecular Genetics," *Annual Review of Genetics* 33 (1999): 313–349; Jean-Louis Fischer and William H. Schneider, eds., *Histoire de la génétique: pratiques, techniques et théories* (Paris: ARPEM, 1990); Jean Paul Gaudillière, "Biologie moléculaire et biologistes dans les années soixante: la naissanace d'une discipline: le cas français" (Ph.D. diss., Université de Paris VII, 1991).

49. T. H. Morgan, *Embryology and Genetics* (New York: Columbia University Press, 1934).

50. Boris Ephrussi, "The Cytoplasm and Somatic Cell Variation," *Journal of Comparative Physiology* 52 (1958): 35–54, at 36.

51. Ephrussi, *Nucleo-Cytoplasmic Relations*, 4.

52. On the development of the idea of the steady states to account for cell heredity, see Sewall Wright, "Genes as Physiological Agents," *American Naturalist* 79 (1945): 289–303, at 198; Max Delbrück, discussion following T. M. Sonneborn and G. H. Beale, "Influence des gènes, des plasmagènes et du milieu dans le déterminisme de caractères antigèniques chez *Paramecium aurelia* (variété 4)," in *Unités biologiques douées de continuité génétique* (Paris: Edition du Centre National de la Recherche Scientifique, 1949), 25–36, at 35; G. H. Beale, *The Genetics of Paramecium aurelia* (New York: Cambridge University Press, 1954); D. L. Nanney, "Epigenetic Control Systems," *Proceedings of the National Academy of Sciences* 44 (1958): 712–717.

53. For the various possible roles of the cytoplasmic genetic system, see T. M. Sonneborn, "Beyond the Gene: Two Years Later," in *Science in Progress*, 7th ser., ed. George H. Baitsell (New Haven: Yale University Press, 1951), 167–202, at 199. In his book *Nucleo-Cytoplasmic Relations in Microorganisms*, Ephrussi summarized the mechanisms cytoplasmic inheritance offered for understanding cellular differentiation: "The non-living environment can induce changes of the concentration of Kappa particles and of antigenic type in Paramecia, and loss of cytoplasmic particles in yeast. . . . Lastly, we find that nucleus and cytoplasm affect each other's activity. The cytoplasmic particles of yeast are activated by a nuclear gene. In turn, in Paramecia, definite cytoplasmic states permit the expression of definite nuclear genes. Here is a set of facts that ought to help explain development." Ephrussi, *Nucleo-Cytoplasmic Relations*, 100.

54. T. M. Sonneborn, "The Role of Genes in Cytoplasmic Inheritance," in *Genetics in the Twentieth Century*, ed. L. C. Dunn (New York: Macmillan, 1951), 291–314, at 311.

55. Ephrussi, *Nucleo-Cytoplasmic Relations*, 101.

56. He argued that "the fundamental problem of genetics in relation to development becomes that of the origin of the specific molecular pattern of the cytoplasm which confers to the egg its vectorial properties." Ephrussi, *Nucleo-Cytoplasmic Relations*, 104.

57. André Lwoff, *Problems of Morphogenesis in Ciliates: The Kinetosomes in Development, Reproduction, and Evolution* (New York: John Wiley and Sons, 1950), 15.

58. Ibid., 2.

59. Ibid., 15.

60. Ibid.

61. Ibid., 7, 29.

62. Ibid., 32.

63. Fauré-Fremiet and Vance Tartar had also argued that differentiation and morphogenesis in ciliates were essentially cortical phenomena. Tartar wrote that "ciliates present us with a living fiber system having morphogenetic capacities." Fauré-Fremiet compared the cortical system, made up of rows of kinetosomes (kineties), to a crystalline network or the complex mesh of a supramolecular structure of the crystalline type, and suggested that the cortex or ciliary system commands morphogenesis. Vance Tartar, "Intracellular Patterns: Facts and Principles Concerning Patterns Exhibited in the Morphogeneis and Regeneration of Ciliate Protozoa," *Growth* 3 (1941): 23–48; E. Fauré-Fremiet, "Les mécanismes de la morphogenèse chez les cilies," *Folia Biotheoretica* 111 (1948): 25–58.

64. Lwoff, *Problems of Morphogenesis*, 28, 62, 78. In Lwoff's view, Sinnot offered a good definition of the morphogenetic field that was adequate because it was very general: "A field is the sum of the reactions which an entire protoplasmic system makes with its external and internal environment, reactions which are determined by the specific physiological activities of the living material of which the organism is composed" (29).

65. The Chicago biologist Paul Weiss, among others, had argued that "a primordial system of spatially organized conditions" was required to set the frame for the later differential settlement of dispersed molecules." Weiss, "The Problem of Specificity in Growth and Development," *Yale Journal of Biology and Medicine* 19 (1947): 235–278. See Lwoff, *Problems of Morphogenesis*, 28.

66. "I should like to say that an orderly or organized asymmetry, like that of an egg or of a ciliate, may only be the reflection of cortical properties. A constantly flowing endoplasm cannot be asymmetrical. The building blocks of different organelles may be asymmetrical; the organelles may be asymmetrical. But if we consider the ciliate as an organism, we reach the conclusion that organized asymmetry or simply organization can belong only to a more or less rigid, or more or less permanent, system, that is to say, to the cortex." Lwoff, *Problems of Morphogenesis*, 86.

67. Ibid., 84. See E. E. Just, *The Biology of the Cell Surface* (Philadelphia: P. Blakison's Son, 1939). On the life of Just, see K. R. Manning, *Black Apollo of Science: The Life of Ernest Everett Just* (New York: Oxford University Press, 1983).

68. Ephrussi, *Nucleo-Cytoplasmic Relations,* 121.

69. See Sapp, "Jean Brachet, *L'hérédité générale*, and the Origins of Molecular Embryology."

Notes to Chapter 16

1. James D. Watson, *The Double Helix* (London: Readers Union, 1969); Gunther Stent, *Molecular Genetics: An Introductory Narrative* (New York: Freeman, 1971); Robert Olby, *The Path to the Double Helix* (London: Macmillan, 1974); Anne Sayer, *Rosalind Franklin and DNA* (New York: Norton, 1975); François Jacob, *The Logic of Life: A History of Heredity*, trans. Betty E. Spillmann (New York: Vintage Books, 1976); H. R. Judson, *The*

Eighth Day of Creation: The Makers of the Revolution in Biology (New York: Touchstone, 1979); Francis Crick, *What Mad Pursuit* (New York: Basic Books, 1988); Lily E. Kay, *Who Wrote the Book of Life? A History of the Genetic Code* (Stanford: Stanford University Press, 2000).

2. F. Miescher, "On the Chemical Composition of Pus Cells," from *Hoppe-Seyler's medizinische–cheische Untersuchungen*, 4 (1871): 441–460; abridged and translated by Mordecai L. Gabriel as "On the Chemical Composition of Pus Cells," in *Great Experiments in Biology*, ed. Mordecai L. Gabriel and Seymour Fogel (Englewood Cliffs, N.J.: Prentice-Hall, 1955), 233–239, at 239.

3. The belief that DNA could only be some sort of structural stiffening was held with tenacity. As one commentator put it, DNA was considered to be "the laundry cardboard in the shirt, the wooden stretcher behind the Rembrandt." See H. F. Judson, *The Eighth Day of Creation: The Makers of the Revolution in Biology* (New York: Touchstone, 1979), 30.

4. F. Griffith, "The Significance of Pneumococcal Types," *Journal of Hygiene* 27 (1928): 113–159.

5. Avery was an expert on the specific carbohydrate molecule that was the characteristic antigen defining the different serotypic variants of the pneumococcus, a necessary basis for the development of serum therapy.

6. O. T. Avery, C. M. MacLeod, and M. McCarty, "Studies on the Chemical Nature of the Substance Inducing Transformation of Pneumococcal Types," *Journal of Experimental Medicine* 79 (1944): 137–158.

7. Oswald Avery to Roy Avery, May 13, 1943, Oswald T. Avery Collection, Profiles in Science, National Library of Medicine, 2.

8. See Joshua Lederberg, "The Transformation of Genetics by DNA: An Anniversary Celebration of Avery, MacLeod, and McCarty, 1944," *Genetics* 136 (1994): 423–426.

9. G. W. Beadle, "Genes and Biological Enigmas," *American Scientist* 36 (1948): 71–74.

10. Salvador Luria and Max Delbrück, "Mutations of Bacteria from Virus Sensitivity to Virus Resistance," *Genetics* 28 (1943): 491–511.

11. Oswald Avery to Roy Avery, May 13, 1943, Oswald T. Avery Collection, Profiles in Science, National Library of Medicine.

12. Joshua Lederberg, "Genetic Recombination in Bacteria: A Discovery Account," *Annual Review of Genetics* 21 (1987): 23–46.

13. Joshua Lederberg, "Cell Genetics and Hereditary Symbiosis," *Physiological Reviews* 32 (1952): 403–430; Lederberg, "Genetic Transduction," *American Scientist* 44 (1956): 264–280.

14. A. D. Hershey and M. Chase, "Independent Functions of Viral protein and Nucleic Acid in Growth of Bacteriophage," *Journal of General Physiology* 36 (1952): 39.

15. Erwin Chargaff, "Chemical Specificity of the Nucleic Acids and Mechanism of Their Enzymatic Degradation," *Experientia* 6 (1950): 201.

16. L. Pauling, H. A. Itano, S. J. Singer, and I. C. Wells, "Sickle-Cell Anemia: A Molecular Disease," *Science* 109 (1949): 443. See also Linus Pauling, "Fifty Years of Progress on Structural Chemistry and Molecular Biology," *Daedelus* 4 (1970): 988–1014.

17. See, e.g., L. Pauling and R. B. Corey, "The Structure of Synthetic Polypeptides," *Proceedings of the National Academy of Sciences* 37 (1951): 3235–3240.

18. There are several biographies of Linus Pauling. See Anthony Serafini, *Linus Pauling: A Man and His Science* (New York: Paragon House, 1989); Thomas Hager, *Force of Nature: The Life of Linus Pauling* (New York: Simon and Schuster, 1995); Hager, *Linus*

Pauling and the Chemistry of Life. (New York: Oxford University Press, 1998); Clifford Mead and Thomas Hager, eds., *Linus Pauling Scientist and Peacemaker* (Corvallis: Oregon State University Press, 2001).

19. L. Pauling and R. B. Corey, "A Proposed Structure for the Nucleic Acids," *Proceedings of the National Academy of Sciences* 39 (1953): 84–97.

20. Watson, *The Double Helix*; Sayers, *Rosalind Franklin and DNA*. See also Anne Piper, "Light on a Dark Lady," *Trends in Biochemical Sciences* 23 (1998): 121–156.

21. Crick, *What Mad Pursuit*, 86.

22. J. D. Watson and F. H. C. Crick, "A Structure for Deoxyribosenucleic Acid," *Nature* 171 (1953): 737–738.

23. Stent, *Molecular Genetics*, 213. See also Kay, *Who Wrote the Book of Life?*

24. Linus Pauling recalled, many years later, that in the 1940s he and his colleagues suspected that there must be some sort of code, probably a linear arrangement of "letters." But at that time the genetic material was usually considered to be protein, rather than nucleic acid, with the "letters" being the different amino acid residues in a polypeptide chain constituting the gene. See Linus Pauling, "Schrödinger's Contributions to Chemistry and Biology," in *Schrödinger: Centenary Celebration of a Polymath,* 225–233, at 230.

25. George Gamow, "Possible Relation between Deoxyribonucleic Acids and Protein Structure, " *Nature* 173 (1954): 318; Gamow, "Information Transfer in the Living Cell," *Scientific American* 193 (1955): 70–78. See also Kay, *Who Wrote the Book of Life?* 131–155.

26. Sydney Brenner, "On the Impossibility of All Overlapping Triplet Codes in Information Transfer from Nucleic Acid to Proteins," *Proceedings of the National Academy of Sciences* 43 (1957): 687–694.

27. See Kay, *Who Wrote the Book of Life?* 156–192.

28. F. H. C. Crick, L.Barnett, S. Brenner, and R. J. Watts-Tobin, "General Nature of the Genetic Code for Proteins," *Nature* 192 (1961): 1227–32.

29. See Kay, *Who Wrote the Book of Life?* 235–293.

30. See, e.g., Richard Dawkins, *River out of Eden* (London: Weidenfeld and Nicolson, 1995). On the relationships between concepts of molecular biology and computer technology, see Evelyn Fox Keller, *Refiguring Life* (New York: Columbia University Press, 1995); Kay, *Who Wrote the Book of Life?*

31. See Stent, *Molecular Genetics*, 218–220.

32. See S. Brenner, F. Jacob, and M. Meselson, "An Unstable Intermediate Carrying Information from Genes to Ribosomes for Protein Synthesis," *Nature* 190 (1961): 576.

33. Stent, *An Introduction to Molecular Genetics*, 469–470.

34. See Bernardino Fantini, "Jacques Monod et la biologie moléculaire," *La recherche* 218 (1990): 180–187; Jean-Paul Gaudillière, "J. Monod, S. Spiegelman, et l'adaptation enzymatique: programme de recherche, cultures locales, et traditions disciplinaires," *History and Philosophy of the Life Sciences* 14 (1992): 23–71; Benno Müller-Hill, *The lac Operon: A Short History of a Genetic Paradigm* (Berlin: Walter de Gruyter, 1996).

35. Jacques Monod and François Jacob, "General Conclusions: Teleonomic Mechanisms in Cellular Metabolism, Growth, and Differentiation," *Cold Spring Harbor Symposia on Quantitative Biology* 26 (1961): 389–401.

36. Ibid., 397.

37. The operon was demonstrated in bacteria; it had not been shown to occur in other cells. Extrapolation to other organisms was quickly made, even though genes in plants and animals could have completely different mechanisms of genetic regulation. Button,

Jacob, and Monod warned their colleagues to be cautious and "refrain from assuming that operons occur in higher forms until the necessary genetic and biochemical tests have been performed." G. Button, F. Jacob, and J. Monod, "The Operon: A Unit of Coordinated Gene Action," in *Heritage from Mendel*, ed. R. A. Brink (Madison: University of Wisconsin Press, 1967), 155–178.

38. F. Jacob and J. Monod, "Elements of Regulatory Circuits in Bacteria," in *Biological Organization at the Cellular and Subcellular Level*, ed. R. J. C. Harris (London: Academic Press, 1963), 1–24.

39. Ibid., 1.

40. See Jacob, *The Logic of Life*, 275.

41. Stent, *Molecular Genetics*, 128.

42. These doctrines are also expressed well in Jacob, *The Logic of Life*, 273, 277.

43. Francis Crick, "On Protein Synthesis: The Biological Replication of Macromolecules," in *Symposia of the Society for Experimental Biology* (Cambridge: Cambridge University Press, 1958), 138–163, at 153.

44. Jacob, *The Logic of Life*, 313.

45. Ibid., 254, 280–281.

Notes to Chapter 17

1. See Elizabeth Pennisi, "Finally, the Book of Life and Instructions for Navigating It," *Science* 288 (2000): 2304–2307.

2. See Alex Mauron, "Is the Genome the Secular Equivalent of the Soul?" *Science* 291 (2001): 831–832.

3. See Carl F. Canor, ed., *Are Genes Us? The Social Consequences of the New Genetics* (Piscataway, N.J.: Rutgers University Press, 1994); Daniel Kevles and Leroy Hood, eds., *The Code of Codes: Scientific and Social Issues in the Human Genome Project* (Cambridge, Mass.: Harvard University Press, 1992); Tom Wilkie, *Perilous Knowledge: The Human Genome Project and Its Implications* (Berkeley: University of California Press, 1993); Timothy F. Murphy and Marc A. Lappé, eds., *Justice and the Human Genome Project* (Berkeley: University of California Press, 1994); Robert Cook-Deegan, *The Gene Wars: Science, Politics, and the Human Genome* (New York: W. W. Norton, 1994); Lois Wingerson, *Unnatural Selection: The Promise and the Power of Human Gene Research* (New York: Bantam Books, 1998); Richard Lewontin, *It Ain't Necessarily So: The Dream of the Human Genome and Other Illusions* (New York: New York Review of Books, 2000); Phillip R. Sloan, ed., *Controlling Our Destinies: Historical, Philosophical, Ethical, and Theological Perspectives on the Human Genome Project* (Notre Dame: University of Notre Dame Press. 2000).

4. See Dean Hamer and Peter Copeland, *Living with Our Genes* (New York: Anchor Books, 1998); Richard Lewontin, *The Triple Helix: Gene Organism and Environment* (Cambridge, Mass.: Harvard University Press, 2000).

5. See, e.g., the excellent study by Dorothy Nelkin and Susan Lindee, *The DNA Mystique: The Gene as a Cultural Icon* (New York: W. H. Freeman, 1995).

6. See Barry Commoner, "Unraveling the DNA Myth," *Harper's*, Feb. 2002, 39–49. See also Jan Sapp, "Cytoplasmic Heretics," *Perspectives in Biology and Medicine* 41 (1998): 225–290; Sapp, "Concepts of Organization: The Leverage of Ciliate Protozoa," in *Developmental Biology: A Comprehensive Synthesis*, ed. S. Gilbert (New York: Plenum Press, 1991), 229–258.

7. See David Adam, "Draft Human Genome Sequence Published," *Nature News*, Feb. 12, 2001; Henry Gee, "A Journey into the Genome: What's There," *Nature News*, Feb. 12, 2001; International Human Genome Sequencing Consortium, "Initial Sequencing and Analysis of the Human genome," *Nature* 409 (2001): 860–920; J. Craig Venter et al., "The Sequence of the Human Genome," *Science* 291 (2001): 1304–1345.

8. Ju Yu et al., "A Draft Sequence of the Rice Genome (*Oryza sativa L.* Ssp. *Indica*)," *Science* 296 (2002): 79–100.

9. See Lynda Hurst, "Genome Just a Start for Maverick Scientist," *Toronto Star*, Apr. 24, 2002.

10. See Sharp, "Split Genes and RNA Splicing."

11. S. Ohno, "So Much Junk DNA in Our Genome," *Brookhaven Symposia in Biology* 23 (1972): 366 ff. See W. F. Loomis and M. E. Gilpin, "Multigene Families and Vestigial Sequences," *Proceedings of the National Academy of Sciences* 83 (1986): 2143; R. Lewin, "'Computer Genome' Is Full of Junk DNA," *Science* 232 (1986): 577–578.

12. On the idea that junk DNA may not be junk, see, e.g., "Hints of a Language in Junk DNA," *Science* 226 (1994): 1320; M. J. Beaton and T. Cavalier-Smith , "Eukaryotic Non-Coding DNA Is Functional: Evidence from the Differential Scaling of Cryptomonal Genomes," *Proceedings of the Royal Society of London*, series B, 266 (1999): 2053–2059.

13. For a discussion of these issues, see Phillip A. Sharp, "Split Genes and RNA Splicing" (Nobel Lecture), *Cell* 77 (1994): 805–815. For an early general review and discussion of the significance of alternative splicing, see B. Lewin, "Alternatives for Splicing: Recognizing the Ends of Introns," *Cell* 22 (1980): 324–326.

14. Athena Andreadis, M. E. Gallego, and B. Nadal-Ginard, "Generation of Protein Isoform Diversity by Alternative Splicing: Mechanistic and Biological Implications," *Annual Review of Cell Biology* 3 (1987): 207–242. See also Bruce Alberts, Dennis Bray, Julian Lewis, Martin Raft, Keith Roberts, James P. Watson, *Molecular Biology of the Cell*, 2d ed. (New York: Garland, 1989), 536, 588.

15. Sharp, "Split Genes and RNA Splicing."

16. See Athena Andreadis, M. E. Gallego, and B. Nadal-Ginard, "Generation of Protein Isoform Diversity by Alternative Splicing: Mechanistic and Biological Implications," *Annual Review of Cell Biology* 3 (1987): 207–242. See also Alberts et al., *Molecular Biology of the Cell*, 2d ed., 536, 588.

17. See Venter et al., "The Sequence of the Human Genome," 1346; International Human Genome Sequencing Consortium, "Initial Sequencing and Analysis of the Human Genome," 898, 901.

18. See Michael E. Zwick, David J. Cutler, and Aravinda Chakravarti, "Patterns of Genetic Variation in Mendelian and Complex Traits," *Annual Review of Genomics and Human Genetics*. 1 (2000): 387–407.

19. See Tom Bethell, "A Map to Nowhere," *American Spectator*, Apr. 2001, 51–56.

20. See Sharp, "Split Genes and RNA Splicing," 807–808. On the history of the gene concept, see P. Portin, "The Concept of the Gene: Short History and Present Status," *Quarterly Review of Biology* 68 (1993): 173–223; Portin, "Historical Development of the Concept of the Gene," *Journal of Medicine and Philosophy* 27 (2002): 257–286; Peter Beurton, Raphael Falk, and Hans-Jörg Rheinberger, eds., *The Concept of the Gene in Development and Evolution: Historical and Epistemological Perspectives* (Cambridge: Cambridge University Press, 2000); Evelyn Fox Keller, *The Century of the Gene* (Cambridge, Mass.: Harvard University Press, 2000).

21. Raphael Falk, "The Gene: A Concept in Tension," in *The Concept of the Gene in Development and Evolution: Historical and Epistemological Perspectives*, ed. Peter Beurton, Raphael Falk, and Hans-Jörg Rheinberger (Cambridge: Cambridge University Press, 2000), 317–348, at 340.

22. F. Jacob, *The Logic of Living Systems: A History of Heredity*, trans. Betty Spillmann (London: Allen Lane, 1974).

23. M. Csikszentmihalyi, *The Evolving Self* (New York: Harper Collins, 1993), 1.

24. Matt Ridly, *Genome* (New York: Harper Collins, 2000), 1.

25. Venter et al., "The Sequence of the Human Genome," 1348.

26. For a philosophical discussion on the meaning of genetic reductionism see, e.g., Sahotra Sarkar, *Genetics and Reductionism* (Cambridge: Cambridge University Press, 1998).

27. F. Harold, "From Morphogenes to Morphogenesis," *Microbiology* 141: 2765–2778, 1995. See also Harold, *The Way of the Cell* (New York: Oxford University Press, 2001).

28. M. A. Corner, "Reciprocity of Structure-Function Relations in Developing Neural Networks: The Odyssey of a Self-Organizing Brain through Research Fads, Fallacies, and Prospects," in *The Self-Organizing Brain: From Growth Cones to Functional Networks*, Progress in Brain Research 102, ed. J. van Pelt, M. A. Corner, H. B. M. Uylings, and F. H. Lopes da Silva (New York: Elsevier, 1994), 3–31.

29. See Franklin Harold, "From Morphogenes to Morphogenesis," *Microbiology* 141 (1995): 2765–2778. See also Harold, *The Way of the Cell* (New York: Oxford University Press, 2001).

30. This is displayed on the large poster in the lobby of the biology department in which I work.

31. See the excellent discussion by Richard Lewontin, "The Dream of the Human Genome," *New York Review of Books*, May 28, 1992.

32. Francis Crick, "On Protein Synthesis: The Biological Replication of Macromolecules," in *Symposia of the Society for Experimental Biology* (Cambridge: Cambridge University Press, 1958), 138–163, at 153.

33. See S. G. Burston and A. R. Clarke, "Molecular Chaperones: Physical and Mechanistic Properties," *Essays in Biochemistry* 29 (1995): 125–136. On the possible importance of chapreones for promoting evolutionary change, see Suzanne L. Rutherford and Susan Lindquist, "Hsp 90 as Capacitor for Morphological Evolution," *Nature* 396 (1998): 336–342.

34. S. V. Paushkin, V. V. Kushnirov, V. N. Smirnov, and M. D. Ter-Avanesyan, "In Vitro Propagation of the Prion-Like State of Yeast Sup35 Protein," *Science* 277 (1997): 381–383; R. B. Wickner, H. K. Edskes, M. L. Maddelein, K. L. Taylor, H. Moriyama, "Prions of Yeast and Fungi: Proteins as Genetic Material," *Journal of Biological Chemistry* 274 (1999): 555–558.

35. See Stanley Prusiner, "The Prion Diseases," *Scientific American*, January 1995, 48–57.

36. As Joshua Lederberg has commented, one implication is that other natural forces could cause the prion to be converted from normal to infectious, and from benign to disease-causing. It is known for example, that proteins adopt new conformations under physical and chemical stress, denaturation, and renaturation. Maltreatment of normal prion protein might then generate infectious protein *de novo*—what we would recognize as "spontaneous" Creutzfeldt-Jakob disease. The rendering of sheep offal used for cattle feed might have made it more infectious in triggering mad cow disease in Britain. See Joshua Lederberg, "Infectious Agents, Hosts, in Constant Flux," *ASM News* 65, no. 1 (1999): 18–

22, part 1; Lederberg, "Parasites Face a Perpetual Dilemma," *ASM News* 65, no. 2 (1999): 77–80, part 2. See also Lederberg, "Some Implications of the Prion Paradigm: Caveat Denaturor," *Journal of the American Medical Association* 282 (1999): 1332–1333; Wickner et al., "Prions of Yeast and Fungi."

37. The endoplasmic reticulum, the major site of the synthesis of membrane lipids and proteins, seems to be derived from preexisting structures, and there is no apparent mechanism to assemble these structures *de novo*. David T. Shima and Graham Warren, "Inheritance of the Cytoplasm During Cell Division," in *Dynamics of Cell Division*, ed. S. A. Endow and D. M. Glover (New York: Oxford University Press, 1998), 248–269.

38. Jan Sapp, "Freewheeling Centrioles," *History and Philosophy of the Life Sciences* 20 (1998): 255–290.

39. Lewis Wolpert, Rosa Beddington, Jeremy Brockes, Thomas Jessell, Peter Lawrence, Elliot Meyerowitz, *Principles of Development* (London: Oxford University Press, 1998), 15.

40. G. Albrecht-Buehler, "In Defense of 'Nonmolecular' Cell Biology," *International Review of Cytology* 20 (1990): 191–241. See also Franklin Harold, "To Shape a Cell: An Inquiry into the Causes of Morphogenesis of Microorganisms," *Microbiological Reviews* 54 (1990): 381–431.

41. S. Luria, discussion following P. Lengyel, "Problems in Protein Synthesis," *Journal of General Physiology* 49 (1966): 305–330.

42. See R. Paton, "The Organisation of Hereditary Information," *Biosystems* 40 (1997): 245–255. See also E. Jablonka and M. Lamb, *Epigenetic Inheritance and Evolution: The Lamarckian Dimension* (Oxford: Oxford University Press, 1994).

43. A. D. Hershey, "Genes and Hereditary Characteristics," *Nature* 226 (1970): 697–700, at 700.

44. I. B. Heath, *Tip Growth in Plant and Fungal Cells* (San Diego: Academic Press, 1990); Heath, "The Cytoskeleton in Hyphal Growth, Organelle Movements, and Mitosis," in *The Mycota*, vol .1, ed. J. G. H. Wessels and F. Meinhardt (Berlin: Springer-Verlag, 1994), 43–65.

45. The inheritance of polarity seems to be specified by preestablished cell structure in some magnetotropic bacteria as well; see R. W. Hedges, "Inheritance of Magnetosome Polarity in Magnetotropic Bacteria," *Journal of Theoretical Biology* 112 (1985): 607–608. One other nonconventional hereditary mechanism based on preestablished structures should be mentioned. It involves the perpetuation of the so-called teeth of the silica-based shell of the protist *Difflugia corona*. See H. S. Jennings, "Formation, Inheritance, and Variation of the Teeth in *Difflugia corona*: A Study of the Morphogenetic Activities of Rhizopod Protoplasm," *Journal of Experimental Zoology* 77 (1937): 287–336.

46. A. Lwoff, *Problems of Morphogenesis in Ciliates: The Kinetosomes in Development, Reproduction, and Evolution* (New York: John Wiley and Sons, 1950).

47. V. Tartar, *The Biology of Stentor* (Oxford: Paramon Press, 1961).

48. T. M. Sonneborn, "Does Preformed Cell Structure Play an Essential Role in Cell Heredity?" in *The Nature of Biological Diversity*, ed. J. M. Allen (New York: McGraw-Hill, 1963), 165–221.

49. J. Beisson and T. M. Sonneborn, "Cytoplasmic Inheritance of the Organization of the Cell Cortex in *Paramecium aurelia*," *Proceedings of the National Academy of Sciences* 53 (1965): 275–282, at 282.

50. D. L. Nanney, *Experimental Ciliatology* (New York: Wiley 1980); J. Frankel, *Pattern Formation: Ciliate Studies and Models* (New York: Oxford University Press, 1989).

51. See Sonneborn, "Does Preformed Cell Structure Play an Essential Role in Cell Heredity?" 213.

52. D. L. Nanney, "Cortical Patterns in Cellular Morphogenesis," *Science* 160 (1968): 496–502.

53. G. W. Grimes, "Pattern Determination in Hypotrich Ciliates," *American Zoologist* 22 (1982): 35–46.

54. See Frankel, *Pattern Formation.*

55. J. Frankel, "The Patterning of Ciliates," *Journal of Protozoology* 38 (1992): 519–525.

56. Scott Gilbert, *Developmental Biology* (Sunderland, Mass.: Sinauer, 1994), 691–692.

57. See R. A. Raff and T. C. Kaufman, *Embryos, Genes, and Evolution* (New York: Macmillan, 1983); Brian Hall, *Evolutionary Developmental Biology* (London: Chapman and Hall, 1992).

58. See Brian K. Hall, "Evo-Devo or Devo-Evo—Or Does It Matter?" *Evolution and Development* 2 (2000): 177–178; B. C. Goodwin, "Development and Evolution," *Journal of Theoretical Biology* 97 (1982): 43–55; S. F. Gilbert, J. M. Ortiz, and R. A. Raff, "Resynthesizing Evolutionary and Developmental Biology," *Developmental Biology* 173 (1996): 357–372.

59. See, e.g., Gary Webster and Brian Goodwin, "The Origin of Species: A Structuralist Approach," *Journal of Social and Biological Structures* 5 (1982): 15–47. See also Anton Markoš, *Readers of the Book of Life: Contextualizing Developmental Biology* (New York: Oxford University Press, 2002), 94–106.

60. Webster and Goodwin, "The Origin of Species," 38.

61. See Lewis Wolpert, Rosa Beddington, Jeremy Brockes, Thomas Jessell, Peter Lawrence, Elliot Meyerowitz, *Principles of Development*, 2d ed. (New York: Oxford University Press, 2002).

62. See S. F. Gilbert, J. M. Oritz, and R. A. Raft, "Resynthesizing Evolutionary and Developmental Biology," *Developmental Biology* 173 (1996): 357–372, at 366–368.

63. D'Arcy Thompson, *On Growth and Form* (Cambridge: Cambridge University Press, 1917).

64. See Gary Webster, "The Relations of Natural Forms," in *Beyond Neo-Darwinism: An Introduction to the New Evolutionary Paradigm*, ed. Mae-Wan Ho and Peter Saunders (London: Academic Press, 1984), 193–217; G. Webster, "The Nature and Scope of Structuralist Analysis in Biology," *Biology Forum* 80 (1987): 173–177.

65. Richard Dawkins, *The Extended Phenotype* (Oxford: Freeman, 1982), 42. Critics of Kauffman and of other complexity theorists have not been kind. As E. O.Wilson remarked, "Their conclusions thus far are too vague and general to be more than rallying metaphors, and their abstract conclusions tell us very little that is really new." E. O. Wilson, *Consilience: The Unity of Knowledge* (New York: Knopf, 1998), 87–89.

66. Stuart Kauffman, *The Origins of Order: Self-Organization and Selection in Evolution* (New York: Oxford University Press, 1993), 5.

67. Brian Goodwin, "A Relational or Field Theory of Reproduction and Its Evolutionary Implications," in *Beyond Neo-Darwinism: An Introduction to the New Evolutionary Paradigm*, ed. Mae-Wan Ho and Peter Saunders (London: Academic Press, 1984), 219–241.

68. Ibid., 226.

69. Joseph Frankel, "Global Patterning in Single Cells," *Journal of Theoretical Biology* 99 (1982): 119–134.

70. D. L. Nanney, review of B. C. Goodwin, N. Horder, and C. C. Wylie, eds., *Development and Evolution, Journal of Protozoology* 31 (1984): 365.

71. A. Lwoff, "L'organisation du cortex chez les cilés: un exemple d'hérédité de caractère acquis," *Comptes rendus de l'Académie des Sciences Paris*, 310 (1990): 109–111; O. E. Landman, "The Inheritance of Acquired Characteristics," *Annual Review of Genetics* 25 (1991): 1–20.

72. See Frankel, *Pattern Formation*, 17; G. W. Grimes and K. J. Aufderheide, *Cellular Aspects of Pattern Formation: The Problem of Assembly,* Monographs in Developmental Biology 22 (Basel: Karger, 1991), 67.

73. J. Maynard Smith, "Evolution and Development," in *Development and Evolution*, ed. B. C. Goodwin, N. Holder, and C. C. Wylie (Cambridge: Cambridge University Press, 1983, 33–46).

74. See also the discussion in Boris Ephrussi, *Hybridization of Somatic Cells* (Princeton, N.J.: Princeton University Press, 1972).

75. See Denis Bray, *Cell Movements: From Molecules to Motility*, 2d ed. (New York: Garland, 2001), 56–58. See also K. K. Hjelm, "Is Non-Genic Inheritance Involved in Carcinogenesis? A Cytotactic Model of Transformation," *Journal of Theoretical Biology* 119 (1986): 89–101.

76. The terms "epinucleic" and "extranucleic" inheritance were suggested by Joshua Lederberg. See J. Lederberg, "Genetic Approaches to Somatic Cell Variation," *Journal of Cellular and Comparative Phsyiology* 52 (1958): 383–392.

77. See Jablonka and Lamb, *Epigenetic Inheritance and Evolution.*

78. A. Levine, G. L. Cantoni, and A. Razin, "Inhibition of Promoter Activity by Methylation: a Possible Involvement of a Protein Mediator," *Proceedings of the National Academy of Sciences* 88 (1991): 6515–6518; A. Levine, G. L. Cantoni, and A. Razin, "Methylation in the Preinitiation Domain Affects Gene Transcription by an Indirect Mechanism," *Proceedings of the National Academy of Sciences* 89 (1992): 10119–10123.

79. See Jablonka and Lamb, *Epigenetic Inheritance and Evolution*, 133–159.

80. See S. W. Brown and H. S. Chandra, "Chromosome Imprinting and the Differential Regulation of Homologous Chromosomes," in *Cell Biology: A Comprehensive Treatise*, ed. L. Goldstein and D. M. Prescott, vol. 1 (New York: Academic Press, 1977).

81. M. Monk, "Genetic Imprinting," *Genes and Development* 2 (1988): 921–925; J. G. Hall, "Genomic Imprinting: Review and Relevance to Human Diseases," *American Journal of Human Genetics* 46 (1990): 857–873; W. Reik, "Genomic Imprinting," in *Transgenic Animals*, ed. F. Grosveld and G. Kollias (London: Academic Press, 1992), 99–126.

82. J. McGrath and D. Solter, "Completion of Mouse Embryogenesis Requires Both the Maternal and Paternal Genomes," *Cell* 37 (1984): 179–183; M. A. H. Surani, S. C. Barton, and M. L. Norris, "Development of Reconstituted Mouse Eggs Suggests Imprinting of the Genome During Gametogenesis," *Nature* 308 (1984): 548–550.

83. It has been suggested that imprinting occurs during gametogenesis due to the different ways in which chromosomes are packaged in eggs and sperm. See the discussion in Jablonka and Lamb, *Epigenetic Inheritance and Evolution*, 111–130.

84. See John Maynard Smith and Eörs Szathmáry, *The Major Transitions in Evolution* (New York: W. H. Freeman, 1995), 247–250.

85. Ibid.; see also John Maynard Smith, "Models of a Dual Inheritance System," *Journal of Theoretical Biology* 143 (1990): 41–53.

86. The term "epigenetic" was coined by Waddington to refer to the study of the processes by which genotype gives rise to phenotype. C. H. Waddington, "The Epigenotype," *Endeavour* 1 (1942): 18–20.

Notes to Chapter 18

1. See J. William Schopf, *Cradle of Life: The Discovery of Earth's Earliest Fossils* (Princeton: Princeton University Press, 1999), 10–11.

2. In *The Origin* Darwin considered the lack of Precambrian fossils to be a grave difficulty for his theory of evolution:

> Consequently, if my theory be true, it is indisputable that before the lowest Silurian stratum was deposited, long periods elapsed . . . and that during these vast, yet quite unknown, periods of time, the world swarmed with living creatures. . . . To the question why we do not find records of these vast primordial periods, I can give no satisfactory answer. . . . The case at present must remain inexplicable; and may be truly urged as a valid argument against the views here entertained. (Charles Darwin, *On the Origin of Species*, with an introduction by Ernst Mayr, facsimile ed. of 1859 [Cambridge, Mass.: Harvard University Press, 1964], 307).

This problem remained for a century. Nonetheless, there were some bold pronouncements, and bitter controversy erupted over Precambrian fossil evidence; see C. F. O'Brien, "Eozoön Canadense, 'The Dawn Animal of Canada,'" *Isis* 61 (1970): 206–223. Walcottt is acknowledged today to have discovered fossilized bacteria and algae from the Precambrian of Montana, but his Precambrian fossils were regarded with skepticism by his contemporaries. Schopf, *Cradle of Life*, 23–34.

3. See Stephen Jay Gould, *Wonderful Life: The Burgess Shale and the Nature of History* (London: Hutchison Radius, 1989), 25.

4. Ibid. Precambrian paleobiology got a big boost in the 1950s from such geologists and paleontologists as Stanley A. Tyler (1906–1963) at the University of Wisconsin, the paleobotanist Elso S. Barghoorn (1915–1984) of Harvard, Vasil'evich Timofeev (1916–1982) and colleagues at the Institute of Precambrian Geochrononology in Leningrad, and Martin Glaessner (1906–1989) at the University of Adelaide in Australia. See Schopf, *Cradle of Life*, 35–70.

5. Ibid.

6. See Lynn Margulis, *Symbiosis in Cell Evolution: Microbial Communities in the Archean and Proterozoic Eons*, 2d ed. (New York: W. H. Freeman, 1993), 118–130. In 1993, J. William Schopf at the University of California reported startling evidence for the oldest fossil-bearing rocks, 3.465 billion years old, from northwestern Australia. J. W. Schopf, "Microfossils of the Early Archean Apex Chert: New Evidence for the Antiquity of Life," *Science* 260 (1993): 640–646. But this evidence has been seriously questioned recently. See Richard Kerr, "Reversals Reveal Pitfalls in Spotting Ancient and E.T. Life," *Science* 296 (2002): 1384–1385.

7. H. D. Holland, "Evidence for Life on Earth More than 3850 Million Years Ago," *Science* 275 (1997): 38–39.

8. In his classic text *Tempo and Mode in Evolution*, the American paleontologist George Gaylord Simpson laid out some of the basic rules of evolution (speciation, specialization, and extinction), and he categorized three different rates of evolution. Fast-evolving species

went extinct after about 1 million years, species evolving at a moderate pace lived for about 10 million years, and slowly evolving species lived for 100 million years. Such slow evolvers are alive today: horseshoe crabs, coelacanths, crocodiles, and opossums. See George Gaylord Simpson, *Tempo and Mode in Evolution* (New York: Columbia University Press, 1944). Schopf suspected that cyanobacteria had changed very little over a huge span of billions of years. See Schopf, *The Cradle of Life*, 210–211.

9. See George Wald, "The Origin of Life," *Scientific American* 192 (1954): 44–53; Iris Fry, *The Emergence of Life on Earth: A Historical and Scientific Overview* (New Brunswick, N.J.: Rutgers University Press, 2000); Robert Shapiro, *A Skeptic's Guide to the Creation of Life on Earth* (New York: Penguin Books, 1986).

10. R. Stanier, M. Douderoff, and E. A. Adelberg, *The Microbial World*, 2d ed. (Englewood Cliffs, N.J.: Prentice-Hall, 1963).

11. Emile Zuckerkandl and Linus Pauling, "Molecules as Documents of Evolutionary History," *Journal of Theoretical Biology* 8 (1965): 357–366. See also Gregory L. Morgan, "Emile Zuckerkandl, Linus Pauling, and the Molecular Evolutionary Clock, 1959–1965," *Journal of the History of Biology* 31 (1998): 155–178; Michael Dietrich, "The Origins of the Neutral Theory of Molecular Evolution," *Journal of the History of Biology* 27 (1994): 21–59; Dietrich, "Paradox and Persuasion: Negotiating the Place of Molecular Evolution within Evolutionary Biology," *Journal of the History of Biology* 31 (1998): 85–111.

12. Zuckerkandl and Pauling, "Molecules as Documents of Evolutionary History."

13. Emile Zuckerkandl and Linus Pauling, "Evolutionary Divergence and Convergence in Proteins," in *Evolving Genes and Proteins*, ed. V. Bryson and H. Vogel (New York: Academic Press, 1965), 97–166, at 148.

14. See, e.g., E. Margoliash, "Primary Structure and Evolution in Cytochrome c," *Proceedings of the National Academy of Sciences* 50 (1963): 672–679.

15. A. M. Maxam and W. Gilbert, "A New Method for Sequencing DNA," *Proceedings of the National Academy of Sciences* 74 (1977): 1258; K. B. Mullis, F. Faloona, S. Scharf, R. K. Saika, and G. T. Horn, "Specific Enzymatic Amplification of DNA in Vitro: The Polymerase Chain Reaction," *Cold Spring Harbor Symposia on Qunatitative Biology* 51 (1987): 263–273.

16. H. F. Judson "A History of the Science and Technology behind Gene Mapping and Sequencing," in *The Code of Codes: Scientific and Social Issues in the Human Genome Project* (Cambridge, Mass.: Harvard University Press, 1992), 37–80; Paul Rabinow, *Making PCR: A Story of Biotechnology* (Chicago: University of Chicago Press).

17. C. R. Woese, *The Genetic Code: The Molecular Basis of Genetic Expression* (New York: Harper and Row, 1967). See also Lily E. Kay, *Who Wrote the Book of Life? A History of the Genetic Code* (Stanford: Stanford University Press, 2000).

18. Oparin first proposed his theory in 1918, less than a year after the Bolsheviks seized power. The czarist censors had not yet been cast aside. Prior to the Russian Revolution, the czar had ruled over both the state and the Orthodox Church. To Russian Orthodoxy, the origin of life was sacrosanct. Oparin's manuscript was rejected for publication. In 1924, his ideas reemerged in a more developed form in a small book, translated into English in 1938. See A. I. Oparin, *The Origin of Life*, trans. Sergius Morgulis, 2d ed. (New York: Dover, 1965).

19. J. B. S. Haldane, "The Origin of Life," *Rationalist Annual* (1929): 148–169.

20. S. L. Miller, "A Production of Amino Acids under Possible Primitive Earth Conditions," *Science* 117 (1953): 528–529; Miller, "Production of Some Organic Compounds

under Possible Primitive Earth Conditions," *Journal of the American Chemical Society* 17 (1955): 2351–2361.

21. In the 1980s, other leading researchers reasoned that the early atmosphere of the earth may not have been a chemically reducing one but may have been neutral or even mildly oxidizing. Under such conditions, no significant amount of organic compounds would be produced. See Joel Levine, Tommy Augustsson, and Murali Natarajan, "The Prebiological Paleoatmopshere: Stability and Composition," *Origins of Life* 12 (1982): 245–259.

22. Freeman Dyson, *Origins of Life*, 2d ed. (Cambridge: Cambridge University Press, 1999), 7. See also Jeremy Pickett-Heaps, "The Cytoplast Concept in Dividing Plant Cells: Cytoplasmic Domains and the Evolution of Spatially Organized Cell Division," *American Journal of Botany* 86 (1999): 153–172. See also the experiments of Sonneborn et al. on the ciliate cortex (chapter 17).

23. See A. G. Cairns-Smith, *Genetic Takeover and the Mineral Origins of Life* (Cambridge: Cambridge University Press, 1982). Also note the experiments showing that when altered, structures made of chitin and silica are inherited in the new form. See H. S. Jennings, "Formation, Inheritance, and Variation of the Teeth in *Difflugia corona*: A Study of the Morphogenic Activities of Rhizopod Protoplasm," *Journal of Experimental Zoology* 77 (1937): 287–336. The simplest cells today are mycoplasmas, small bacteria-like organisms less than a billionth the size of a single-celled protist. Almost all mycoplasmas are parasites, scaled-down descendants of free-living larger microbes, and they grow and reproduce only inside other cells.

24. See Woese, *The Genetic Code*, 179–195.

25. Ibid., 194.

26. Ibid., 176.

27. C. R. Woese, "Molecular Mechanics of Translation: A Reciprocating Ratchet Mechanism," *Nature* 226 (1970): 817–820; G. E. Fox and C. R. Woese, "5S RNA Secondary Structure," *Nature* 256 (1975): 505–507.

28. Virginia Morell, "Microbiology's Scarred Revolutionary," *Science* 276 (1997): 699–702, at 700.

29. In 1966, of course, none of today's efficient methods for sequencing genetic material existed. Instead, Woese relied on a tedious, labor-intensive technique known as oligonucleotide cataloging. In this method, an rRNA molecule, which is a long string of four nucleotides (adenine, cytosine, uracil, and guanine, or ACUG) was broken into small fragments by cutting it at every G residue. Each of these fragments or oligonucleotides was then broken into subfragments with enzymes that sliced at different residues. This allowed Woese to reconstruct the sequence of the original rRNA fragment.

30. C. Woese, G. E. Fox, L. Zablen, T. Uchida, L. Bonen, K. Pechman, B. J. Lewis, and D. Stahl, "Conservation of Primary Structure in 16s Ribosomal RNA," *Nature* 254 (1975): 83–86; C. R. Woese and G. E. Fox, "The Concept of Cellular Evolution," *Journal of Molecular Evolution* 10 (1977): 1–6; C. R. Woese, "Bacterial Evolution," *Microbial Reviews* 51 (1987): 221–271.

31. G. E. Fox et al., The Phylogeny of Prokaryotes," *Science* 209 (1980): 457–463.

32. See M. L. Sogin, "Evolution of Eukaryotic Microorganisms and Their Small Subunit Ribosomal RNAs," *American Zoologist* 29 (1989): 487–499.

33. Carl R. Woese and George E. Fox, "Phylogenetic Structure of the Prokaryotic Domain: The Primary Kingdoms," *Proceedings of the National Academy of Sciences* 74 (1977): 5088–5090, at 5089.

34. C. R. Woese, O. Kandler, and M. L. Wheelis, "Towards a Natural System of Organisms: Proposal for the Domains Archae, Bacteria, and Eucarya," *Proceedings of the National Academy of Sciences* 87 (1990): 4576–4579.

35. Woese and Fox, "Phylogenetic Structure," 5089.

36. See John L. Howland, *The Surprising Archaea: Discovering Another Domain of Life* (New York: Oxford University Press, 2000).

37. Charles Darwin to Joseph Hooker, 1871, in *Life and Letters of Charles Darwin*, ed. F. Darwin, vol. 3 (London: John Murray, 1887), 18.

38. Morell, "Microbiology's Scarred Revolutionary."

39. Edouard Chatton, "*Pansporella Perplexa*. Reflexions sur la biologie et la phylogénic des protozoaires," *Annales des sciences naturelles* 10e series vii (1925): 1–84, 26.

40. R. Stanier, M. Douderoff, and E. Adleberg, *The Microbial World*, 2d ed. (Engelwood Cliffs, N.J.: Prentice-Hall, 1963).

41. R. H. Whittaker, "New Concepts of Kingdoms of Organisms," *Science* 163 (1969): 150–160; R. H. Whittaker and L. Margulis, "Protist Classification and the Kingdoms of Organisms," *Biosystems* 10 (1978): 3–18; L. Margulis and K. V. Schwartz, *Five Kingdoms Illustrated Guide to the Phyla of Life on Earth* (San Francisco: W. H. Freeman, 1988).

42. G. Olsen and C. Woese, "Lessons from an Archael Genome: What Are We Learning from *Methanococcus jannaschii*?" *Trends in Genetics* 12 (1996): 377–379.

43. Norman Pace, "A Molecular View of Microbial Diversity and the Biosphere," *Science* 276 (1997): 734–740, at 734.

44. William B. Whitman, David C. Coleman, and William J. Wiebe, "Prokaryotes: The Unseen Majority," *Proceedings of the National Academy of Sciences* 95 (1998): 6578–6583.

45. Carl Woese, "A Manifesto for Microbial Genomics," *Current Biology* 8, no. 22 (1999): R781–R783.

46. Pace, "A Molecular View of Microbial Diversity and the Biosphere."

47. N. R. Pace, D. A. Stahl, D. J. Lane, and G. J. Olsen, "Analyzing Natural Microbial Populations by rRNA Sequences," *ASM News* 51 (1985): 4–12.

48. Pace, "A Molecular View of Microbial Diversity and the Biosphere."

49. See Carl Woese, "The Archaea: Their History and Significance," in *The Biochemistry of Archae (Archaebacteria)*, ed. M. Kates, D. J. Kushner, and A. T. Matheson (Dordrecht: Elsevier, 1993), vii–xxxix, at vii. See Virginia Morell, "Microbiology's Scarred Revolutionary."

50. See Ernst Mayr, "A Natural System of Organisms," *Nature* 348 (1990): 491; T. Cavalier-Smith, "Bacteria and Eukaryotes," *Nature* 356 (1992): 570.

51. See Michael Dietrich, "Paradox and Persuasion: Negotiating the Place of Molecular Evolution within Evolutionary Biology," *Journal of the History of Biology* 31 (1998): 87–111.

52. See Pnina Abir-Am, "The Politics of Macromolecules: Molecular Biologists, Biochemists, and Rhetoric," *Osiris* 7 (1992): 164–191.

53. Theodosius Dobzhansky, "Evolutionary and Population Genetics," *Science* 142 (1963): 1131–1135.

54. Ernst Mayr, "The New versus the Classical in Science," *Science* 141 (1963): 763.

55. See M. Kimura, *The Neutral Theory of Molecular Evolution* (Cambridge: Cambridge University Press, 1983); G. L. Morgan, "Emile Zuckerkandl, Linus Pauling, and the Molecular Evolutionary Clock, 1959–1965," *Journal of the History of Biology* 31 (1998): 155–178. See also Michael Dietrich, "The Origins of the Neutral Theory of Mo-

lecular Evolution," *Journal of the History of Biology* 27 (1994): 21–59; Dietrich, "Paradox and Persuasion"; William Provine, "The Neutral Theory of Molecular Evolution in Historical Perspective," in *Population Biology of Genes and Molecules*, ed. B. Takahata and J. Crow (Tokyo: Baifukan, 1990), 17–31.

56. Simpson and Mayr protested that evolution worked on the whole phenotype and could not single out genes unless they have phenotypic effects separable both phenotypically and genetically from other genes. Dietrich, "Paradox and Persuasion," 97. William Provine has suggested that underlying the confrontation was substantial confusion about what neutral theorists were arguing. Neutral theorists were not arguing that functional proteins emerged without selection, but only that one could detect amino acid sequence changes unaffected by selection. In other words, drift and natural selection were not mutually exclusive alternatives. Provine, "The Neutral Theory of Molecular Evolution in Historical Perspective."

57. G. G. Simpson, *Concession to the Improbable: An Unconventional Autobiography* (New Haven: Yale University Press, 1978), 269.

58. Ernst Mayr, "Two Empires or Three?" *Proceedings of the National Academy of Sciences* 95 (1998): 9720–9723, at 9722.

59. Ibid. Protistologists counted 200,000 species of protists to date. See John Corliss, "Toward a Nomenclatural Protist Perspective," in *Illustrated Glossary of Protista*, ed. L. Margulis, H. I. McKhann, L. Olendzenski (Boston: Johns and Bartlett, 1993), xxvii–xxxii.

60. Mayr, "Two Empires or Three?" 9723.

61. C. R. Woese, "Default Taxonomy: Ernst Mayr's View of the Microbial World," *Proceedings of the National Academy of Sciences* 95 (1998): 11043–11046, at 11045.

62. See W. Ford Doolittle, "Phylogenetic Classification and the Universal Tree," *Science* 284 (1999): 21124–21128.

63. For an excellent review of the biology of horizontal gene transfer, see Frederic Bushman, *Lateral DNA Transfer: Mechanisms and Consequences* (New York: Cold Spring Harbor Laboratory Press, 2002).

64. See Doolittle, "Phylogenetic Classification."

65. See Howard Ochman, Jeffery G. Lawrence, and Eduardo Groisman, "Lateral Gene Transfer and the Nature of Bacterial Innovation," *Nature* 405 (2000): 299–304; Jonathan Eisen, "Horizontal Gene Transfer among Microbial Genomes: A New Insight from Complete Genome Analysis," *Current Opinion in Genetics and Development* 10 (2000): 606–611. See also Lynn Margulis and Dorion Sagan, *Acquiring Genomes* (New York: Basic Books, 2002).

66. See S. Sonea and P. Panisset, *The New Bacteriology* (Boston: Jones and Bartlett, 1983); S. Sonea and L. G. Mathieu, *Prokaryotology* (Montreal: Presses de l'Université de Montréal, 2000).

67. See William Martin, "Mosaic Bacterial Chromosomes: A Challenge en Route to a Tree of Genomes," *Bioessays* 21 (1999): 99–104; W. Ford Doolittle, "Uprooting the Tree of Life," *Scientific American*, Feb. 2000, 90–95.

68. Maria Rivera and James Lake, "Evidence that Eukaryotes and Eocyte Prokaryotes are Immediate Relatives," *Science* 257 (1992): 74–76.

69. See, e.g., Radhey Gupta, "Protein Phylogenies and Signature Sequences: A Reappraisal of Evolutionary Relationships among Archaebacteria, Eubacteria, and Eukaryotes," *Microbiology and Molecular Biological Reviews* 62 (1998): 1435–1491. See also E. V. Koonin, A. R. Musegian, M. Y. Galperin, and D. R. Walker, "Comparison of Archaeal and Bacterial Genomes: Computer Analysis of Protein Sequences Predicts Novel Func-

tions and Suggests a Chimeric Origin for the Archaea," *Molecular Microbiology* 25 (1997): 619–637.

70. But see Tsuneaki Asai, Dmitry Zaporojets, Craig Squires, and Catherine L. Squires, "An *Escherichia coli* Strain with All Chromosomal rRNA Operons Inactivated: Complete Exchange of rRNA Genes between Bacteria," *Proceedings of the National Academy of Sciences* 96 (1999): 1971–1976.

71. See Martin, "Mosaic Bacterial Chromosomes"; Doolittle, "Phylogenetic Classification and the Universal Tree"; Woese and Olsen, personal communication.

72. See Carl Woese, "The Universal Ancestor," *Proceedings of the National Academy of Sciences* 95 (1998): 6854–6859. See also Otto Kandler, "The Early Diversification of Life and the Origin of the Three Domains: A Proposal," in *Thermophiles: The Keys to Molecular Evolution and the Origin of Life?* ed. Juergen Wiegel and Michael Adams (New York: Taylor and Francis, 1998), 19–31.

73. Darwin had reasoned that "probably all organic beings which have ever lived on this earth have descended from one primordial form, into which life was first breathed." Charles Darwin, *On the Origin of Species*, with an introduction by Ernst Mayr, facsimile ed. of 1859 (Cambridge, Mass.: Harvard University Press, 1964), 484.

74. Woese, "The Universal Ancestor," 6854.

Notes to Chapter 19

1. Jan Sapp, *Evolution by Association: A History of Symbiosis* (New York: Oxford University Press, 1994).

2. See John Lear, *Recombinant DNA: The Untold Story* (New York: Crown, 1978); Ted Howard and Jeremy Rifkin, *Who Should Play God?* (New York: Dell, 1977); June Goodfield, *Playing God* (New York: Random House, 1977); Nicholas Wade, *The Ultimate Experiment: Man-Made Evolution* (New York: Walker, 1977); Sheldon Krimsky, *Genetic Alchemy: The Social History of the Recombinant DNA Controversy* (Cambridge, Mass.: MIT Press, 1982); Raymond A. Zilinskas and Burke K. Zimmerman, eds., *The Gene-Splicing Wars: Reflections on the Recombinant DNA Controversy* (New York : Macmillan, 1986). See also Philip Kitcher, *The Lives to Come: The Genetic Revolution and Human Possibilities* (New York: Touchstone, 1996); Tom Wilkie, *Perilous Knowledge* (Berkeley: University of California Press, 1993); Lois Wingerson, *Unnatural Selection* (New York: Bantam Books, 1998).

3. Stephen Jay Gould, *Wonderful Life: The Burgess Shale and the Nature of History* (London: Hutchison Radius, 1989), 46–47.

4. As an illustration of how the null hypothesis over horizontal gene transfer has changed, in 2001, human genome researchers published suggestive, though controversial, evidence that humans have acquired hundreds of genes directly from bacteria. See Nicholas Wade, "Link between Human Genes and Bacteria Is Hotly Debated," *New York Times*, May 18, 2001. See also Steven L. Salzberg, Owen White, Jeremy Peterson, and Jonathan A. Eisen, "Microbial Genes in the Human Genome: Lateral Transfer or Gene Loss?" *Science* 292 (2001): 1903–1906; M. J. Stanhope, A. Lupes, M. J. Italia, K. N. Koretke, C. Volker, and J. R. Brown, "Phylogenetic Analyses Do Not Support Horizontal Gene Transfers from Bacteria to Vertebrates," *Nature* 411 (2001): 940–944.

5. Joshua Lederberg suggested the term "symbiome" to signify "the whole symbiotic complex, namely the superorganism." Personal correspondence, May 22, 2000. Lederberg

also coined the term "microbiome" "to signify the ecological community of commensal, symbiotic, and pathogenic microorganisms that literally share our body space and have been all but ignored as determinants of health and disease." J. Lederberg and Alexa McCray, "Ome Sweet 'Omics: A Genealogical Treasure of Words," *Scientist* 15 (Feb. 4, 2001): 8.

6. See Theodor Rosebury, *Life on Man* (London: Paladin, 1969); Frederic Bushman, *Lateral DNA Transfer: Mechanisms and Consequences* (New York: Cold Spring Harbor Laboratory Press, 2002).

7. See L. V. Hooper, L. Bry, P. G. Falk, and J. I. Gordon, "Host-Microbial Symbiosis in the Mammalian Intestine: Exploring an Internal Ecosystem," *BioEssays* 20 (1998): 336–343; L. V. Hooper, M. H. Wong, A. T. Lennart Hansson, P. G. Falk, and J. I. Gordon, "Molecular Analysis of Commensal Host-Microbial Relationships in the Intestine," *Science* 291 (2001): 881–884.

8. These bacteria convert nitrates to nitrites, which, in the acid environment of our stomachs, produce the pathogen-fighting chemical nitric oxide. See Elizabeth Pennisi, "Integrating the Many Aspects of Biology," *Science* 287 (2000): 419–421.

9. R. D. Berg, "Probiotics, Prebiotics, or Conbiotics?" *Trends in Microbiology* 6 (1998): 89.

10. Sapp, *Evolution by Association.*

11. See, e.g., W. Schwemmler, *Symbiogenesis, a Macro-Mechanism of Evolution: Progress toward a Unified Theory of Evolution* (Berlin: Walter de Gruyter, 1989); Mary Beth Saffo, "Symbiosis in Evolution," in *The Unity of Evolutionary Biology: Proceedings of the Fourth International Congress of Systematics and Evolutionary Biology*, ed. E. C. Dudley (Portland, Ore.: Dioscorides Press, 1991), 674–680. See also essays in Lynn Margulis and René Fester, eds., *Symbiosis as a Source of Evolutionary Innovation: Speciation and Morphogenesis* (Cambridge, Mass.: MIT Press, 1991); Sapp, *Evolution by Association*; Lynn Margulis and Dorion Sagan, *Acquiring Genomes: A Theory of the Origins of Species* (New York: Basic Books, 2002).

12. See, e.g., Ernst Mayr, *The Growth of Biological Thought* (Cambridge, Mass.: Harvard University Press, 1982), 13; Peter Bowler, *Evolution: The History of an Idea* (Berkeley: University of California Press, 1984); David Kohn, ed., *The Darwinian Heritage* (Princeton: Princeton University Press 1985); David Depew and Bruce H. Weber, *Darwinism Evolving: Systems Dynamics and the Genealogy of Natural Selection* (Boston: MIT Press, 1995); Stephen Jay Gould, *The Structure of Evolutionary Theory* (Cambridge, Mass.: Harvard University Press, 2002).

13. Anton de Bary, "Die Erscheinung der Symbiose," *Vortrag auf der Versammlung der Naturforscher und Aertze zu Cassel* (1879): 1–30, at 21–22.

14. A. B. Frank, "Ueber die auf Wurzelsymbiose beruhende Ernährung gewisser Bäume durch unterirdische Pilze," *Berichte der Deutschen Botanischen Gesellschaft* 3 (1885): 128–145.

15. See Sapp, *Evolution by Association.*

16. See M. F. Allen, *The Ecology of Mycorrhizae* (New York: Cambridge University Press, 1991).

17. David H. Lewis, "Mutualistic Symbiosis in the Origin and Evolution of Land Plants," in *Symbiosis as a Source of Evolutionary Innovation: Speciation and Morphogenesis*, ed. Lynn Margulis and René Fester (Cambridge, Mass.: MIT Press, 1991), 288–300.

18. Kris Pirozynski, "Galls, Flowers, Fruits, and Fungi," in *Symbiosis as a Source of Evolutionary Innovation: Speciation and Morphogenesis*, ed. Lynn Margulis and René Fester (Cambridge, Mass.: MIT Press, 1991), 364–379, at 371–72.

19. See Anton Quispel, "Evolutionary Aspects of Symbiotic Adaptations: Rhizobium's Contribution to Evolution by Association," in *The Rhizobiaceae: Molecular Biology of a Model Plant-Associated Bacteria*, ed. H. P. Spaink, A. Kondorosi, and P. J. Hooykaas (Dordrecht: Kluwer, 1998), 487–507. Horizontal gene transfer (that which is not derived from parents) of bacterial DNA to plant chromosomes is well known to be involved in the formation of crown galls, the tumorous responses to gene transfer from *Agrobacterium tumefaciens*. Crown gall disease affects about a hundred families of plants and causes millions of dollars worth of damage each year to agricultural and ornamental plants. See, e.g., C. I. Kado, "Molecular Mechanisms of Crown Gall Tumorigenesis," *Critical Reviews in Plant Sciences* 10 (1991): 1–32.

20. M. J. McFall-Ngai and E. G. Ruby, "Sepiolids and Vibrios: When First They Meet," *Bioscience* 48 (1998): 257–265.

21. See Paul Buchner, *Endosymbiosis of Animals with Plant Microorganisms*, trans. Bertha Mueller (New York: Interscience, 1965).

22. See Angela Douglas, "Nutritional Interactions in Insect-Microbial Symbioses: Aphids and Their Symbiotic Bacteria *Buchnera*," *Annual Review of Entomology* 43 (1998): 17–37.

23. See J. H. Werren, "Biology of Wolbachia," *Annual Review of Entomology* 42 (1997): 587–609.

24. Sapp, *Evolution by Association*.

25. See C. Merezhkowsky, "Über Natur und Ursprung der Chromatophoren im Pflanzenreiche," *Biologisches Centralblatt* 25 (1905): 593–604; English translation in W. Martin and K. V. Kawallik, "Annoted English Translation of Merezkowsky's 1905 paper 'Über Natur und Ursprung der Chromatophoren im Pflanzenreiche,'" *European Journal of Phycology* 34 (1999): 287–295; C. Mérejkovsky, "La plante considérée comme un complexe symbiotique," *Bulletin de la Société des Sciences Naturelles de l'ouest de la France* 6 (1920): 17–98. For an account of Merezhkowsky's life, Jan Sapp, Francisco Carrapico, and Mikhail Zolotonosoz, "The Hidden Face of Constantin Merezhkowsky," *History and Philosophy of the Life Sciences* (in press).

26. C. Mereschkowsky, "Theorie der zwei Plasmaarten als Grundlage der Symbiogenesis: einer neuen Lehre von der Entstehung der Organismen," *Biologisches Centralblatt* 30 (1910): 277 ff.

27. P. Portier, *Les symbiotes* (Paris: Masson, 1918).

28. Ibid., vii.

29. See Sapp, *Evolution by Association*.

30. Ibid. See also William Summers, *Félix d'Herelle and the Origins of Molecular Biology* (New Haven: Yale University Press, 1999).

31. Félix d' Herelle, *The Bacteriophage and Its Behavior*, trans. George H. Smith (Baltimore: Williams and Wilkins, 1926), 320.

32. On biologists' myths about the discovery of lysogeny, see Sapp, *Evolution by Association*, 107–108, 158–160. See also Summers, *Félix d'Herelle and the Origins of Molecular Biology*.

33. I. E. Wallin, *Symbionticism and the Origin of Species* (Baltimore: Williams and Wilkins, 1927).

34. Sapp, *Evolution by Association*. See also Donna Mehos, "Ivan E. Wallin's Theory of Symbionticism," in *Concepts of Symbiogenesis: History of Symbiosis as an Evolutionary Mechanism*, ed. L. N. Khakhina, L. Margulis, and M. McMenamin (New Haven: Yale University Press, 1993), 149–163.

35. See Jan Sapp, "Paul Buchner (1866–1978) and Hereditary Symbiosis in Insects," *International Journal of Microbiology* 5 (2002): 145–160.

36. Buchner, *Endosymbiosis of Animals with Plant Microorganisms.*

37. See Sapp, "Paul Buchner (1866–1978) and Hereditary Symbiosis in Insects."

38. See Buchner, *Endosymbiosis of Animals with Plant Microorganisms*, 69.

39. See Sapp, *Evolution by Association.*

40. For more detailed discussion, see ibid.

41. Portier, *Les symbiotes*, 294.

42. Wallin, *Symbionticism and the Origin of Species.*

43. See Frank Ryan, *Tuberculosis: The Greatest Story Never Told* (Bromsgrove: Swift, 1992). Some 40 million people are reportedly infected with the human immunodeficiency virus (HIV). See also Michael B. Oldstone, *Viruses, Plagues, and History* (New York: Oxford University Press, 1998).

44. John Harley Warner, "Rethinking the Reception of the Germ Theory of Disease: Comparative Perspectives," introduction to special issue of *Journal of the History of Medicine and Allied Sciences* 52 (1997): 7–16.

45. See Stephen Morse, ed., *Emerging Viruses* (New York: Oxford University Press, 1993); Paul Ewald, *Evolution of Infectious Disease* (New York: Oxford University Press, 1994); Joshua Lederberg, ed., *Biological Weapons: Limiting the Threat* (Cambridge, Mass.: MIT Press, 1999).

46. H. J. Muller, "Artificial Transmutation of the Gene," *Science* 46 (1927): 84–87.

47. E. M. East, "The Nucleus-Plasma Problem," *The American Naturalist* 68 (1934): 289 ff.

48. H. J. Muller summarized the classical geneticists' viewpoint in 1951:

> In mitigation of the current conception that cytoplasmically located genes or gene-complexes form an essential part of the genetic constitution of animals, the following points should be noted: 1) the extreme rarity with which illustrations of such inheritance have been found in animal material . . . 2) the dispensability of the cytoplasmically located particles in the cases studied . . . 3) the fact that, in these same cases, the agents have been proved to be able to pass as infections . . . 4) the lack of a fundamental basis for distinguishing between these and cases of undoubtedly parasitic or symbiotic microorganisms or viruses of exogenous derivation. These are points which, taken together, would appear to argue for most or all of these agents in animals having at one time arrives as invaders; for their tenure usually being insecure, as compared with that of the native chromosomes. This conclusion is, moreover, reinforced by a consideration of their mode of distribution and aggregation, since it is not only rather precarious but apparently far less suitable than that of the chromosomal genes for the simultaneous retention and the accumulation of numerous different types within the same germ plasm. (H. J. Muller, "The Development of the Gene Theory," in *Genetics in the Twentieth Century*, ed. L. C. Dunn [New York: Macmillan], 77–100, 117)

49. H. G. Wells, J. Huxley, and G. P. Wells, *The Science of Life*, vol. 3 (New York: Doubleday, Doran, 1930).

50. One of the real exceptions to this was the work of W. C. Allee and colleagues, who focused principally on intraspecific cooperation. See W. C. Allee, A. E. Emerson, O. Park, and K. P. Schmidt, *Principles of Animal Ecology* (Philadelphia: Saunders, 1949). See also

G. Mitman, *The State of Nature: Ecology, Community and American Social Thought* (Chicago: University of Chicago Press, 1992); Sapp, *Evolution by Association.*

51. F. G. Gregory, "A Discussion of Symbiosis Involving Micro-Organisms," *Proceedings of the Royal Society of London*, series B, 139 (1951): 202–203.

52. P. S. Nutman and Barbara Mosse, editors' preface, in *Symbiotic Associations: Thirteenth Symposium of the Society for General Microbiology*, ed. P.S. Nutman and Barbara Mosse (Cambridge: Cambridge University Press, 1963), ix–xx.

53. On the historical development of the ecosystem concept, see Donald Worster, *Nature's Economy: A History of Ecological Ideas* (Cambridge: Cambridge University Press, 1977); Joel Hagen, *An Entangled Bank: The Origins of Ecosystem Ecology* (New Brunswick, N.J.: Rutgers University Press, 1992); Frank B. Golley, *A History of the Ecosystem Concept in Ecology* (New Haven: Yale University Press, 1993).

54. In the preface to his well-known book of 1982, Kenneth Mann explained that he had seriously considered writing the entire text without naming a single organism. K. H. Mann, *Ecology of Coastal Waters: A Systems Approach*, Studies in Ecology 9 (Berkeley: University of California Press, 1982), ix.

55. This phenomenon is responsible for the very rapid spread of antibiotic resistance among widely different microbial species, which has heightened concerns about the wide use of antibiotics in medicine as well as in feed supplements in the meat industry.

56. See Sapp, *Evolution by Association*, 157–163. Tracy Sonneborn insisted that these infectious or "migratory" plasmagenes must be considered as part of heredity, and that mitochondria and chloroplasts that are not infectious today may have originated as symbionts. See T. M. Sonneborn, "The Cytoplasm in Heredity," *Heredity* 4 (1950): 11–36, at 22.

57. C. D. Darlington, "Mendel and the Determinants," in *Genetics in the Twentieth Century*, ed. L. C. Dunn (New York: Macmillan, 1951), 315–332, at 320.

58. Ibid., 331.

59. J. Lederberg, "Genetic Studies in Bacteria," in *Genetics in the Twentieth Century*, ed. L. C. Dunn (New York: Macmillan, 1951), 263–289; Lederberg, "Cell Genetics and Hereditary Symbiosis," *Physiological Reviews* 32: 403–430, 1952.

60. Lederberg, "Cell Genetics," 413.

61. Ibid., 424.

62. Ibid., 425–426.

63. R. Dubos, "Integrative and Creative Aspects of Infection," in *Perspectives in Virology*, vol 2, ed. M. Pollard, (Minneapolis: Burgess, 1961), 200–205.

64. Ibid., 204.

65. On the origins of recombinant DNA technology, see Susan Wright, "Recombinant DNA Technology and Its Social Transformation, 1972–1982," *Osiris* 2 (1986): 303–360; On "breaching the species barrier," see Sheldon Krimsky, *Genetic Alchemy: The Social History of the Recombinant DNA Controversy* (Cambridge, Mass.: MIT Press, 1982). See also Raymond A. Zilinskas and Burke K. Zimmerman, eds., *The Gene-Splicing Wars: Reflections on the Recombinant DNA Controversy* (New York: Macmillan, 1986).

66. Erwin Chargaff, 'Recombinant DNA Research: A Debate on the Benefits and Risks," *Chemical and Engineering News* 55 (1977): 35; Chargaff, "Profitable Wonders," *Sciences* 15 (1975): 21–26.

67. George Wald, "The Case against Genetic Engineering," *Sciences* 16 (1976): 6–11.

68. Robert Sinsheimer, lecture at UCLA, Nov. 1976, in *The Recombinant DNA Debate*, ed. David A. Jackson and Stephen P. Stich (Englewood Cliffs, N.J.: Prentice-Hall, 1979), 86.

69. Lynn Sagan, "On the Origin of Mitosing Cells," *Journal of Theoretical Biology* 14 (1967): 225–274. See also Lynn Margulis, *Symbiosis in Cell Evolution* (San Francisco: W. H. Freeman, 1981).

70. Hans Ris and Walter Plaut, "Ultrastructure of DNA-Containing Areas in the Chloroplast of Chlamydomonas," *Journal of Cell Biology* 13 (1962): 383–391, at 388–390; Sylan Nass and Margit M. K. Nass, "Intramitochondrial Fibers with DNA Characteristics," *Journal of Cell Biology* 19 (1963): 613–628; F. J. R. Taylor, "Implications and Extensions of the Serial Endosymbiosis Theory of the Origin of Eukaryotes," *Taxon* 23 (1974): 229–258.

71. See Ruth Sager, *Cytoplasmic Genes and Organelles* (New York: Academic Press, 1972); Nicoholas Gillham, *Organelle Heredity* (New York: Raven Press, 1978).

72. See Margulis, *Symbiosis in Cell Evolution*.

73. S. Watasē, "On the Nature of Cell Organization," *Woods Hole Biological Lectures* (1893), 83–103. See also Sapp, *Evolution by Association*.

74. L. Sagan, "On the Origin of Mitosing Cells."

75. See, e.g., Michael Gray, Gertraud Burger, and B. Franz Lang, "Mitochondrial Evolution," *Science* 283 (1999): 1476–1481.

76. Roger Stanier, "Some Aspects of the Biology of Cells and Their Possible Evolutionary Significance," in *Organization and Control in Prokaryotic Cells: Twentieth Symposium of the Society for General Microbiology*, ed. H. P. Charles and B. C. Knight (Cambridge: Cambridge University Press, 1970), 1–38, at 31.

77. C. R. Woese and G. E. Fox, "Phylogenetic Structure of the Prokaryote Domain: The Primary Kingdoms," *Proceedings of the National Academy of Sciences* 75 (1977): 5088–5090; C. R. Woese, "Endosymbionts and Mitochondrial Origins," *Journal of Molecular Evolution* 10 (1977): 93–96.

78. See M. W. Gray and W. F. Doolittle, "Has the Endosymbiont Hypothesis Been Proven?" *Microbial Reviews* 46 (1982): 1–42. See also Michael Gray, "The Endosymbiont Hypothesis Revisited," *International Review of Cytology* 141 (1992): 233–257.

79. See J. L. Hall. and D. J. Luck, "Basal Body–Associated DNA: *In situ* Studies in *Chlamydomonas reinhardtii*," *Proceedings of the National Academy of Sciences* 92 (1995.): 5129–5133. One of the biological lessons from this is that cell structures do not have to have nucleic acids in order to be inherited. See Jan Sapp, "Freewheeling Centrioles," *History and Philosophy of the Life Sciences* 20 (1998): 255–290.

80. See Michael Chapman, Michael Dolan, and Lynn Margulis, "Centrioles and Kinetosomes: Form, Function, and Evolution," *Quarterly Review of Biology* 75 (2000): 409–429.

81. See Sapp, *Evolution by Association*.

82. Watasē, "On the Nature of Cell Organization"; Theodor Boveri, "Ergebnisse über die Konstitution der chromatischen Kernsubstance," *Jena* (1904): 90; C. Mereschkowsky, "Theorie der zwei Plasmaarten als Grundlage der Symbiogenesis." For historical discussion, see Sapp, *Evolution by Association*.

83. Nucleus, chloroplasts, mitochondria, and centrioles were not the only organelles that were proposed to have originated by symbiosis. In 1982, the Nobel Prize–winning cell biologist Christian de Duve at the Rockefeller Foundation suggested that another organelle, peroxisomes, arose from aerobic bacteria, which were adopted as endosymbionts before mitochondria. Peroxisomes are as widely distributed throughout the plant and animal world as mitochondria but are much simpler in structure and composition. Although today they carry out various metabolic activities, de Duve argued that their original bene-

fit was to rescue their anaerobic hosts from the toxic effects of oxygen, which greatly accumulated in the primitive atmosphere some 2 billion years ago after photosynthetic cyanobacteria arose. Unlike mitochondria and chloroplasts, peroxisomes, like centrioles, have no remnants of an independent genetic system. See Christian de Duve, "The Birth of Complex Cells," *Scientific American*, Apr. 1996, 50–57.

84. Jeremy Picket Heaps, "The Evolution of Mitosis and the Eukaryotic Condition," *BioSystems* 6 (1974): 37–48.

85. James Lake, "Mapping Evolution with Ribosome Structure: Intralineage Constancy and Interlineage Variation," *Proceedings of the National Academy of Sciences* 79 (1982): 5948–5952, at 5951; James Lake and Maria Rivera, "Was the Nucleus the First Symbiont?" *Proceedings of the National Academy of Sciences* 91 (1994): 2880–2881.

86. See Radhey Gupta and G. Brian Golding, "The Origin of the Eukaryotic Cell," *Trends in Biology* 21 (1996): 166–170; Radhey Gupta, "Protein Phylogenies and Signature Sequences: A Reappraisal of Evolutionary Relationships among Archaebacteria, Eubacteria, and Eukaryotes," *Microbiology and Molecular Biology Reviews* 62 (1998): 1435–1491.

87. See William Martin, and Miklós Muller, "The Hydrogen Hypothesis for the First Eukaryote," *Nature* 392 (1998): 37–41; T. Marin Embley and William Martin, "A Hydrogen-Producing Mitochondrion," *Nature* 396 (1998): 517–519.

88. Chapman, Dolan, and Margulis, "Centrioles and Kinetosomes."

89. Arguing that the genes affecting the cytoskeleton, which allowed phagocytosis, are found in no existing bacterial lineages, Mitchel Sogin and Russel Doolittle have suggested that the cytosol arose from a merger involving a now-extinct kind of bacteria. See W. Ford Doolittle, "Uprooting the Tree of Life," *Scientific American* 2000, 90–95, at 94. See also H. Hartman and A. Federov, "The Origin of the Eukaryotic Cell: A Genomic Investigation," *Proceedings of the National Academy of Sciences* 99 (2002): 1420–1425.

90. See, e.g., W. Schwemmler, *Symbiogenesis: A Macro-Mechanism of Evolution*; Saffo, "Symbiosis in Evolution." See essays in Margulis and Fester, *Symbiosis as a Source of Evolutionary Innovation*. See also Sapp, *Evolution by Association.*

91. See, e.g., Lederberg, "Cell Genetics and Hereditary Symbiosis"; Taylor, "Implications and Extensions"; K. W. Jeon and J. F. Danelli, "Microsurgical Studies with Large Free-Living Amoebas," *International Review of Cytology* 30 (1971): 49–89; K. W. Jeon, "Amoeba and X-Bacteria: Symbiont Acquisition and Possible Species Change," in *Symbiosis as a Source of Evolutionary Innovation: Speciation and Morphogenesis*, ed. Lynn Margulis and René Fester (Cambridge, Mass.: MIT Press, 1991), 118–134.

92. Paul Nardon and Anne-Marie Grenier, "Serial Endosymbiosis Theory and Weevil Evolution: The Role of Symbiosis," in *Symbiosis as a Source of Evolutionary Innovation: Speciation and Morphogenesis*, ed. Lynn Margulis and René Fester (Cambridge, Mass.: MIT Press, 1991), 153–169.

93. See K. W. Jeon, *Intracellular Symbiosis* (New York: Academic Press, 1983).

94. C. M. Cavanaugh, S. L. Gardiner, M. L. Jones, H. W. Jannasch, and J. B Waterbury, "Prokaryotic Cells in the Hydrothermal Vent Tube-Worm *Riftia pachiptila* Jones: Possible Chemoautotrophic Symbionts," *Science* 213 (1981): 340–342; C. M. Cavanaugh, "Symbiotic Chemoautotrophic Bacteria in Marine Invertebrates from Sulfide-Rich Habitats," *Nature* 302 (1983): 58–61; H. Felbeck, "Chemoautotrophic Potential of the Hydrothermal Vent Tube-Worm *Riftia Pachyptila* Jones (Vestimentifera)," *Science* 213 (1981): 336–338; H. Felbeck, "Sulfide Oxidation and Carbon Fixation in the Gutless Clam *Solemya reidi*: An Animal-Bacterial Symbiosis," *Journal of Comparative Physiology* 152 (1983): 3–11.

95. Russel D. Vetter, "Symbiosis and the Evolution of Novel Trophic Strategies: Thiotrophic Organisms at Hydrothermal Vents," in *Symbiosis as a Source of Evolutionary Innovation: Speciation and Morphogenesis*, ed. Lynn Margulis and René Fester (Cambridge, Mass.: MIT Press, 1991), 219–248.

96. J. H. Werren,, "Biology of Wolbachia," *Annual Review of Entomology* 42 (1997): 587–609. See also James Higgins and Abdu F. Azad, "Use of Polymerase Chain Reaction to Detect Bacteria in Arthropods: A Review," *Journal of Medical Entomology* 32 (1995): 13–22; J. H. Werren, "Invasion of the Gender Benders," *Natural History*, February 2003: 58–63; John H. Werren and Donald Windsor, "Wolbachia Infection Frequencies in Insects: Evidence of a Global Equilibrium," *Proceedings of the Royal Society of London*, series B, 267 (2000): 1977–1995.

97. See Carl Zimmer, "Wolbachia: A Tale of Sex and Survival," *Science* 292 (2001): 1093–1096.

98. H. Abdelaziz, A-M. Grenier, C. Khatchadourian, H. Charles, and P. Mardon, "Four Intracellar Genomes Direct Weevil Biology: Nuclear, Mitochondrial, Principal Endosymbiont, and *Wolbachia*," *Proceedings of the National Academy of Sciences* 96 (1999): 6814–6819.

99. Charles B. Beard, Ravi V. Durvasula, and Frank F. Richards, "Bacterial Symbiosis in Arthropods and the Control of Disease transmission" *Emerging Infectious Diseases* 4 (1998): 581–591.

100. See S. K. Pierce, T. K. Mausel, M. E. Rumpho, J. J. Hanten, and W. L. Mondy, "Annual Viral Expression in a Sea Slug Population: Life Cycle Control and Symbiotic Chloroplast Maintenance," *Biological Bulletin* 197 (1999): 1–6.

101. See Elizabeth Pennisi, "Evo-Devo Devotees Eye Ocular Origins and More," *Science* 296 (2002): 1010–1011.

102. Jan O. Andersson, W. Ford Doolittle, and Camilla L. Nesbø, "Are There Bugs in Our Genome?" *Science* 292 (2001): 1848–1850; Steven L. Salzberg, Owen White, Jeremy Peterson, and Jonathan A. Eisen, "Microbial Genes in the Human Genome: Lateral Transfer or Gene Loss?" *Science* 292 (2001): 1903–1906.

103. See Roswitha Löwer, Johannes Löwer and Reinherd Kurth, " The Viruses in All of Us: Characteristics and Biological Significance of Human Endogenous Retrovirus Sequences," *Proceedings of the National Academy of Sciences* 93 (1999): 5177–5184.

104. J. Robin Harris, "Placental Endogenous Retrovirus (ERV): Structural, Functional, and Evolutionary Significance," *BioEssays* 20 (1998): 307–316.

105. See N. Eldredge and S. J.Gould, "Punctuated Equilibria: An Alternative to Phyletic Gradualism," in *Models in Paleontology*, ed. T. J. M. Schopf (San Francisco: Freeman, Cooper, 1972), 82–115; S. M. Stanley, "A Theory of Evolution above the Species Level," *Proceedings of the National Academy of Sciences* 72 (1975): 646–650; Niles Eldredge, *The Unfinished Synthesis* (New York: Oxford University Press, 1985), 128–130.

106. See Niles Eldredge, *Macro-Evolutionary Dynamics, Species, Niches, and Adaptive Peaks* (New York: McGraw-Hill 1989).

107. See Jeffrey H. Schwartz, *Sudden Origins: Fossils, Genes, and the Emergence of Species* (New York: John Wiley and Sons, 1999).

108. Stephen Jay Gould, *Wonderful Life: The Burgess Shale and the Nature of History* (London: Hutchison Radius, 1989), 310. Symbiosis is not mentioned (and bacteria are allotted 3 pages out of 1200) in Stephen Jay Gould's treatise *The Structure of Evolutionary Thought* (Cambridge, Mass.: Harvard University Press, 2002).

109. John Maynard Smith and Eors Szathmáry, *The Origins of Life: From the Birth of Life to the Origin of Language* (New York: Oxford University Press 1999), 107. See also Maynard Smith and Szathmáry, *The Great Transitions in Life* (New York: W. H. Freeman, 1995), 195.

Notes to Chapter 20

1. See Connie Barlow, ed., *From Gaia to Selfish Genes: Selected Writings in the Life Sciences* (Cambridge, Mass.: MIT Press, 1991). Robert N. Brandon and Richard M. Burian, eds., *Genes, Organisms, Populations: Controversies over the Units of Selection* (Cambridge, Mass.: MIT Press, 1984).

2. Charles Darwin, *The Decent of Man*, 2d ed. (New York: Hurst, 1874), 146.

3. Ibid.

4. Ibid., 146–147.

5. Ibid.,148.

6. Darwin wrote that "Ultimately our moral sense or conscience becomes a highly complex sentiment, originating in the social instincts, largely guided by the approbation of our fellow men, ruled by reason, self-interest, and in later times by deep religious feeling, and confirmed by instruction and habit." Ibid.

7. See D. Worster, *Nature's Economy: A History of Ecological Ideas* (Cambridge: Cambridge University Press, 1977), 328–331. G. Mitmann, *The State of Nature: Ecology, Community, and American Social Thought* (Chicago: University of Chicago Press, 1992).

8. W. C. Allee, "Where Angels Fear to Tread: A Contribution from the General Sociology to Human Ethics," *Science* 97 (1943): 517–525. Reprinted in *The Sociobiology Debate*, ed. A. L. Caplan (New York: Harper and Row, 1978), 41–56, at 55.

9. Worster, *Nature's Economy*, 329.

10. See V. C. Wynne-Edwards, *Animal Dispersion in Relation to Social Behaviour* (New York: Hafner, 1962); Wynne-Edwards, *Evolution through Group Selection* (Oxford: Blackwell Scientific Publications, 1986).

11. See, e.g., Robert C. Paehlke, *Environmentalism and the Future of Progressive Politics* (New Haven: Yale University Press, 1989). Garret Hardin, *Living Within Limits: Ecology, Economics and Population Taboos* (New York: Oxford University Press, 1993).

12. Wynne-Edwards, *Animal Dispersion in Relation to Social Behaviour*, 19.

13. Ibid., 131.

14. See V. C. Wynne-Edwards, "Ecology and the Evolution of Social Ethics," in J. W. S. Pringle ed., *Biology and the Human Sciences*, New York: Oxford University Press, 1972. For the contrary view of the times—of "man as a manmade species"defined and molded as much by human culture as by biological evolution—see J. Lederberg, "The Genetics of Human Nature," *Social Research* 40 (1973): 375–406.

15. "For the non-rational creature, on the other hand, there appears to be no choice: conformity with the social code is unquestioned, and virtue is automatic." In regard to limiting human population growth, Wynne-Edwards pointed to primitive tribes which had adopted conventional customs of one form or another to restrict fertility and reduce family size by several means: head hunting, sacrifice cannibalism, and by three main methods: 1) taboos that prevented women to become pregnant throughout lactation of the previous child; 2) Abortions; and 3) infanticide. Wynne-Edwards, "Ecology and the Evolution of Social Ethics," 492. He lamented that "The evident loss by man, almost within

the historic period, of means of limiting population growth, which he formerly possessed like other animals, stands out with disturbing clarity." Ibid., v.

16. See Richard Dawkins, *The Selfish Gene* (Oxford: Oxford University Press, 1976), 112.

17. J. B. S. Haldane, "Population Genetics," *New Biology* 18 (1955): 34–51.

18. J. Maynard Smith, "Groups Selection and Kin Selection," *Nature* 201 (1964): 1145–1147.

19. W. D. Hamilton, "The Genetic Evolution of Social Behaviour (I and II)," *Journal of Theoretical Biology* 7 (1964): 1–16, 17–52.

20. See John Tyler Bonner, *The Evolution of Complexity* (New Jersey: Princeton University Press, 1988), 65–116. The general consensus is that multicellularity evolved independently among diverse organisms many times in the course of life on earth: plants, animals, fungi, as well as some protists and some bacteria.

21. Leo Buss, *The Evolution of Individuality* (Princeton: Princeton University Press, 1987), 20–25.

22. Ibid., 31.

23. Ibid. 29–30.

24. See also J. Maynard Smith, "Evolutionary Progress and Levels of Selection," In M. Nitecki, ed., *Evolutionary Progress*, (Chicago: University of Chicago Press,1988), 219–230. John Maynard Smith and Eörs Szathmáry, *The Major Transitions in Evolution* (New York: W. H. Freeman, 1995), 244–246.

25. Ibid, 244. See also Eva Jablonka and Marion Lamb, *Epigenetic Inheritance and Evolution*, (New York: Oxford University Press, 1995), 206. Maynard Smith compared epigenetic development of a plant or an animal made up of genetically identical somatic cells to cooperation in an insect colony See John Maynard Smith, *A Darwinian View of Symbiosis*, in René Fester and Lynn Margulis eds., *Symbiosis as a Source of Evolutionary Innovation*, 26–39, 32.

26. See Lewis Wolpert, "Gastrulation and the Evolution of Development," *Development* 1992 Supplement (1992): 7–13; at 7; Wolpert, "The Evolutionary Origin of Development: Cycles, Patterning, Privilege, and Continuity," *Development* 1994 Supplement (1994): 79–84.

27. Maynard Smith and Szathmáry, *The Major Transitions in Evolution*, 245.

28. Robert Trivers, "The Evolution of Reciprocal Altruism," *Quarterly Review of Biology* 46 (1971): 35–39, 45–47; reprinted in Arthur L. Caplan, *The Sociobiological Debate* (New York: Harper and Row, 1978), 213–226, at 220.

29. E. O. Wilson, *Sociobiology: The New Synthesis* (Cambridge, Mass.: Harvard University Press, 1975).

30. Richard Dawkins, *The Selfish Gene*, new ed. (Oxford: Oxford University Press, 1989), 19–20.

31. Dawkins, *The Extended Phenotype*, 237. See also Garrett Hardin, "The Tragedy of the Commons," *Science* (1968): 1243–1248.

32. Dawkins, *The Selfish Gene*, 3.

33. See, e.g., Maynard Smith and Szathmáry, *The Great Transitions in Evolution.*

34. There are a vast number of books and articles on the sociobiology debate. I will mention here Arthur Caplan, ed., *The Sociobiology Debate* (New York: Harper and Row, 1978); Joe Crocker, "Sociobiology: The Capitalist Synthesis," *Radical Science Journal* 13 (1983): 55–71; W. R. Albury, "The Politics of Truth: A Social Interpretation of Scientific Knowledge, With an Application to the Case of Sociobiology," in *Nature Animated*, ed. Michael Ruse

(Dordrecht: D. Reidel, 1983), 115–129; S. Rose, L. J. Kamin, and R. C. Lewontin, *Not in Our Genes: Biology and Human Nature (*Harmondsworth: Penguin, 1984); Philip Kitcher, *Vaulting Ambition: Sociobiology and the Quest for Human Nature* (Cambridge, Mass.: MIT Press, 1987); Ullica Segerstråle, *Defenders of the Truth: The Battle for Science in the Sociobiology Debate and Beyond* (New York: Oxford University Press, 2000).

35. Lewontin, *Biology as Ideology*, 67.

36. See Ernst Mayr, "The Objects of Selection," *Proceedings of the National Academy of Sciences* 94 (1997): 2091–2094.

37. Stephen Jay Gould and Richard Lewontin, "The Spandrels of San Marco and the Panglossian Paradigm: A Critique of the Adaptationist Program," *Proceedings of the Royal Society of London*, series B, 205 (1978): 581–598.

38. On insect flight see J. H. Marden and M. G. Kramer, "Plecopteran Surface-Skimming and Insect Flight Evolution: Reply," *Science* 270 (1995): 1685; Marden and Kramer "Locomotor performance of insects with rudimentary wings," *Nature* 377 (1995): 332–334; J. H. Marden, M. R. Wolf, and K. E. Weber, "Aerial Performance of *Drosophila melanogaster* from Populations Selected for Upwind Flight Ability," *Journal of Experimental Biology* 200 (1997): 2747–2755; M. G. Kramer and J. H. Marden, "Almost Airborne," *Nature* 385 (1997): 403–404.

39. Voltaire, *Candide or Optimism*, trans. John Butt (London: Penguin Books, 1947), 20.

40. William Bateson, "Address of the President of the British Association for the Advancement of Science," *Science* 40 (1914): 287–302, at 293.

41. See Massimo Pigliucci and Jonathan Kaplan, "The Fall and Rise of Dr. Pangloss: Adaptionism and the Spandrels Paper Twenty Years later," *TREE* 15 (2000): 66–69.

42. Dawkins wrote, "If an individual dies in order to save ten close relatives, one copy of the kin-altruism gene may be lost, but a larger number of copies of the same gene is saved."*The Selfish Gene*, 90.

43. S. J. Gould, "Sociobiology: The Art of Storytelling," *New Scientist* 80 (1978): 530–533.

44. See, e.g., Segerstråle, *Defenders of the Truth;* Alcock, *The Triumph of Sociobiology*.

45. See Daniel C. Dennet, *Darwin's Dangerous Idea: Evolution and the Meanings of Life* (New York: Touchstone, 1996).

46. Richard Dawkins, *The Blind Watchmaker* (Harlow: Longman 1987), 296; Dawkins, Dawkins, *River out of Eden: A Darwinian View of Life* (New York: Basic Books, 1995).

47. See, e.g., John Alcock, *The Triumph of Sociobiology* (New York: Oxford University Press, 2001).

48. Sarah B. Hrdy, *The Woman That Never Evolved* (Cambridge, Mass.: Harvard University Press, 1983), 21.

49. See Maynard Smith and Szathmary, *The Major Transitions in Evolution.*

50. See H. Bernstein, G. S. Byers, and R. E. Michod, "Evolution of Sexual Reproduction: Importance of DNA Repair, Complementation, and Variation," *American Naturalist* 117 (1981): 537–549. See also H. Bernstein, F. A. Hopf, and R. E. Michod, "Is Meiotic Recombination and Adaptation for Repairing DNA Producing Genetic Variation, or Both?" in *Evolution of Sex: An Examination of Current Ideas*, ed. R. E. Michod and B. R. Levin (Sunderland, Mass.: Sinauer, 1988), 106–125.

51. Reg Morrison, *The Spirit of the Gene: Humanity's Proud Illusion and the Laws of Nature* (Ithaca: Cornell University Press, 1999).

52. Ibid., 173–174.

53. See Sapp, *Evolution by Association.*

54. See Noël Bernard, "Etudes sur la tubérisation," *Revue générale botanique* 14 (1902): 5–25; Félix d'Herelle, *The Bacteriophage and Its Behavior*, trans. George H. Smith (Baltimore: William and Wilkins, 1926), 320; P. S. Nutman and Barbara Moss, "Editors' Preface," in *Symbiotic Associations: Thirteenth Symposium of the Society for General Microbiology*, ed. P. S. Nutman and Barbara Mosse (Cambridge: Cambridge University Press, 1963), x.

55. See James Lovelock, *Gaia: A New Look at Life on Earth* (Oxford: Oxford University Press, 1979); Lovelock, "Hands Up for the Gaia Hypothesis," *Nature* 344 (1990): 100–102; John Postgate, "Gaia Gets Too Big for Her Boots," *New Scientist* 7 (1988): 60. In 1993, Lynn Margulis insisted that "Gaia is not an individual, it is an ecosystem." Lynn Margulis, *Symbiosis in Cell Evolution* (New York: W. H. Freeman, 1993). See also Sapp, *Evolution by Association*.

56. Roscoe Pound, "Symbiosis and Mutualism," *American Naturalist* 27 (1893): 509–520, at 519.

57. H. G. Wells, Julian Huxley, and G. P. Wells, *The Science of Life*, vol. 3 (New York: Doubleday, Doran, 1930), 932.

58. See K. F. Meyer, "The 'Bacterial Symbiosis' in the Concretion Deposits of Certain Operculate Land Mollusks of the Families Cyclostomatidae and Annularidae," *Journal of Infectious Diseases* 36 (1925): 1–107, at 94–95; L. R. Cleveland, "Symbiosis Among Animals with Special Reference to Termites and Their Intestinal Flagellates," *Quarterly Review of Biology* 1 (1926): 51–56.

59. D. C. Smith and A. E. Douglas, *The Biology of Symbiosis* (London: Edward Arnold, 1987), 262.

60. Sapp, *Evolution by Association.*

61. Maurice Caullery, *Parasitism and Symbiosis*, trans. Averilm Lysaght (London: Sidgwick and Jackson, 1952), 2.

62. Gunther S. Stent, *An Introduction to Molecular Genetics* (San Francisco: W. H. Freeman, 1971), 622.

63. Lewis Thomas, *The Lives of a Cell: Notes of a Biology Watcher* (New York: Viking Press, 1974), 71.

64. Ibid., 73–74.

65. R. Axelrod and W. D. Hamilton, "The Evolution of Cooperation," *Science* 211 (1981): 1390–1396.

66. See Robert Axelrod, *The Evolution of Cooperation* (New York: Basic Books, 1984),103.

67. Ibid., 100.

68. Ibid., 143.

69. G. C. Williams, *Adaptation and Natural Selection: A Critique of Some Current Evolutionary Thought* (Princeton: Princeton University Press, 1966); Robert May, *Stability and Complexity in Model Ecosystems* (Princeton University Press, 1973). May considered mutualisms to be more prevalent in stable nonseasonal, nonfluctuating environments, such as the tropics than in temperate regions. R. May, "Mutualistic Interactions among Species," *Nature* 296 (1982): 803–804; May, "A Test of Ideas about Mutualism," *Nature* 307 (1984): 410–411.

70. For a discussion, see Robert May, "Ecology and Evolution of Host-Virus Associations," in *Emerging Viruses*, ed. Stephen Morse (New York: Oxford University Press, 1993), 58–68, at 62–67. See René Dubos, *Man Adapting* (New Haven: Yale University Press, 1965), 196.

71. See R. M. May and R. M. Anderson, "Coevolution of Hosts and Parasites," Parasitology 85 (1982): 411–426; May and Anderson, "Epidemiology and Genetics in the Coevolution of Parasites and Hosts," *Proceedings of the Royal Society of London*, series B, 219 (1983): 281–313; Paul Ewald, "Host-Parasite Relations, Vectors, and the Evolution of Disease Severity," *Annual Review of Ecology and Systematics* 14 (1983): 465–485.

72. See Maynard Smith and Szathmáry, *The Major Transitions in Evolution*, 197.

73. Ibid., 141.

74. Ibid., 9–10, 23.

75. See Michael Gray, Gertraud Burger and Franz Lang, "Mitochondrial Evolution," *Science* 283 (1999): 1476–1481.

76. Charles Kurland and Siv Andersson have argued that most of the nuclear genes affecting mitochondrial structure and function were not derived from the symbiont at all: See C. G. Kurland and S. G. E. Andersson, "Origin and Evolution of the Mitochondrial Proteome," *Microbiology and Molecular Biology Reviews* 64 (2000): 786–820. Still other researchers have suggested that the ancestor of mitochondria may not have been captured by a nucleated cell in the remote past, but was a partner in a fusion event with another kind of bacterium that resulted in the nucleated cell itself. See William Martin and Miklós Muller, "The Hydrogen Hypothesis for the First Eukaryote," *Nature* 392 (1998): 37–41; T. Marin Embley and William Martin, "A Hydrogen-Producing Mitochondrion," *Nature* 396 (1998): 517–519; William Martin, Meike Hoffmeister, Carmen Rotte, and Katrin Henze, "An Overview of Endosymbiotic Models for the Origins of Eukaryotes, Their ATP-Producing Organelles (Mitochondria and Hydrogenosomes), and Their Heterotrophic Lifestyle," *Biological Chemistry* 382 (2001): 1521–1539.

77. See J. H. Werren, "Biology of Wolbachia," *Annual Review of Entomology* 42 (1997): 587–609; J. H. Werren, "Invasion of the Gender Benders," *Natural History*, Ferbruary 2003, 58–63; Carl Zimmer, "Wolbachia: A Tale of Sex and Survival," *Science* 292 (2001): 1093–1096.

Index

Abiogenesis, 80
Adaptationism, 64, 66–68, 149–150, 154, 259
Adelberg, Edward, 220, 226
Agassiz, Louis, 19, 65
Agnosticism, 5, 33, 34, 87
Agriculture, 130–132, 135–137, 234, 249, 269
Albrecht-Buehler, Guenter, 208
Albumin, 165
Algae, 238
Alkaptonuria, 158–160
Allee, Warder, 253
Allelomorph, origin of term, 134
Alpha-proteobacteria, 246
Altman, Sidney, 223
Altmann, Richard, 88, 188
Altruism, 57, 69, 255. *See also* Mutalism
American Civil Liberties Union, 4
Amoeba proteus, 247
Anarchism, 43, 55–60, 179
Ancestral inheritance, 125, 131, 135
Andersson, Siv, 345
Animals. *See also* Organism; Experimental embryology; Genetics
  chloroplasts in, 249
  cooperation among, 55–59, 253–258
  origins of, 98–99, 156, 218–230, 256–257, 296, 342
  populations, 254, 255
  sex and, 260, 261
  as superorganisms, 236
Anthropomorphism, 60–62, 99, 252–253, 260, 262–263
Antibiotic resistance, 337
Antibodies, 169
Antigenetics 171–173, 179–180
American Breeder's Association, 137–138
American Genetics Society, 178
American Society of Naturalists, 138
Amino Acids, 188–189, 191, 222
Amoeba proteus, 247
Antoinette, Marie, 6
Aphids, 238, 248–249, 264
Appel, Toby, 13–14, 276
Applied science. *See* Science
Archaea, 225–233
Archaebacteria, 147, 225–233
Archaeopteryx, 35
Archetype, 35–36
Aristotle, 4, 38, 61, 110, 274
Arthropods, 218, 248. *See also* Insects
Ascaris, 89, 96–97, 269
Astbury, William T., 192
Atheism, 41. *See also* Agnosticism
Autocatalytic, 196
Avery, Oswald, 189–190
Avery, Roy, 190
Axelrod, Robert, 263–264

Bacteria, 156, 190–191, 196, 233, 269. *See also* Archaebacteria; Bacterial genetics; Eubacteria; Symbiosis; Lateral gene transfer
  biochemists and, 220
  classification, 220, 224
  conjugation, 167
  diversity, 216, 227–228, 238

Bacteria (*continued*)
DNA, 228
environments of, 227, 237, 248
first organisms, 219–220
gut, 200, 235–236, 264
history of biology and, xi, 108
luminescent, 237
nitrogen-fixing, 237, 335
phylogenetics, 220–33
species concept and, 231
species number, 229
sulfur, 248
as superorganism, 231
transduction, 167, 189–191
transformation, 167, 189–190
Bacterial genetics, origins, 166–167
Bacterial sex, 166–167
Bacteriophage, 167–169, 181, 190– 191, 195, 243, 244. *See also* Viruses
Balfour, Francis, 40
Bateson, William
on adaptationism, 124–126, 136, 131, 259, 307
biometricians and, 124–125
criticism of chromosome theory, 113
on embryology, 133
on Garrod, 159–160
on gene mutations, 160
genetic lexicon and, 124, 134
on genetics' aims, 130, 131, 133, 136
on Mendel's neglect, 125–126
on Mendel's paper, 127
saltationism of, 71 124–126, 144
on telelogy, 134
Barnes, Barry, 283–284
Baur, Erwin, 146
Beadle, George
on bacterial transformations, 190
and corn genetics, 137
on cytoplasmic inheritance, 178
on Garrod, 157–58, 160–162
on human gene number, 169
Ephrussi and, 163–164
on Moewus, 165
Neurospora and, 163–165
one-gene: one enzyme theory, 157, 161–162
Rockefeller Foundation and, 168–170
Tatum and, 157, 164
Beagle, *H.M.S.*, voyage of the, 18, 20, 24, 52
Beisson, Jane, 211
Benda, C., 90
Benevolence, 69. *See also* Altruism; Mutualism
Benzer, Seymour, 170
Bergson, Henri, 180
Bernal, John Desmond, 192
Bernard, Claude, 110
Bible, 4, 18. *See also* Natural theology
Big Bang, 226
Biochemical genetics, 157–170
Biochemistry, 157, 159, 181, 188, 191–193, 220. *See also* Biochemical genetics; Molecular biology
Biogenetic law, 83, 98–99. *See also* Recapitulation theory
Biology
coined, 5
division of labor in, 84
and social sciences, 84, 270
and specialization, 84, 130–132, 138–140, 143–144, 155, 240, 270–271
Biological species concept, 151
Bioluminescence, 237
Biometricians, 124–125, 131, 133–135
Biosphere, 262
Biotechnology, 187, 202, 234
Birds
reptiles and, 67
species number, 229
Bismark, Otto von, 180
Blastula, 104–105
Blending inheritance, 64–65, 117, 267
Blyth, Edward, 22–23
Bohr, Niels, 169
Bolsheviks, 172
Bonaparte, Napoleon, 6, 11, 14, 60, 180
Bonner, David, 164
Botany
cell theory and, 76, 83, 98
chloroplasts and, 90
chromatography and, 191
classification, 5, 9
evolutionary synthesis and, 150–151
genetics and, 113, 137–146, 172–177
Mendel's laws and, 118–119
Boveri, Theodor, 89, 90, 91, 96, 106, 112, 246
Bowler, Peter, 308
Boyle, Robert, 303–304
Brachet, Albert, 107, 139
Brachet, Jean
Communist Party and, 177
meeting with Lysenko, 177–178
on microsomes, 184
plasmagene theory and, 175–176, 184, 196
on RNA, 184

Bragg, William Henry, 192
Bragg, William Lawrence, 192
Brannigan, Augustine, 121–122
Brave New World, 32
Breeders, 125, 130, 132, 133, 135–137
Bridges, C. O., 136, 160
Brücke, Ernst von, 82
Buchner, Paul, 240
Burgess Shale, 218
Buss, Leo, 256–257
Bussy Institution, 137
Butler, Samuel, 93
Brown, Robert, 76
Bridgewater Treatises, 61
Buffon, Georges
  evolutionary views of, 5, 9–10
  and natural history, 9, 274
  species definition, 65
Butler Act, 41

California Institute of Technology, 157, 163, 168–170
Callender, A., 123
Cambrian explosion, 156, 218–226
Cambridge University, 15, 18, 19, 147, 148, 192–193
Cancer, 193
Capitalism, 261. *See also* Laissez-faire
Carlsberg Laboratories, 166
Carnegie, Andrew, 46–47
Carroll, Lewis, 148
Caspersson, Torbjörn, 196
Castle, W. E., 137, 149
Catastrophism, 12, 18, 19
Caullery, Maurice, 113, 263
Cech, Thomas, 223
Cell, *See also* Egg; Cell state
  as bag of enzymes, 182, 183
  central control and, 263, 264
  as chamber, 86–87
  as chemical factory, 110
  as empire, 263
  form, 210–214
  as information society, 198
  as misnomer, 86
  nuclear monopoly, 174
  origin of term in science, 87
  polarity, 106, 110–11, 209–214, 299, 325
  soup, 200
Cell cortex. *See also* Cortical inheritance
  inheritance, 209–214, 299, 319
  polarity and, 110–111, 299
  morphogenetic field and, 182–184, 209, 212
Cell heredity, 109, 139–140, 181–184. *See also* Cellular differentiation; Epigenetic inheritance, Structural inheritance; Field heredity
Cell lineage studies, 104–105
Cell state, 82–84
  criticisms of, 98–100
Cell structure
  inheritance, 207–214
  morphogenetic role of, 182–184, 210–216
Cell theory, xi, 75–91, 296–297. *See also* Cell state
Cellular differentiation, 109–112, 181–184. *See also* Epigenetic inheritance; Structural inheritance; Field heredity
  and Mendelian genetics, 109, 140
  paradox resolved, 198, 204
Cellular pathology, 75, 78, 80
Centre national de la recherche scientifique, 179
Centrifuge, 110–111
Centrioles, 175, 207, 210, 239
  in cell division, 90–91, 245
  DNA and, 246
  kinetosomes and, 91
  reproduction of, 208
  symbiosis and, 238, 245–247
Centrosomes, 90–91, 95, 110. *See also* Centrioles
Chain, Ernst, 241
Chambers, Robert, 17
Chance and evolution, 63. *See also* Natural selection
Chaperonins, 207, 217
Chargaff, Erwin, 191–192, 245
Chase, Martha, 190–191
Chatton, Edouard, 226, 296
Chemists, 111, 187, 200. *See also* Biochemistry; Biochemical genetics; Chromatography; Molecular biology
Chetverikov, Sergie, 149–150
Chimera, 234, 235
Chlamydomonas genetics, 164–166, 296
Chloroplasts, 207
  in animals, 249
  DNA, 245
  genome, 207–08
  inheritance, 114, 177
  reproduction of, 95, 110, 175
  symbiotic origin, 90, 235, 239, 263 338

Chromatin, 88. *See also* Chromosomes
Chromosomes, *See also* Genes; Genetics; Nucleus; Weismannism
  in cell division, 88, 89, 91, 94, 104, 166
  heredity and, 107–110, 113–114, 197–198, 236, 242
  origin of word, 88–89
  reproduction of, 95, 136
Cilia, 175. *See also* Centrioles; Kinetosomes
Ciliates, 210–214
Clark, Terrance, 180
Classification, *See also* Phylogeny; Comparative anatomy; Comparative embryology; Microbial phylogeny; Molecular evolution
  of animals, 10, 11, 227, 275
  Aristotelian, 38
  of bacteria, 85–86, 227, 220–233
  of fungi, 227
  genealogical, 26
  hierarchical nature of, 231, 250
  of plants, 5, 9, 227
  of protists, 36, 85–86, 98–99, 227
Cleland, Ralph, 169
Clements, Frederick, 83
Cloning, 109, 298
Coevolution, 264
Cold Spring Harbor Symposium, 161, 190
Cold War, xii, 170–184
Colloid chemistry, 200
Columbia University, 100, 136–137, 191
Commensalism, 58–59, 236
Common descent, 26–27, 231, 250
Communist Party, 173, 177, 178, 179
Comparative anatomy, 11–12, 37. *See also* Comparative embryology
  analogies, 36
  functionalist, 13, 35
  genetics and, 133
  homologies, 35–36
  structuralist, 13–14, 35
  transcendental, 13–15, 35–36
Comparative embryology, 37–40
Competition, *See* Darwinism; Malthusianism; Natural selection; Social Darwinism; Sociobiology; Struggle for existence
Condorcet, Jean Marie, 52
Conklin, Edwin G.
  on cell theory, 75, 100
  cellular differentiation, 109
  on cytoplasmic inheritance, 103, 106–107
  on development and heredity, 131–134, 139
  on Mendelian inheritance, 103, 109
  on reductionism, 299
Convoluta roscoffensis, 238
Cooperation, x, 254. *See also* Mutualism; Altruism
  among biologists, 270–271
  among cells, 255–256
  among deer, 252
  among genes, 256, 265
  among humans, 253–255, 264, 341
  among insects, 254
Cope, Edward Drinker, 68–70
Copernicanism, 40, 155
Copernicus, Nicolas, 40
Corals, 238
Corey, Robert, 193
Corn genetics, 137, 146
Correlative characters, 71
Correns, Carl, 180, 294
  cytoplasmic inheritance, 113, 174–175
  de Vries and, 121–122
  Mendel's laws and, 118–122
Cortical Inheritance. *See* Cell cortex; Cell organization; Polarity; Morphogenetic field
Creationism, 4–5, 23, 33, 34, 41–42, 155, 268
Creator, 13, 23, 26
Creutzfeldt-Jakob disease, 207
Crichton, Michael, 205
Crick, Francis
  central dogma, 199
  frozen accident, 223
  genetic code and, 195
  structure of DNA, 168, 194
  Watson and, 168, 192, 193, 194
Crown Galls, 244, 335
Crystal Palace Exhibition, 35
Crystallography. *See* X-ray crystallography
Currie, Pierre, 64
Cuvier, Georges, 77
  catastrophism and, 12, 18
  comparative anatomy of, 11, 25–26, 275
  embranchements and, 11, 19
  fixist views of, 11–12
  Geoffroy and, 13–14
  Lamarck and, 11–15, 275
  Napoleon and, 11, 275
Cyanobacteria, 219, 235, 246
Cystic fibrosis, 202
Cytoarchitecture. *See* Cytoskeleton; Cortical inheritance

Cytogenetics, 136–137, 144, 150
Cytology, 72, 86–91, 103–105, 110–111, 131, 133, 209–210. *See also* Cells; Microscopy; Protists
Cytoplasm. *See also* Egg
  nucleus and, 93, 97
  origin of word, 88
  as symbiont, 238, 246–247, 339
Cytoplasmic inheritance. *See also* Cortical inheritance; Mitochondria; Chloroplasts; Symbiosis
  cellular differentiation and, 109, 139–140, 181–182
  criticisms of, 107–108 112–114, 214, 242, 251, 336
  embryology and, 103–114
  evolution and 103, 106–108, 112–114
  genetic evidence for, 174–179, 181–184
  inheritance of acquired characteristics and, 171, 176–177
Cytoskeleton, 210–214
Cytosol, 339
Cytotaxis, 210

D'Arcy Thompson, 213
Darlington, Cyril D., 150, 166, 243
Darrow, Clarence, 41
Dart, Raymond, 68
Darwin, Charles, ix, xi, 3, 16, 45, 96, 144, 218
  agnosticism of, 33–34
  on altruism, 253
  Beagle voyage, 18, 20–22
  comparative embryology and, 39
  on classification, 26
  early education of, 17–18
  evolutionary theory of, 26–30, 39, 66
  Durkheim and, 284–285
  gemmules, 92
  genetics and, 117
  on inheritance of acquired characteristics, 7, 18
  journal of, 21
  Lamarck and, 7, 8, 17
  Lyell and, 18–20, 22
  Malthus and, 52–54
  Marx and, 49
  Mendel and, 303
  on morality, 69
  on mutualisms, 58
  Mythical precursors, 22–23
  natural theology and, 23
  The Origin and, 15, 22, 277–278
  on origin of life, 226
  pangenesis theory of, 92
  on precambrian fossils, 328
  recapitulation theory and, 39
  social Darwinism and, 49–50
  species concept of, 28, 65
  on struggle for existence, 22, 29–30, 49, 50–51, 54, 58
  switch to evolution, 20–21
  Wallace and, 24–25
Darwin, Erasmus, 5, 7, 17, 24
Darwinism, *See also* Evolution; Natural selection; Social Darwinism; Struggle for existence
  eclipse of, 154
  gradualism and, 27–28, 32–33, 124–126, 131, 152, 236
  Judeo-Christian theology and, 3–5
  Mendelism and, 143–156
  natural theology and, 22–23, 31
  naturalism and, 31, 43
  sociopolitical theory and, 43–51, 270
  tenets of 4, 26–30
Dawkins, Richard, 213, 257, 258, 255
de Bary, Anton, 98, 236, 262
de Beer, Gavin, 123
de Duve, Christian, 338–339
Delbrück, Max, 170, 190
de Vries, Hugo
  Correns and, 121–122
  Mendel's laws and, 118, 121
  mutation theory, 144, 145, 309
  saltationist views of, 71
Demeric, Milislav, 137
Derham, William, 23
Descartes, René, 299
Descent of Man, 33, 50
Desmond, Adrian, 15
Development. *See also* Cellular differentiation; Comparative embryology; Egg; Experimental embryology
  evolution and, 38–39, 103, 106–108, 112–114, 156, 212–213, 256–257
  Mendelian genetics and, 135–136
Developmental symbiosis, 235–236
d'Herelle, Félix, 239
Dialectical materialism, 172
Difflugia corona, 325
Dimitrieff, Elizabeth, 60
Dinosauria coined, 35
Dinosaurs, 70, 219, 229–230

Divergence, 27, 50–51, 66
Division of labor, 284–285, 292
  among cells, 83–85, 255, 270
  ecological, 51, 83
  physiological, 51, 85
  in society, 51, 284–285, 292
  among specialities, 84, 244, 270, 292
  among species, 27, 50–51
DNA. *See also* Genes; Genomics; Molecular evolution
  as basis of the gene, 167, 184, 187–191
  as blueprint, 201, 205
  as book of life, 2001
  centrioles and, 246
  in chloroplasts, 207–208, 245
  in chromosomes, 188
  as code, 201
  as computer program, 196, 261
  fingerprints, 221
  junk, 203
  as master molecule, 187, 199, 205
  medicine and, 201
  methylation, 214–215
  in mitochondria, 245
  as orchestra conductor, 205–206
  proteins and, 194–200
  as recipe, 205
  recombinant, 187
  as secret of life, 201
  as self-replicating, 206
  sex and, 261
  as soul, 202
  structure, 187
  transcription, 196–197
Dobbell, Clifford, 98–99
Dobzhansky, Theodosius, xi, 307
  on genetic drift, 150
  on macroevolution, 152
  on molecular evolution, 228–229
  on pneumococcus transformations, 190
  population genetics and, 150
  species concept, 151
Dodge, B. O., 164
Doolittle, W. Ford, 227, 232, 246
Double helix, 194. *See also* DNA
Douderoff, Michael, 220, 226
Douglas, Angela, 262
Driesch, Hans, 100–102, 104, 110
Drift. *See* Genetic drift
Drosophila genetics, 72, 112–114, 136–138, 141, 150, 161–164, 269
  mutations, 112, 141–142
Dubos, René, 244, 247
Dumortier, Barthélemy C., 77
Dunn, Leslie C., 108, 139
Durkheim, Emile, 284–285
Dutrochet, Henri, 76

E. Coli, 197, 206
Earth. *See also* Geology
  age of, 4, 64, 217
  pre-life conditions, 217–219, 222, 330
East, E. M., 137, 146, 175, 242
Echinoderms, 86, 97, 100, 269
Ecology, 51. *See also* Malthusianism
  superorganism and, 83
  trends in, 243 254, 255
  word coined, 36
Ediacaran biota, 218–219
Edwards, Henri Milne, 51
Egerton, Frank, 23
Egg. *See also* Experimental embryology
  organization of 104–106, 110–111, 182, 207–210, 299, 318–319
  role in heredity, 103, 106–108
Egoism, 45, 56, 253–255, 258. *See also* Selfish genes; Cooperation; Mutualism.
Ehrlich, Paul, 80
Eidos, 28
Eimer, Theodor, 69–71
Einsteinian Atsrophysics, 155
Eisley, Loren, 119, 258
Eldredge, Niles, 249–250
Electron microscope, 196
Elementary organisms, 82, 95–96. *See also* Cell state
Elk, 70, 289
Elysia chlorotica, 249
Embranchements, 11
Embryology. *See* Comparative embryology, Experimental embryology
Emergence, 111
Emerson, R. A., 137, 163
Endoplasmic reticulum, 325
Engels, Freidrich, 49
Entelechy, 110
Entwicklungsmechanik, 97, 140, 144. *See also* Experimental embryology
Enzymes
  DNA reproduction and, 223
  gene function and, 140–141, 157–163, 199, 203–204
  RNA and, 223
Eocyte, 322

Eohippus, 67
Ephrussi, Boris, 181
  Beadle and, 163–164
  on cellular differentiation, 181–182
  cytoplasmic inheritance, 175
  on egg organization, 182, 318
  Lysenkoism and, 178, 179
  McClintock and, 157
  Morgan and, 168, 181
  on protein synthesis, 184
  Rockefeller Foundation and, 168
  yeast genetics and, 166
Epigenesis, 100–102
Epigenetic, coined, 174
Epigenetic inheritance, 200, 214–216
Epigenetic landscape, 174
Epigenotype, 316
Epinucleic inheritance, 215
Eubacteria, 225–233
Eugenics, 132–133, 137, 147, 203
Euglena, 177, 245
Eukarya, 233, 227–229
Eukaryotes, 269
  evolutionary significance of, 226–227
  genome, 229
  origin of term, 226, 229, 230
  species numbers, 227, 229
  as superorganisms, 235
  on symbiotic nature of, 238–240, 245–247, 263, 264
Evolution. *See also* Comparative embryology; Darwinism; Fossils; Microbial evolution; Molecular evolution; Mutualism; Natural selection; Social Darwinism; Symbiosis
  and chance, 4–5, 63
  early evidence for, 10–11, 20–22, 67–68, 219
  gradualism, 27–28, 147, 149, 152
  liberalism and, 44–45
  neutral theory, 229, 332
  origins of term, 15, 39, 276
  as population genetics, 150
  progress and, 11–13, 20, 39–54, 58, 62, 63, 68–69
  religion and, 4–5, 41, 63, 68, 258
  saltationism, 68, 71–72, 112, 124, 144–146, 152, 251
  society and, x, 42–54
  tempo and mode, 153–154
Evolutionary synthesis, xi, 218
  embryology and, 156
  genetics, and, 142–151
  microbes and, 156, 216, 250
  scientific change and, 154–155
  symbiosis and 156, 241–242
Exons, 204
Experimental embryology, 207, 209. *See also* Cellular differentiation
  cell theory and, 95–99
  cytoplasmic inheritance and, 103, 106–108
  evolution and, 112–114
  evolutionary synthesis and, 156
  genetics and, 109–114, 138–140
  heredity and, 131–133
  origins of, 95
  techniques of, 96–98, 105–106, 108, 110–111
Extinction, 11, 20
Eye, evolution of, 66, 249

Falk, Raphael, 205
Fascism, 169
Fertilization, 72, 83
Field. *See* Morphogenetic field
First World War, 241, 243
Fisher, Ronald A.
  eugenics and 147
  on Mendel's data, 126–127
  on Mendel's paper, 117, 123, 126–127
  populations genetics and, 147–149
  Wright and, 149
Fitness, 132, 148
Fitzroy, Robert, 18
Fleming, Alexander, 241
Flemming, Walther, 88
Flory, Howard, 241
Forman, Paul, 302
Fossil record, 10, 18–20, 67–68, 70–71, 218–220, 238, 248–249, 288
Founding Father Fables, 123–24, 157–163
Foucault, Michel, 12
Fox, George, 225
Frank, Albert Bernard, 237
Frankel, Joseph, 213–214
Franklin, Rosalind, 192–194
Fraud
  Kammerer, 178
  Lysenko, 171–173
  Mendel, 120, 126–127
  Moewus, 164–166
French Revolution, 6, 52
Functionalism, 11, 25–26, 267

Gaertner, C. F., 122–123
Gaia hypothesis, 262

Galapagos archipelago, 20–22
Galileo, 40
Galton, Francis, 125
Galtonian theory, 131, 132
Gametogenesis, 327
Gamow, George, 195
Garrod, Archibald, xi, 157–163, 271
Garstang, Walter, 40
Gastrulation, 104–105
Genes 135–137. *See also* DNA
  as abstract entities 135, 140–141, 204–205, 259
  as beads on string, 136
  and cellular differentiation, 109, 140, 251
  of classical genetics, 140–142
  of classical molecular biology, 198–200
  coined, 135
  enzymes and, 140–141, 157–163
  of eukaryotes, 201, 203–204
  homeobox, 251
  mapping 112, 136 137
  morphogenesis and, 106–112, 207–214, 268
  mutations of, 112–114, 146, 160
  of population genetics, 148, 149
  of prokaryotes, 201, 203
  proteins and, 169, 184, 190–191, 199, 204, 215–216
  regulation, 182, 197–198
  regulatory genes, 251
  selection, 332
  selfish, x, 255–261
  as self-replicating, 196–200
  as sentence, 198–199
  of sociobiology, 255–261
  splicing, 187
  split, 204
  as viruses, 141, 168
Genealogy, 26, 37, 231. *See also* Phylogeny
Genesis. *See also* Abiogenesis; Gametogenesis; Heterogenesis; Morphogenesis; Mutagenesis; Ontogenesis; Orthogenesis; Pangenesis; Parthenogenesis; Phytogenesis; Symbiogenesis
  biblical, 4, 18
Genetic code, 195–196, 203–204
  evolution of, 221–224
Genetic determinism, 136, 202, 257–258, 261 304
Genetic drift, 149–150, 250
Genetic program, 200, 205–206
Genetic recombination, 136, 137, 142, 146
Genetics. *See also* Biochemical genetics; Molecular biology; Population genetics; Mendel; Microbial genetics
  agriculture and, 172
  biometrics and, 124–125, 133
  breeders and, 130, 137–138
  in Britain,124–125, 136, 160, 177
  coined, 124
  comparative anatomy and, 133
  cytology and, 72
  Darwinism and, 117, 124–127, 143–156
  embryology and, 103, 104, 106–114, 133, 135–136, 138–140, 144, 181–184, 197–198, 208, 212–214
  eugenics and, 135, 137–138
  evolutionary synthesis and, 142–152
  in France, 175, 179, 181–184, 197–198
  in Germany, 146, 164–166, 174–175, 180
  naturalists and, 133, 138
  physiology and, 158–159
  saltationism and, 71–72, 144,146
  in Soviet Union, 149–150, 172–173
  in the United States, 112–114, 136–140, 150, 161–164, 166–170, 176, 182
  Weismannism and, 94, 109, 135
Genetics Society of America, 138
Genome
  bacterial, 227, 229
  chloroplast, 207–208, 264
  defined, 174, 200
  human, 203–204
  mitochondrial, 207–208, 264–265
Genomic imprinting, 215, 327
Genomics, 202–205, 227, 230
Genotype concept, 134–137
  blending inheritance and, 117
  development and, 135–136
Geoffroy Saint Hilaire, Etienne, 13, 17, 35, 213
  Cuvier and, 14–15
  Lamarckism and, 14, 15
Geographical isolation, 64
Geology, 209
  age of the earth, 4, 64
  catastrophism, 12, 19
  Darwin and, 16, 18–20
  Lamarck and, 6
  precambrian fossils and, 218–219
  uniformitarianism,18–20
Germ-free mice, 236
Germ layers, 105
Germ plasm, 92–93, 137, 176

Germ theory, 80, 240–241
Gillham, Nicholas, 166, 245
Glass, Bentley, 158
God, 23, 41, 62, 268. *See also* Creationism; Creator; Natural theology
Godwin, William, 52, 55
Goethe, Johann Wolfgang von, 112
Goldschmidt, Richard, 152, 180, 310
Golgi, Camillo, 294
Golgi bodies, 207
Goodwin, Brian, 212
Gorilla, 35
Gould, Stephen J., 258–260
 on evolutionary change, 249–250
 on the evolutionary synthesis, 153–154
 on panselectionism, 258–260
 on symbiosis, 251
Grant, Robert, 15–17
Gray, Asa, 25, 65
Gray, Michael, 227, 246
Great Chain of Being, 4, 8–9, 11, 38
Greene, John C., 49
Gregory, W. K., 138
Griffith, Frederick, 189–190
Group selection, 254–256
Guinea pigs, 160
Gupta, Radhey, 232, 247
Guyer, Michael, 107

Haeckel, Ernst, 31
 on the clergy and aristocracy, 40–41
 Hobbesian views of, 56
 on idealism, 37
 Lamarckism and 68
 Monist philosophy, 41,47
 National Socialism and, 47
 ontogeny and phylogeny, 36–39
 protistology and, 85
Haldane, John Burdon S.
 the evolutionary synthesis and, 148
 on the inheritance of acquired characteristics, 177
 on kin selection, 255
 on the origin of life, 222
 on physiological genetics, 160, 161
Hamilton, William, 255, 263–264
Hardy, G. H., 308
Harold, Franklin, 206
Harrison, Ross, 100, 140
Hartmann, Max, 180
Harvard, 19, 137, 258
Harvey, Ethel B., 97
Hegel, George Wilhelm Friedrich, 19, 37
Hemoglobin, 192, 221
Henneguy, Félix, 91
Henslow, Joseph S., 18
Heredity. *See also* Inheritance; Genetics; Experimental embryology
 breeders' view of, 132
 early field of, 130–134
 embryologists' definition, 108, 139
 geneticists' definition, 139
 genotype concept of, 134–136
 metaphor, x, 134
 unified view of, 216
Hereditary symbiosis. *See* symbiosis
Hershey, Alfred, 169, 190–191, 209
Heterocatalytic, 196
Heterogenesis, 80
Hippocratic writers, 7
Hitler 47, 254. *See also* Nazism
Hobbes, Thomas, 56, 57, 62, 253, 258
Hogg, John, 85
Holism defined, 111
Holley, Robert, 196
Holy grail, 202, 217. *See also* DNA
Hominids, origins, 4
Homo sapiens, 33, 217
Homology, 35–36. *See also* Comparative anatomy
Homozygote, 134
Hooke, Robert, 86–87
Hooker, Joseph, 8, 25, 26, 31, 226
Hopkins, Frederick G., 201
Horizontal gene transfer. *See* Lateral Gene transfer; Infectious heredity; Symbiosis
Horowitz, Norman, 164
Horse, 69–70
Hrdy, Sarah Blaffer, 260
Human evolution, 8, 33–36, 57, 69, 218, 274. *See also* Social Darwinism; Sociobiology
Human nature, 57, 69, 202, 252–255, 257–261. *See also* Social Darwinism; Sociobiology
Human genome
 bacterial symbiosis and, 249
 disease and, 203–204
 as holy grail, 202
 regulation of, 204
 size,169, 203
 viruses and, 249
Human Genome Project, 202, 203, 217–221, 227, 269, 333
Hutton, James, 19

Huxley, Aldous, 32
Huxley, Andrew, 32
Huxley, Julian, 2, 136, 143, 151, 154, 166
Huxley, Leonard, 32
Huxley, Thomas Henry
  agnosticism of, 87
  on amoral nature, 57
  on birds and reptiles, 67
  on cell theory, 98
  Darwin and, 31–33
  on human evolution, 33–36
  Owen and, 34–36
  materialism of, 45, 80, 87
  saltationist views of, 32–33
  Wallace and, 25–26
  Wilberforce and, 34
Hyatt, Alpheus, 69, 70
Hybridists tradition, 120–121, 126
Hydra, 238
Hydro-thermal vents, 248

Inborn Errors of Metabolism, 158–160. *See also* Garrod, Archibald
Idealism, 36, 37, 45, 81. *See also Naturphilosphie*
Ideology, 258
Idioplasm, 92
Indiana University, 166, 168, 169
Industrial Revolution, 44
Infective heredity, 175, 243. *See also* Lateral gene transfer; Symbiosis
Infusoria, 8
Inheritance, 134. *See also* Ancestral inheritance; Blending inheritance; Cell cortex; Chloroplasts; Chromosomal inheritance; Epigenetic inheritance; Infectious heredity; Kinetosomes, Mitochondria; Structural inheritance; Symbiosis
Inheritance of acquired characteristics, 68–71. *See also* neo-Lamarckism; Lysenkoism; Symbiosis
  cortical inheritance and, 214
  Darwin and, 7, 29
  epigenetic inheritance and, 215, 267
  fraud and, 178
  genotype and, 135
  Lamarck and, 7–9
  Lysenkoism and, 171–174
  morality and, 69
  plasmagenes and, 171, 176–179
  Weismannism and, 92, 94
Insects
  colonies, 58, 98, 255, 256
  species numbers, 227
  symbiosis in, 227, 238, 240, 248–249, 264, 265
Instinct, 253. *See also* Human nature
Insulin, 169, 234
Introns, 203
Invertebrates, coined, 6, 248

Jacob, François, 12, 181, 187
  on the gene, 198
  on gene regulation, 197–198
  on genetic program, 200, 205
Jacobs, Walter, 188
Jardin du Roi, 5, 6
Jenkin, Fleeming, 65
Jenkinson, J.W., 107
Jennings, Herbert Spencer, 165–166, 320
Jeon, Kwang, 247
Johannsen, Wilhelm, 71, 131, 144, 307, 308
  genotype concept of heredity, 134–137
  on inheritance of acquired characteristics 135
  non-Mendelian views of, 112–113
  purelines, 145–146
  on Weismannism, 135
John Innes Horticultural Institute, 124
Johns Hopkins University, 164, 166
Jost, L., 110
Judeo-Christian theology, 3–5, 41
Jukes, Thomas, 229
Jumping genes. *See* Transposable elements
Jurassic Park, 205
Just, Ernest Everett, 183–184
Just-so-stories, 258–260

Kaiser Wilhelm Institute for Biology, 180
Kaiser Wilhelm Institute for Chemistry, 169
Kaiser Wilhelm Institute for Medical Research, 165
Kandler, Otto, 227
Kappa, 176, 243
Kammerer, Paul, 178
Kant, Immanuel, 19, 37, 292
Kauffman, Stuart, 213
Kelvin, William Thomson, Lord, 64, 71
Kendrew, John, 193
Khorana, Har Gobind, 196
Kimura, Mooto, 229
Kin selection, 255

Kinetosomes. *See also* Centrioles; Cortical inheritance
  centrioles and, 91
  morphogenesis and, 182–184, 319
  plasmagenes, and, 175, 183
  symbiosis and, 245–246
King, Jack, 229
Kingdoms, 85, 227
Kings College, 192
Koch, Robert, 80, 240
Kolliker, Rudolf, 88
Kol'tsov Institute, 150
Kropotkin, Peter, 55–59, 253
Krushschev, Nikita, 173
Kuhn, Richard, 165
Kuhn, Thomas, 154–155
Kurland, Charles, 232, 345

Laissez-faire economic theory, 44, 50–51, 270
Lake, James, 227, 232, 247
Lamarck, ix, 3, 29. *See also* neo-Lamarckism
  botany and, 5–6
  cell theory and, 290
  Cuvier and, 11–13
  evolutionary theory of, 9–11
  Geoffroy and, 13–15
  human evolution, 8, 274
  on life, 274
  myths, 6–8
Lamarckism. *See* neo-Lamarckism
Lambda phage 244
Lateral gene transfer, 249–250, 270, 333, 335
  bacterial evolution and, 230–233, 243–244
  biotechnology and, 234, 245
  gene theory and, 167, 168, 189–190
  individual and, 231, 243–244
  neo-Darwinism and, 231, 249–25
  species concept and, 231, 245
Lederberg, Joshua, 157
  bacteria recombination, 166–167
  on gene theory, 168, 313–314
  on hereditary symbiosis, 243–244, 247
  on the individual, 244, 333–334
  on infective heredity, 243
  on plasmids, 167
  on prions, 324
  on transduction, 190, 243–244
  on transformation, 190
Leeuwenhoek, Anthony, 87
Lenhosseék, Mihaly, 91
Les Forces Français Libres, 164
Levene, Phoebus, 188
Leviathan, 56
Lewontin, Richard
  on adaptationism, 259
  on human diversity, 304
  on ideology in science, 25
  on macro-evolution, 310
  on sociobiology, 258–260
L'Héritier, Philippe, 175
Lichens, 236, 238, 263
Life. *See also* Cell; DNA; Protoplasm; Vitalism
  enzyme theory of, 160–161
  gene theory of, 141
  germ-plasm theory and, 92–93
  materialistic conception of, 6, 8–9, 87
  origin of, 4, 91, 110, 217–226, 232–233
  physiological view of, 96, 110, 299
Lillie, Frank R., 100, 136, 139, 140
Limoges, Camille
  on the division of labor, 284–285
  on mythical precursors of Darwin, 278
Lindegren, Carl, 166
Linnaeus, Carolus
  Darwin and, 29
  classification and, 9
  evolutionary views of, 9–10, 120
  hybridists and, 120
  natural theology and, 23, 61
Lister, Joseph, J., 79
Loeb, Jacques, 97, 110
  on cytoplasmic heredity, 103, 107
The Logic of Life, 198, 200, 205
Lovelock, James, 262
Luria, Salvador, 169, 190, 193, 209
Lwoff, André
  on cortical inheritance, 214, 319
  on morphogenesis, 181, 198, 210, 296
  plasmagene theory and, 175, 176
Lyell, Charles, 22, 25, 26, 28, 31, 39, 44
  evolution and, 19–20
  steady state theory of, 20
  uniformitarian views of, 18, 19, 277
Lysenko, Trofim Denisovich, 171–173, 177–179, 269
Lysogeny, 239

Macleod, Colin, 189–190
Macroevolution, 152–154, 247–251, 310
Mad Cow disease, 207
Malthus, Thomas, 44, 57, 254
  Darwin and, 24, 52–54
  Wallace and, 54

Mangold, Hilde, 101
Man's Place in Nature, 33
Margoliash, Emmanuel, 221
Margulis, Lynn, 227, 245–247
Marsh, Othniel, 67
Martin, William, 247, 345
Marx, Karl, 48–49
Marxism, 43
Materialism, 3, 41, 81, 87
Maternal Inheritance, 106–108
Mathews A. P., 110
Matthaei, J. Heinrich, 196
Matthew effect, 306
May, Robert, 344
Mayr, Ernst, 28–29, 151, 152, 274
  on Archaebacteria, Eubacteria, Eukarya, 228–230
  early neo-Lamarckian views of, 149
  on the evolutionary synthesis, 155
  on the neutral theory, 229, 332
  on population genetics, 149
  on typology and nominalism, 65
  on speciation, 15
McCarthyism, 178
McCarty, Maclyn, 189–190
McClintock, Barbara, 137, 157, 163
McFall Ngai, Margaret, 237
Mechanistic-materialism, 3, 101, 110
Meckel-Serrés Law, 38
Medawar, Peter, 127–128, 296
Medicine and biology, 75, 169–170, 187, 269
Meiosis, 89, 94, 104, 166
Meischer, Johann, Friedrich, 88, 188
Mendel, Gregor
  Darwin and, 120
  fraud and, 126–127
  Garrod and, 158
  legend of, xi, 117–122, 271, 294
Mendelian-chromosome theory, 112–114, 136–137, 140–142. *See also* Genetics
Mendel's laws, ix, 114, 118, 268
Merezhkowsky, Constantin, 57, 239, 247
Merogonic hybridization, 97–98, 106, 108, 298
Merton, Robert, 306
Metchnikoff, Elii, 86
Michaelis, Peter, 174
Michurin, Ivan Vladimirovich, 172
Microbial Evolution, xi, 224–233
Microbial fossils, 219. *See also* Precambrian
Microbial genetics, 161–170, 184
Microbial genome initiative, 227
Microbiology, 163, 181, 220–223, 270. *See also* Microbial genetics
Microbiome, 334
Microscopy, 79–80, 86–91, 196, 269, 294
Microsomes, 184
Militarism, 47–48
Mill, John Stuart, 128
Miller, Stanley, 222
Mimicry, 67
Minchin, Edward, 86
Mind, 69
Mirbel, Brisseau, 76
Mitchell, M. B., 164
Mitochondria, 90, 95, 110, 175, 238
  genome, 207–208, 265
  symbiotic origin of, 235, 238, 239, 245 263–265, 338
Mitosis, 88, 89, 90–91, 166, 245–246
Mitotic spindles, 175
Mivart, Saint-George Jackson, 70, 116
Model organisms, 269
  for biochemical genetics,161–168
  for classical genetics, 137
  for embryology, 97–98
  for molecular biology, 167, 190–191, 201, 203, 209
  for structural inheritance, 210–214
Moewus, Franz, 164–166
Mohl, Hugo, 77
Molecular biology, ix, x. *See also* Biochemical Genetics; Microbial phylogeny; Molecular evolution
  coined, 168
  doctrines, 198–201, 203–209, 215–216, 260
  origins, 167–170, 184, 187–198
Molecular clocks, 220–221, 224, 229
Molecular evolution, 156, 220–221
Monera, 37, 85
Monism, 41
Monist League, 47–48
Monod, Jacques, 181, 197–198
Monotheism, 41
Morality, 41, 42, 55–58, 69, 253, 258, 261. *See also* Altruism; Mutualism; Natural theology
Morgan, Thomas Hunt, 100, 134, 150. *See also Drosophila* genetics
  at Cal Tech, 163, 168–169
  on cellular differentiation, 140
  at Columbia University, 112, 114, 136
  on cytoplasmic inheritance 113–114
  on embryology, 136, 140, 181

on evolution, 71–72, 112–113, 131, 144–146
on gene mutations, 113–114, 146
on gene theory, 141–142, 160
on genetics' aims, 112, 114, 136–137, 164
Lysenkoism and, 173
on Mendel, 302
Morphogenesis. *See also* Cell structure; Ciliates; Experimental embryology
cell cortex and, 182–184, 210–216
Mendelian genetics and, 103–111
molecular biology and, 205–211
Morphogenetic field
defined, 111, 212
and cell cortex, 111, 182–184
inheritance of 183, 211–214
nature of, 213–214
Mosaic embryos, 102
Moscow University, 172
Müller, Johannes, 77
Muller, Hermann J., 136, 169, 241
on cytoplasmic inheritance, 178, 242, 336
on gene theory, 141–142, 160, 168
on physiologists, 159
X-rays and, 141, 241
Muller, Miklós, 345
Muséum d'histoire naturelle, 6, 11
Mutagenesis, 141–142
Mutualism, x, 65, 262
and anarchism, 55–60
and natural theology, 23, 61–62
Mycoplasmas, 330
Mycorrhiza, 237, 262

Nägeli, Carl von, 70, 77,92
Nanney, David, 211
Nardon, Paul, 247
NASA, 222, 225
National Institutes of Health, 168, 227
Naturphilosophie, 19, 36– 38, 56, 81, 112, 291
Natural selection, 148, 174, 262, 268. *See also* Adaptationism; Darwinism; Evolution; Evolutionary synthesis; neo-Darwinism
criticism of, 63–72
Darwin and, 22–24, 26–30
targets of, 252–262
Wallace and, 24–26
Natural Theology, 4, 22, 23, 31, 56, 61–62
Naturalism, 3, 15, 31, 43, 44
Naturalists, 9, 17, 56, 71, 131, 134–135, 155. *See also* Classification; Paleontology
Nazism, 48, 132, 133, 152, 165, 269
Needham, Joseph, 138
Neo-Darwinism, ix–xi, 68, 103, 142, 146, 214, 249–250. *See also* Darwinism; Evolutionary Synthesis; Weismannism
Neo-Lamarckism, 112, 131, 153. *See also* Inheritance of acquired characteristics
in France, 68–69, 179–181
genetics and, 113, 149
in Germany, 68
Malthusianism and, 68
in the USA, 68
Neurospora genetics 161–165, 194
Neutral theory. *See* Evolution
Newton, Isaac, 40
Nilsson-Ehle, H., 146, 307
Nirenberg, Marshall, 196
Noah, 4
Nominalism. *See* Species
Non-Mendelian inheritance. *See* Cytoplasmic inheritance; Structural inheritance; Epigenetic inheritance; Symbiosis
Nordenskiöld, Eric, 63, 274, 275
Nuclear monopoly, 174
Nuclear theory of inheritance, 92–94, 97
Nuclear war, 263
Nuclear weapons, 170, 193
Nucleic acid, coined, 88, 188. *See also* DNA; RNA
Nuclein, 88, 188
Nucleotide, 188
Nucleotide sequencing. *See* Genomics; Molecular evolution; Microbial Phylogeny
Nucleus, 76, 91–95, 110. *See also* Chromosomes; Mitosis; Meiosis; Weismannism
symbiotic theories of, 238, 246–247, 338

Oehlkers, Friedrich, 174
Oenothera Lamarckianna, 145, 175
Oken, L., 19, 38
Oligonucleotide cataloging, 330
Olsen, Gary, 227, 232
On the Origin of Species
concepts in, 26–30
publication of, 26, 32–33
One gene: one enzyme hypothesis
biochemical genetics and, 157, 160–162
criticisms of, 161, 162–163, 169
human genome project and, 203–204
molecular biology and, 199
on refutation of, 203–204, 268
Onslow, Muriel Whedale, 160

Ontogenesis, 36, 156
Ontogeny, coined, 36
Oparin, Aleksandr Ivanovich, 222, 329
Operon theory, 197–198, 321
Orchids, 238
Organism. *See also* Division of labor; Morphogenesis, Symbiome; Superorganism
  as cell state, 82–84
  noncellular, 98–99, 296
  society as, 82–85
  sociobiology and, 255–257
  as a whole, 95, 98–100, 108
Orthogenesis, 69–71, 112, 153, 289
Osborn, Henry Fairfield, 70
Owen, Richard, 34–36, 85, 281
Oxford, 15, 40
Oxytricha fallax, 212

Pace, Norman, 227, 228
Paleobiology, 218–220
Paleontology, 16, 67–70, 133, 153–154, 218–219, 226, 249–250
Paley, William, 23, 61
Pangenesis, 92
Pangloss, Dr., 134, 259
Panselectionism, 154, 229, 259
Paradigm, 143, 155
Paramecium genetics, 166, 175–177, 182–183, 211, 269
Parasitism, 58–58, 244, 262, 264. *See also* Symbiosis; Mutualism
Paris Commune, 60
Paris Revolution, 15
Parthenogenesis, 97, 106, 248
Pasteur, Louis, 80–81, 166, 181 220, 240
Pasteur Institute, 80, 197
Pauling, Linus
  Beadle and, 170
  on genetic code, 321
  molecular evolution, 220–221
  on protein structure, 192–193
Pearson, Karl, 124, 131, 133, 147
Penicillin, 241
Peppered moth, 148–149
Pepsin, 169
Peroxisomes, 338
Phage. *See* Bacteriophage
Phenotype, 117, 134
Phlogiston theory, 113
Phylogeny, coined, 36
  microbial, 85–86, 224–233
Phylum, coined, 36
Physics
  biology and, 102
  evolution and, 213
  malaise in, 170
  molecular biology and, 170, 187, 189, 271
Physiological division of labor. *See* Division of labor
Physiological genetics, 158–161. *See also* Biochemical genetics
Physiologists
  and cell theory, 110
  and genetics, 159
Phytogenesis, 76
Picket-Heaps, Jeremy, 247
Pitt, William, 52
Plants. *See also* Botany; Cell theory; Chloroplasts; Classification; Eukaryotes; Organisms; Symbiosis
  evolution of, 9, 27, 227, 237, 262
  genetics of, 172–177
  origins of, 4, 90, 98–99, 156
  speciation, 150–152
Plasmagene theory, 171, 179, 181–184, 243
Plasmids, defined, 243–244
Plasmon, 174–175
Plato, 4, 19, 35
Pleiotropy, 145, 161
Pneumococcus, 189–190
Polarity. *See* Cell polarity; Cell structure
Polygeny, 145, 161
Polymerase chain reaction, 221, 248
Polypeptide, 192
Polyploidy, 151–152
Poor laws, 52
Population genetics, xi, 147–150, 149–150
Population growth, 24, 29, 45, 254. *See also* Malthus
Population thinking, 28, 142
Portier, Paul, 239, 240, 241, 247
Pouchet, Félix, 80
Pound, Roscoe, 262
Pragmaticism, 267, 271
Precambrian, 218–220, 328
Preformation, 102
Prezent, I. I., 171, 178
Principles of Geology, 18–20
Prions, 207, 216, 324
Prisoner's dilemma, 263
Progenote, 232–233
Prokaryotes, 217, 226 229, 269. *See also* Bacteria; Microbial evolution

Protelarian biology, 172
Proteins. *See also* Enzymes; Gene theory
  chaperon, 207–217
  evolution, 229
  genes and, 156, 157–163, 184
  hereditary, 200, 207, 268, 324
  structure, 170, 191–193
  synthesis, 196–199, 203–207, 215–216
Protista, 85, 100
Protistology, 85–86, 210
Protists, 98, 99, 156. *See also* Ciliates
  on acellular nature of, 296
  colonies, 98, 100, 296
  and origin of metazoa, 342
Protoplasm, 87
Protozoa, 88, 296
Protozoology, 210. *See also* Ciliates; Protists
Proudhon, Pierre-Joseph, 60
Psychic explanations, 66, 112
Punctuated equilibria, 250–251
Purelines, 309
Purkinje, Johannes, E., 87

Rabl, Karl, 88
Race, 132, 133, 304
Racism, 47–48, 132–133, 254
Radioactive isotopes, 187, 219
Rapoport, Anatol, 263
Rattlesnake, H.M.S., 32
Ray, John, 61
Rayleigh, R. J., 64
Raymond, Percy, 153
Recapitulation theory
  cell theory and, 83, 98, 99
  comparative embryology and, 38–40
  experimental embryology and, 40
Recessive mutations, 160
Recombinant DNA, 187. *See also* Biotechnology
Red Queen's dilemma, 148
Reductionism, 102, 111, 203, 267
Reign of terror, 6
Remak Robert, 77–78
Renner, Otto, 174, 175
Rensch, Bernard, 149
Rhizobium, 237
Rhoades, Marcus, 137, 163
Ribosomal RNA phylogenies, 224–233
Ribosomes, 184, 196, 197, 223
Ribozymes, 223
Richards, Robert, 203–204
Rivera, Maria, 247
RNA, 156
  as enzymes, 223
  messenger, 204
  phylogenies, 224–233
  ribosomal, 224–233
  viruses and, 184
RNA World, 223
Rockefeller Foundation, 168–170
Rockefeller Institute, 189–190
Rothschild Institute for Physico-chemical Biology, 166, 168
Roux, Wilhelm, 97, 89, 100–102
Royal Society, 86

Sachs, Julius, 88, 95
Saffo, Mary Beth, 247
Sager, Ruth, 166, 245
Salmonella, 244
Saltationism. *See* Evolution; Symbiosis
Scala naturae, 4. *See also* Great chain of being
Schelling, Freidrich, 19, 37
Schopf, William J., 328
Schimper, Andreas, 90, 239
Schleiden, Matthias Jacob, xi, 75–79
Schleifer, Karl Hans, 232
Schmalhausen, Ivan Ivanovich, 172–173
Schrödinger, Erwin, 170, 189
Schwann Theodor, 75–79, 84
Schwemmler, Werner, 247
Science
  and ideology, 50, 252
  pure and applied, xii, 130, 269–270
  and religion, 23, 258, 329
Science for the People, 258
Scientific paper, 122, 127–129, 303–304
Scientist, word coined, 32
Scopes, John, 41
Scott-Moncrieff, Muriel, 160
Scrabble, 204
Sea anemones, 238
Sea urchin, 97, 100, 269
Second World War, 156, 161, 163, 164, 170, 175, 189, 243
Sedgwick, Adam, 18–19, 28, 98, 124
Self-assembly, 200, 207–212
Selfish genes, 255–257
Seven Years War, 5
Sex, 146
  determination, 136, 305–306
  microbes and, 166–167
  origins of, 260–261
Sexism, 193, 260, 305–306

Sexual selection, 66
Shapin, Stephen, 283–284
Sharp, Leslie, W., 137
Sharp, Philip, 203–205
Shaw, George Bernard, 43
Shifting balance theory, 149–150
Sickle-cell anemia, 192
Sigma, 177
Sinsheimer, Robert, 245
Simpson, George Gaylord
  on macro-evolution, 153–154
  on neutral theory, 228–229, 332
  on rates of evolution, 328–329
Smallpox, 241
Smith, Adam, 44, 50, 270
Smith, David, 262
Smith, John Maynard
  on cortical inheritance, 214
  on development, 256–257
  on kin selection, 255
  on symbiosis, 251, 264
Smuts, Jan C., 111
Social Darwinism, 132
  anarchism and, 55–58
  Darwin and, 49–50
  liberalism and, 44–47
  Marxism and, 48–49
  Militarism and, 47–48
  as misnomer, 50
  symbiosis and, 263
Sociobiology, 255–261
Sogin, Mitchell, 227
Sonneborn, Tracy, M., 210–211
  on cortical inheritance, 210–211
  on inheritance of acquired characteristics, 171, 176
  on Lysenkoism, 171, 178
  on Moewus, 165–166
  on plasmagenes, 175–176
  on symbiosis, 337
Soul, 4, 38, 202
Spandrels of San Marcos, 259–260
Species
  bacteria and, 231
  boundaries, 23
  definitions of, 23, 65–66, 151
  duration of, 250, 328–329
  fixism and, 4–5, 11–12, 18–20, 23
  nominalism and, 28–29, 65
  numbers of, 229
  polymorphic, 288
  symbiosis and, 234–235
  typology and, 28–29
Specialization. *See* Biology
Speciation, 66, 152, 250, 251, 328–329. *See also* Symbiosis
  allopatric, 151
  sympatric, 151
Speigelman, Sol
  on plasmagenes, 175–176
  yeast genetics, 166
Spemann, Hans, 180
  on cellular differentiation 109, 298
  embryonic induction, 101, 109
  on evolution, 112
Spencer, Herbert, 44–46, 92, 96, 253
Sperm, 72, 83, 104, 108. *See also* Egg; Fertilization
Spielberg, Steven, 205
Spiritualism, 112, 180
Spliceosomes, 204
Spontaneous generation, 80–81, 91, 110
Squid, 237
Stadler, L. J., 141
Stalin, Joseph, 173
Stanford, 164, 169
Stanier, Roger, 220, 226, 246
Stebbins, G. Ledyard, 151, 309
Stent, Gunther, 170, 195, 199, 263
Stentor, 210
Stevens, Nettie Maria, 136, 305–306
Stomatolites, 219
Sturtevant, Alfred H., 136, 146, 175
Strasburger, Edouard, 88
Statisticians, 133. *See also* Biometricians
Structural inheritance, 182, 207– 215. *See also* cortical inheritance
Structuralism, 14, 212–213, 267
Struggle for existence. 24, 45. *See also* Darwinism; Hobbes; Mutualism, Natural Selection; Social Darwinism; Sociobiology
  Darwin on, 29–30, 49–50, 54, 58
  Industrial Revolution and, 43
  materialism, and, 54, 56
  mutualism and, 55–62
  neo-Lamarckism and, 68
  use before Darwin, 22–23
Sulfur bacteria, 248
Sullaway, Frank, 275
Sumner, William Graham, 46
Superorganism, *See also* Cell state theory; Division of labor
  biosphere as 262
  ecological succession and, 83

society as, 44
symbiosis and, 235, 244
Survival of the fittest, 45, 69
Sutton, William, 89–89
Symbiodinium, 238
Symbiogenesis, 245
Symbiome, xi, 236
Symbiomics, 234–251
Symbionts, 90
Symbiosis, x, xi, *See also* Chloroplasts, Mitochondria, Nucleus
definitions of, 235, 240, 261–263
developmental, 235–236
evolution of, 264–265
in evolutionary change, 240–241, 244, 247–249
germ theory and, 234, 240–241
hereditary, 175–177, 216, 233, 235, 238, 243–250, 268–269
land plant evolution, and, 237
morphogenesis and, 236, 238
neo-Darwinism and, 250–251
oppositions to, 240–245
parthenogenesis, 248
physiological affects, 236–238, 248
as slavery, 263–265
social Darwinism and, 263
speciation, 248
species concept and, 234, 244–245
Szathmáry, Eörs, 251, 256, 257

Tansley, Arthur, 83
Tartar, Vance, 210, 319
Tatum, Edward, 157–158, 164–165
Taylor, Max, 247
Teleology, 4, 259, 262, 268
Termites, 238
Tessier, Georges, 179, 181
Tetrahymena, 211
Tetranucleotide hypothesis, 188, 190
Theology, 31, 87
Thomas, Lewis, 263
Thomson, William, 64
Tissue culture, 181
Tit-for-tat, 263–264
Totipotent, 102
Transcendental anatomy, 13–14, 35
Transduction. *See* Bacteria
Transgenic organisms, 187, 234
Transposable elements, 163
Tree of life, 4, 13, 27, 231, 250
Treitschke, H. von, 48
Treviranus, 5
Trilobites, 218
Trivers, Robert, 255, 257
Troland, Leonard, 160
Trotsky, Leon, 49
Tschermak, Eric, 119, 121
Tube worms, 248
Tuberculosis, 241
Turpin, J. P. F., 76
Tyndall, John, 80
Typology 65, 29

Ukranian Commissariat of Agriculture, 172
United States Department of Energy, 227
Unger, Franz, 77
Uniformitarianism, 18
University academic structures, 180, 181, 317. *See also*, Specialization
University of Chicago, 149
Urey, Harold, C., 222
Ussher, James, 4

van Benden, Edouard, 88–89, 91, 96
van Beneden, Pierre-Joseph, 58–62
Vavilov, Nikolai Ivanovich, 172, 173
Venter, Craig, 203, 205
Vernalization, 172
Vestiges of Creation, 17, 34
Virchow, Rudolf, 77–78, 82, 84, 110, 209
Viral genetics, 209
Virulence, 204, 264
Viruses. *See also* Bacteriophage
DNA and, 190–193
as genes, 168, 169, 184
human genome and, 249
as plasmagenes, 176, 177
self-assembly, 200, 269
symbiosis and, 177, 216, 239, 244, 249
Vitalism, 9, 70, 80–81, 87, 110, 179–180
Vitamins, 164, 236, 238, 248
Voltaire, 259
von Baer, Karl E., 38
von Wettstein, Fritz, 174

Waddington, Conrad H., 173–174
Wagner, Moritz, 66
Walcott, Charles, D., 218
Wald, George, 245
Waldeyer, Wilhelm, 88
Wallace, Alfred Russell
Darwin and, 26
Malthus and, 54
spirtualism and, 66
voyages, 17, 24, 25

Wallin, Ivan E., 240, 241
Ward, Lester Frank, 282
Watasē, Shôsaburô, 245, 246
Watson, James D., 168, 192–195, 204
Weaver, Warren, 168–169, 314
Webster, Gary, 212
Wedgwood, Emma, 22
Wedwood, Susannah, 17
Weinburg, Wilhelm, 308
Weinstein, Alexander, 121
Weismann, August, 68, 69, 92–94, 100, 101, 104, 144, 176, 213
Weismannism, 95, 200, 268, 294, 298
Weiss, Paul, 319
Weldon, W. F. L. 124, 131
What Is Life?, 170, 189
Wheeler, William Morton, 138, 139
Whewell, William, 18–19, 32
White, Michael, J. D., 150
Whitman, Charles Otis, 82, 84, 245
Wigglesworth, V. B., 108
Wilberforce, Samuel, 34
Wilkins, Maurice, 168, 170, 192, 193
Williams, George C., 255
Wilson, Edmund Beecher
  on cell organization, 91, 131
  on cell state theory, 84–85, 100
  on genes and enzymes, 161
  on history of cell theory, 76, 88
  on sex determination, 136, 305–306
Wilson, Edward O., 255, 257, 261
Winge, Otto, 166
Winkler, Hans, 174
Wittaker, R. H., 227
Woese, Carl
  on genetic code, 221–222
  on genomics, 217
  Mayr and, 229–230
  on molecular clock, 224
  on the progenote, 232–233
  three-domain proposal, 225–228
Wolbachia, 238, 248, 265
Wolpert, Lewis, 208, 213
Woods Hole, 97, 181, 245
Worster, Donald, 254
Wright, Sewall, 147, 149–150, 154, 160–161, 309
Wynne-Edwards, Vero C., 254–255, 341

X-ray crystallography, 187, 191–194, 302
X-rays, 141, 164, 241

Yale, 167, 222
Yeast genetics, 166, 182

Zillig, Wolfram, 227
Zinder, Norton, 190
Zoology. *See* Classification; Comparative embryology; Experimental embryology
Zoocentricism, 176
Zuckerkandl, Emile, 220–221